William Robinson, William Whitman Bailey

Alpine Flowers for English Gardens

William Robinson, William Whitman Bailey

Alpine Flowers for English Gardens

ISBN/EAN: 9783743323513

Manufactured in Europe, USA, Canada, Australia, Japa

Cover: Foto ©berggeist007 / pixelio.de

Manufactured and distributed by brebook publishing software (www.brebook.com)

William Robinson, William Whitman Bailey

Alpine Flowers for English Gardens

CONTENTS.

	PAGE
INTRODUCTION	x

Part I.

CULTURAL AND STRUCTURAL	1
POSITION FOR THE ROCK-GARDEN	7
MATERIALS	7
HIDDEN WEALTH	9
PATHWAYS, ETC.	10
SOIL	13
WATER	14
DETAILS OF CONSTRUCTION	15
CASCADES, ROCKY BRIDGES, ROCKY MARGINS, ISLETS	23
THE ROCK-GARDEN FERNERY	28
ROCKWORK FORMED OF CONCRETE, ETC.	31
MINIATURE ROCK-GARDENS	33
RUIN AND WALL-GARDENS	36
ALPINE FLOWERS IN THE MIXED BORDER	44
ALPINE PLANTS IN SHRUBBERY BORDERS	50
THE NATURAL ROCK-GARDEN	54
ALPINE PLANTS ON WINDOW-SILLS	57
ALPINE PLANTS IN POTS	59
ALPINE PLANTS FROM SEED	65
PLANTING	72
THE BOG-GARDEN	72
A SELECTION OF CHOICE BOG-PLANTS	77
HARDY AQUATIC PLANTS	77
WHAT TO AVOID	84
ON THE GEOLOGICAL ASPECTS OF ROCKWORK	90
PLANT-HUNTING ON THE MOUNTAINS	102
MOUNTAIN VEGETATION IN AMERICA	144

Part II.

	PAGE
A SELECTION OF THE CHOICEST ALPINE FLOWERS ALPHABETICALLY ARRANGED, WITH INSTRUCTIONS FOR THE CULTURE AND POSITION FOR EACH KIND	155

Part III.

A SELECTION OF DWARF ALPINE AND ROCK-PLANTS THAT WILL THRIVE IN ORDINARY SOIL IN LEVEL GROUND	423
A SELECTION OF ALPINE AND ROCK-PLANTS WORTHY OF BEING GROWN IN NURSERIES	425
ALPINE AND ROCK-PLANTS WITH WHITE OR WHITISH FLOWERS	426
ALPINE AND ROCK-PLANTS WITH BLUE, BLUISH, OR PURPLE FLOWERS	427
ALPINE AND ROCK-PLANTS WITH ROSY, CRIMSON, SCARLET, RED, AND PINKISH FLOWERS	428
ALPINE AND ROCK-PLANTS WITH YELLOW FLOWERS	428
A SELECTION OF CHOICE DWARF SHRUBS FOR THE ROCK-GARDEN, ETC.	429
A SELECTION OF ALPINE AND ROCK-PLANTS TO RAISE FROM SEED	430
A SELECTION OF ALPINE PLANTS, ETC., SUITABLE FOR PLANTING ON THE MARGINS OF BEDS AND MASSES OF RHODODENDRONS AND OTHER AMERICAN SHRUBS	431
A SELECTION OF PLANTS FOR FORMING "CARPETS"	431
ALPINE PLANTS THAT WILL GROW WELL IN AND NEAR CITIES	433
ALPINE PLANTS GREEN IN WINTER	433
A SELECTION OF ALPINE PLANTS SUITED FOR CULTURE IN POTS FOR EXHIBITION	434
ALPINE AND HERBACEOUS PLANTS SUITABLE AS FLOWERING EDGINGS FOR BEDS OR BORDERS	435
A SELECTION OF ORNAMENTAL AQUATIC PLANTS	435
A SELECTION OF PLANTS THRIVING IN MARSHY OR BOGGY GROUND	436
TRAILERS, CLIMBERS, ETC.	436
LIST OF FERNS THAT MAY BE GROWN IN THE ROCK-GARDEN	437
SELECTION OF ALPINE AND ROCK-PLANTS FOR GROWING ON OLD WALLS, RUINS, STONY BANKS, ETC.	437
DWARF HARDY PLANTS OF A SILVERY OR VARIEGATED TONE, AND MOSTLY SUITABLE FOR EDGINGS	438
DWARF SHRUBS SUITED FOR THE ROUGHER PARTS OF ROCK-GARDENS, ETC.	438
LIST OF DWARF ALPINE SHRUBS, ETC., FOR THE ROCK-GARDEN	439
SELECTION OF ALPINE AND ROCK-PLANTS OF PROSTRATE OR DROOPING HABIT	439

LIST OF ILLUSTRATIONS.

	PAGE
Alpine Plants on a sloping Ridge	2
Mountain flank in process of degradation	3
Corner of natural Rock covered with Alpine Plants	5
Mound of Earth, with exposed points of Rock	9
Unearthed Rocks in a Sussex Garden	10
Alpine Plants on Vertical Rock	11
Passage in Rock-garden	12
Rude Stair from deep recess of Rock-garden	13
Right and wrong forms of Rockwork	15
Half-buried Stone surrounded by Alpine Plants	16
Well-formed Sloping Ledges	17
Artificial Rock on which Plants do not thrive	17
Horizontal Fissure	18
Right and wrong forms of Oblique Fissure	18
Do. of Steep Rockwork	19
Do. of Vertical Fissure	19
Ledge of Alpine Flowers	21
Alpine Plants growing on level ground	22
Waterfall fringed with Yuccas, etc.	23
Young Plants of Clematis falling over the face of Artificial Rock	24
Bird's-eye view of Islands above the Falls of Niagara	25
Stepping-stone Bridge	26
Stepping-stone Bridge, plan of	27
Rock-garden near Water	27
Margin of Island in Lake Maggiore	28
Rocky Water-margin at Oak Lodge, Kensington	28
Entrance to Cave for Killarney Fern	30
Masses of Artificial Rock	31
View of Artificial Rock at Oak Lodge	32
Scene in the Gardens at Oak Lodge	33
Small rocky bed of Alpine Flowers	35
Rock-garden on margin of Shrubbery	37
Ruined Castle	38
Ruins and Bridge	40
Rock-plants on an old Fort Wall	41
Saxifraga longifolia	41
Pansy on Brick-wall	42
Stone Wall covered with Alpines	43
Alpines in Level Border	47
Mixed Border	48
Mixed Border, plan of	49
Alpine Plants on Border	50
A Natural Rock-garden	55
The Window Rock-garden	58
Alpines in Pots	63
Bed of Alpines plunged in Sand	64
Bed kept saturated by perforated pipes	65
Illustration of right and wrong mode of Planting	72
Alpine Bog-garden	73
White Water Lily	78
Yellow Water Lily	79
Great Water Dock	80
Frontispiece of Book on Alpines	84
Arch (after Loudon)	85
Rockwork in Villa at Hammersmith	86
All the Alps seen from the Hall-door (after Macintosh)	86
Fountain and Rockworks (after Loudon)	87
"Infandi scopuli" (after ——)	87
Rockwork (after Mrs. Loudon)	87
Ground-plan of Rockworks in a London Park	88
Sketch from Kew	89
Sketch from the Botanic Gardens, Regent's Park	89
Granite Tor	91
Chalk	93
Limestone	95
Old Red Sandstone	97
Mica Schist	100
"Excelsior!"	102
An Alpine Lake	103
In the Woody Region	104
Pine Woods, Glacier, and Alpine Village	106
View of a distant range	109
Chillon	110
An Alpine Valley and River-bed	112
An Alpine Pathway	114
An Alpine Village	116
An Alpine Waterfall	117
An Alpine Stream	118
"The glassy Ocean of the Mountain Ice"	123
The Limit of Life	125
Alpine Larch-wood	126
Cascade in a High Wood	128
Alpine Road through Cliff	129
Island in Lake Maggiore	130
Scene in the higher Alps	133
The limit of the Pines	135
View on the Simplon Road	136
A Glacier	136
A glimpse at the Home of the two-flowered Saxifrage	140
Scene in the Rocky Mountains	143
Isolated Rocks in Rocky Mountains	144
Mountain Woods of California	145
Isolated Rocks on Plains eastward of the Rocky Mountains	397

Works by the same Author.

THE GARDEN.
PUBLISHED WEEKLY AND IN HALF-YEARLY VOLUMES.

"That excellent periodical, 'The Garden.'"—PROFESSOR OWEN.

THE PARKS AND GARDENS OF PARIS.
CONSIDERED IN RELATION TO THE WANTS OF OTHER CITIES, AND OF PUBLIC AND PRIVATE GARDENS.

New and Revised Edition, with many Illustrations. 8vo. 18s.

"For a long time we have not read a more interesting and instructive book than this."—*The Times.*

HARDY FLOWERS.
"A minute Encyclopædia."—*Saturday Review.*

THE WILD GARDEN;
OR, OUR GROVES AND SHRUBBERIES MADE BEAUTIFUL BY THE NATURALISATION OF HARDY EXOTIC PLANTS.

With many Illustrations by Alfred Parsons.

THE SUB-TROPICAL GARDEN;
OR, BEAUTY OF FORM IN THE FLOWER GARDEN.

Second Edition, with Illustrations of all the finer Plants used for this purpose

INTRODUCTION.

THIS book is written to dispel a very general but erroneous idea, that the exquisite flowers of alpine regions cannot be grown in gardens. There are few who have not heard of or beheld the beauty and vividness of colour of alpine flowers; but such knowledge is usually accompanied by the conviction that these can only be seen upon the high Alps, and that it is impossible to cultivate them in lowland regions. This erroneous idea is not confined to the general public; it has been propagated by our most famous botanists and horticulturists past and present, whenever they have had to figure or allude to mountain flowers; while almost every Alpine traveller has lugubriously regretted that we could not enjoy in our gardens these most charming of all flowers.

The Duke of Argyll, presiding, two or three years since, at the dinner of the Gardeners' Royal Benevolent Institution, told the company an anecdote about the great interest felt by the Queen in some alpine flowers gathered in a highland excursion, in which he had accompanied her, and he took the opportunity of telling the crowd of assembled horticulturists that, though they had overcome almost every difficulty of cultivation, they were conquered by one—that of growing alpine plants. Any reader of this book may satisfy himself that this idea is as unfounded as it is general, and that intelligent cultivation will prove as successful with the plants of the coldest and most elevated regions in our open gardens as it has already proved with the choicest plants of steaming tropical forests in hothouses. So far from its being true that high mountain plants cannot be cultivated, there is no alpine flower that ever cheered the traveller's eye with its brilliancy that cannot be successfully grown in these islands.

What are alpine plants? The word *alpine* is used in an arbitrary sense to denote the vegetation that grows naturally on the most elevated regions of the earth—on all very high mountain-chains, whether they spring from hot tropical plains or green northern pastures. Above the cultivated land these flowers begin to occur on the fringes of the stately woods; they are seen in multitudes in the vast and delightful pastures with which many great mountain-chains are robed, enamelling their soft verdure with innumerable dyes; and where neither grass nor loose herbage can exist; where feeble world-heat is quenched by mightier powers; where mountains are crumbled into ghastly slopes of shattered rock by the contending forces of heat and cold; even there, amidst the glaciers, they brilliantly spring from Nature's ruined battle-ground, as if the mother of earth-life had sent up her sweetest and loveliest children to plead with the spirits of destruction.

Alpine plants fringe the vast fields of snow and ice of the high hills, and at great elevations have often scarcely time to flower and ripen a few seeds before they are again imbedded in the snow; while sometimes, if the previous year's snow has been very heavy, and the present year's heat below the average, numbers of them may remain beneath the surface for more than a year. Enormous areas of the earth, inhabited by alpine plants, are every year covered by a deep bed of snow. Where the tall tree or shrub cannot exist from the intense cold, a deep soft mass of down-like snow settles upon these minute plants, like a great cloud-borne quilt, under which they safely rest, untortured by the alternations of frost and biting winds with moist and spring-like days.

But let it not for a moment be supposed that these conditions are indispensable for their growth! The reason that they predominate in these very elevated regions is because no taller vegetation can exist there; were these places inhabited by trees and shrubs, we should find few alpine plants among them; on the other hand, were no stronger vegetation found at a lower elevation, these plants would there make their appearance. Many plants found on the high Alps, and popularly believed to grow only near or among fields of snow, are also met with in rocky or bare places at much lower elevations. *Gentiana verna*, for example, one of the loveliest gems in the Flora of the Alps, often flowers very late in summer when the snow thaws on a very high mountain; yet it is also found on comparatively low hills, and occurs in England and Ireland. Numbers of other plants could be mentioned of which the same is true. In the close struggle for existence upon the plains and low tree-clad hills, the more minute species are often overrun by trees, trailers, bushes, and vigorous herbs, but, where in northern and elevated regions these fail from the earth, the choicer jewellery of vegetable-life known as alpine plants prevail.

Alpine plants possess the great charm of endless variety. They include subjects from many widely separate divisions of the vegetable kingdom, embracing endless diversities of form and colour. Among them are tiny orchids, as interesting as their tropical brethren, though so much smaller; Lilliputian trees, and even a tree-like moss (*Lycopodium dendroideum*), that branches and grows into an erect little pyramid, as if in imitation of the mountain-loving Pines, which, in their massy strength, are often tortured and depressed by storms, but rarely submit to become miniatures of what they are in lower regions. There are ferns that peep from narrowest crevices of high rocky places, often so diminutive that they seem to cling to the rocks for shelter, not daring to throw forth their fronds with airy grace as they do in more favourable positions. Numerous too are alpine bulbous plants, from Lilies to Bluebells, which appear to have been refined in Nature's laboratory,—all coarseness and ruggedness eliminated, all preciousness and beauty retained. There are evergreen shrubs, perfect in leaf and blossom and fruit, yet so small that an inverted finger-glass would make a roomy conservatory for them. There are exquisite creeping plants, rarely venturing much above mother earth, yet trailing and spreading freely along it, and, when they creep over the brows of rocks or stones, draping them with curtains of colour as lovely as any afforded by the most vigorous climbers of tropical forests. There are numberless minute plants that scarcely exceed the mosses in size, and quite surpass them in the way in which they mantle the earth with fresh green carpets in the midst of winter; and "succulent" plants in endless variety, which yield not in beauty to those of America or the Cape, though frequently smaller than the mosses of our bogs: in a word, alpine vegetation embraces nearly every type of the plant-life of northern and temperate climes, chastened in tone and diminished in size, and infinitely more attractive to the human eye than any other known, forming "a veil of strange intermediate being; which breathes, but has no voice; moves, but cannot leave its appointed place; passes through life without consciousness, to death without bitterness; wears the beauty of youth without its passion; and declines to the weakness of age without its regret."

With reference to the merits of "alpine" and allied types of gardening, as compared with those commonly in vogue, there can be little doubt in the minds of all who give the subject any thought. On the one hand, we have sweet communion with Nature; on the other, the process which is commonly called "bedding out" presents to us simply the best possible appliance for depriving vegetation of every grace of form, beauty of colour, and vital interest. The genius of cretinism itself could hardly delight in anything more tasteless or ignoble than the absurd daubs of colour that every summer flare in the neighbourhood of most country-houses in Western Europe. Enter the garden of a rich amateur, who spends a small fortune on his flowers, say in the neighbourhood of Liverpool or Lyons. You find orchids from Mexico and the Eastern Archipelago; the beauties of the

Flora of New Holland, as healthy as ever they were in their native homes; tropical fruits perfect in flavour and size, ferns gathered in many climes, and exotics from all parts of the world; but mention the name of some long-discovered native of North Europe or Siberia, hardy as Ivy and beautiful as numbers of the most gorgeous exotics, and in all but extremely rare cases the owner will never even have heard of it! Visit any of our large country gardens, and probably the first thing that will be triumphantly told you is the number of scores of thousands of plants "bedded out" every year, though no system ever devised has had a more miserable effect on our gardens. Even our great botanic gardens, which ought beyond all others to teach us the capabilities of the plants of our own climes, do not exhibit anything better than the gaudiness of great masses of flowers of the same colour on the one hand, and the repulsive formality resulting from so-called "scientific" arrangement of plants on the other.

Numbers of amateurs who cultivate numerous hot-house plants, and who generally have not a dozen of the equally beautiful flowers of northern and temperate regions in their gardens, might grow an abundance of them at a tithe of the expense required to fill a glass-house with costly Mexican or Indian orchids. Our botanical and great public gardens, in which alpine plants are usually found in frames in obscure corners, or which perhaps contain a few dozen of indifferent kinds on some absurdly-formed rockwork (half hidden under trees and shrubs, or a canvas roller-blind, as if very properly ashamed of itself), might each exhibit a beautiful alpine-garden, at half the expense and trouble which are now bestowed on some tropical family displayed in an enormous glass-house. In a word, there is not a garden of any kind, even in the suburbs of our great cities, in which the flowers of alpine lands may not be grown and enjoyed. And every person who makes himself a garden of them may be assured that, more than of any kind of garden he has ever seen, he will say of it,—" A garden is a beautiful book, writ by the finger of God: every flower and every leaf is a letter. You have only to learn them—and he is a poor dunce that cannot, if he will, do that—to learn them and join them, and then to go on reading and reading. And you will find yourself carried away from the earth by the beautiful story you are going through. You do not know what beautiful thoughts grow out of the ground, and seem to talk to a man. And then there are some flowers that seem to me like overdutiful children: tend them but ever so little, and they come up and flourish, and show, as I may say, their bright and happy faces to you."

W. R.

LONDON, *March 17th*, 1870.

Preface to the Second Edition.

Since the publication of the first edition of "Alpine Flowers," the culture of alpine plants has become much more popular in gardens, and, from the large collections which have been brought together in nurseries, they would seem to be growing in favour. More species are now in good health in our gardens than at any former time. The present edition has been considerably altered in the cultural section, and many additional illustrations added, these being for the most part sketched from actual scenes selected by the author. The alphabetical arrangement, or second portion, was stereotyped, and remains for the present as it was first published. The third part, in which selections of the species suited for various purposes are given, has been re-written and made more comprehensive and useful. The acknowledgments of the author are due to Mr. Jas. C. Niven, Curator of the Botanic Gardens at Hull; Mr. Backhouse, of York; Mr. George Maw, of Benthall Hall; Mr. James Atkins, of Painswick; the Rev. H. W. Ellacombe, of Bitton; and to many others, for much useful information as to the species enumerated in the book, and for other kindly help.

24th March, 1875.

ALPINE FLOWERS.

PART I.

CULTURAL AND STRUCTURAL.

In treating of the culture of alpine plants, the first important consideration is that much difference exists among them as regards constitution and vigour. We have, on the one hand, a number of valuable subjects that merely require to be sown or planted in the roughest way to flourish—the common Arabis and Aubrietia for example; but, on the other, there are many kinds, like Gentiana verna, and the Primulas of the high Alps, with many of their beautiful companions near the perpetual snows, which we rarely or never see in good health in these islands, or elsewhere, in gardens. It is as to the less vigorous species that advice is chiefly required. Nearly the whole of the misfortunes which these little plants have met with in our gardens are to be attributed to a false conception of what a rockwork ought to be, and of what the true alpine plant requires. These plants live on high mountains; therefore it is erroneously thought they will do best in our gardens if merely elevated on such tiny heaps of stones and brick rubbish as we frequently see piled together and dignified by the name of "rockwork." Mountains are often "bare," and cliffs are usually devoid of soil; but we must not conclude from this that the choice jewellery of plant life scattered over the ribs of the mountain or the interstices of the crag lives upon little more than the mountain air and the melting

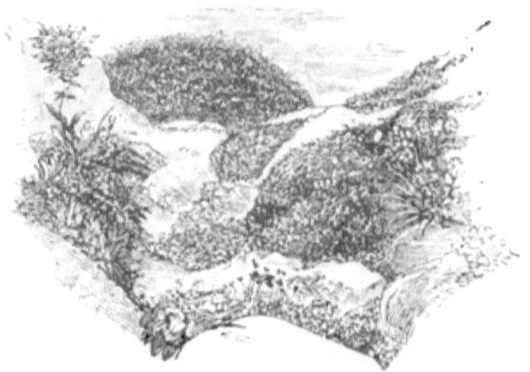

Alpine Plants growing at the bottom of a sloping ridge.

snow. Where will you find such a depth of well-ground stony soil, and withal such perfect drainage, as on the ridges of *débris* flanking some great glacier, stained all over with tufts of crimson Saxifrage? Can you gauge the depth of that narrow chink, from which peep tufts of the diminutive and beautiful Androsace helvetica? No; for ages and ages it has gathered the crumbling grit and scanty soil, into which the roots enter so far that nothing the tourist carries with him can bring out enough of them to enable the plant to exist elsewhere. And suppose we find plants growing apparently from mere cracks without soil; if so, the roots simply search farther into the heart of the flaky rock, so that they are safer from any want of moisture than in the best and deepest soil.

In 1868 I met on the Alps with plants not more than an inch high, and so firmly rooted in crevices of half-rotten slaty rock that any attempt to take them directly out would have proved futile. But, by carefully knocking and peeling away the sides from some isolated bits of projecting rock, I succeeded in laying the roots quite bare, and found them radiating in all directions against a flat rock, some of the largest being more than a yard long. We think it rapacious of the Ash, a towering forest tree, to send its roots under the walls of our gardens and rob the soil therein, and are surprised at finding the roots of a tree more than a hundred feet high descending a fifth or a sixth of that distance into the ground; but here is an instance of a plant one inch high penetrating into the earth to a distance forty times greater than it ventures into the alpine air! And there need be no doubt whatever that even smaller plants descend quite as deep, or even deeper, though it is rare to find the texture and position of the rock such as will admit of tracing them. It is true you occasionally find hollows in fields of flat hard rock, into which moss and leaves have gathered for ages, and where, in a sort of basin, without an outlet of any kind in the hard mountain, shrubs and plants grow freely enough; but in exceptional droughts they are just as liable to suffer from want

Mountain flank in process of degradation.

of water as they would be in our plains. On level or sloping spots of ground in the Alps the earth is of great depth, and if it be not all *earth* in the common sense of the word, it is more suitable to the plants than what we commonly understand by that term. Stones of all sizes broken up with the soil, sand, and grit, greatly tend to prevent evaporation; the roots lap round them and follow them deeply down. While in such positions, they never suffer from want of food and moisture, or vicissitudes of weather. Stone, it need scarcely be remarked, is a great preventive of evaporation, and shattered stone forms the soil as well as the subsoil of the mountain flanks where the rarest alpine plants abound. It should also be taken into account that the degradation so continually effected by melted snow water and heavy rains in summer, serves to earth up, so to speak, many alpine plants. I have torn up tufts of them showing this in a marked manner, for the remains of many generations of the old plants were seen buried and half buried in the soil beneath their descendants. This would, of course, be effected to some extent by the decaying of the plants themselves, but very frequently grit and peat are washed down plentifully among them, and, in cases where these do not come so thickly as to overwhelm them completely, they thrive with unusual luxuriance.

Now, if we consider how dry even our English air becomes in summer, and that no positions in our gardens afford such moist and cool rooting-places as those described, the necessity of giving to alpine plants a treatment quite different from what has hitherto been in vogue will be fully seen. The only sound principle generally employed is that of elevating the plants above the level of the ground. Naturally protected in winter by a dry bed of thick snow, some of them cannot exist on our wet soils in that season, if not raised well above the level. But this principle of elevation should in all cases be accompanied by the more essential one of giving the plants abundant means of rooting deeply into good and perfectly firm soil—sandy, gritty, peaty, or mingled with broken stone, as the case may be. How *not* to do this is capitally illustrated by persons who stuff a little soil into a chink between the stones in a rockery, and insert some minute alpine plant in that. There is usually a vacuum between the stones and the soil beneath them, and the first dry week sees the death of the plant—that of course not being attributed to the right cause. Precisely the same end would have come of it if the experiment had been tried on some alp bejewelled with Gentians and Primulas! Every plant of these two brilliant families should have means of rooting a yard or more into a suitable medium. We should not pay so much attention to the stones or rocks as to the earth from which they protrude. There are certainly alpine plants that do not require a deep soil, or what is usually termed soil at all; but all require a firm roomy medium for the roots.

In numbers of gardens an attempt at "rockwork" of some sort has been made; but in nine cases out of ten, the result is simply ridiculous; not because it is puny when compared with Nature's work in this way, but because it is generally so arranged that rock-plants cannot exist upon it. The idea of rockwork arose at first from a desire to imitate those natural croppings out of rocks which in temperate and cold countries are frequently covered with a dwarf but beautiful vegetation. It is strange that the conditions which surround these, and their texture and position, should rarely be taken into account by those who make rockwork in gardens. Numerous places occur in every county in which a sort of sloping stone or burr wall passes as "rockwork," a dust of soil being shaken in between the stones, and the whole so arranged that, if you do cover it with suitable plants, they perish speedily. In others, made upon a better plan as regards the base, the "rocks" are all stuck up on their ends, and so close that soil, or room for a plant to root, is out of the question. The best thing that usually happens to a structure of this sort is that its nakedness gets covered by a Cotoneaster, or some friendly climbing shrub, or some rampant weed, of course to the exclusion of true rock-plants; but in most cases the attempted rockwork is a standing eyesore.

Corner of a ledge of natural rock covered with Alpine Plants.

In moist and elevated districts, where frequent rains and showers keep porous stone in a continually humid state, this straight-sided, stone-wall-like rockwork may manage to support a few plants; but in by far the larger portion of the British Isles it is quite useless, and always ugly. It is not alone because the mountain air is pure and clear and moist that the Gentians and like plants prefer it, but because the elevation is unsuitable to the coarser-growing vegetation; and the alpines have it all to themselves. Take a healthy patch of Silene acaulis, by which the summits of some of our highest mountains are sheeted over with rosy crimson of various shades, and plant it two thousand feet lower down in suitable soil, keeping it moist enough and free from weeds, and you may grow it to perfection; but leave it to Nature in the same neighbourhood, and the strong grasses and herbage will soon run through and cover it, excluding the light, and finally and quickly killing the hardy and vigorous but diminutive Moss Campion.

Although hundreds of brilliant alpine flowers may be grown without a particle of rock near them, yet the slight elevation given by rockwork is very congenial to numbers of the most valuable kinds. The effect of a tastefully disposed rock-garden is very desirable in garden scenery. It furnishes a home for many pretty native and other interesting plants, which may not safely be put elsewhere; and therefore it is important that the most essential principle to be borne in mind, when making it, should be generally known.

The chief mistake generally made is that of not providing a feeding-place for the roots of the plants that are to embellish the rockwork. On ordinary rockwork even the coarsest British weeds cannot find a resting-place, simply because there is no motherly body of soil or matter into which the descending roots can penetrate, and find nourishment sufficient to keep the plant fresh and bright and well in all weathers. It is not only those who make their "rockwork" out of spoilt bricks, cement, and perhaps clinkers, that err in this respect, but the designers of some of the most expensive works in the country. At Chatsworth, for instance, and also to some extent at the Crystal Palace, you see rockwork not offensive so far as its distant effect in the landscape is concerned; but, when examined closely, it might well be imagined that rockwork and rock plants were never intended for each other's company, so bare are these large works of their proper and best ornaments. They are, for the most part, pavements of small stones, huge masses of rock, or imitation rock, formed by laying cement over brickwork, and in none of these cases are they adapted for the cultivation of high mountain plants.

It is quite possible to combine the most picturesque effects of which rockwork is capable with all the requirements for plant-growing; but, in the case of extensive rockwork-making for the sake of its picturesque effect, the owner must either call to his aid a landscape gardener of some skill in this way, or possess much taste and knowledge of the work himself. It is easy to use the largest stones and make the boldest prominences, and leave at the same time rather level intervening spaces of rocky ground in which rock-plants may luxuriate.

POSITION FOR THE ROCK-GARDEN.

The position selected for the rock-garden should never be near walls; never very near a house; never, if possible, within view of formal surroundings of any kind. It should generally be in an open situation; and of course a diversified spot, or one with bold prominences, should be selected, if available. No efforts should be spared to make all the surroundings, and every point visible from the rockwork, as graceful, quiet, and natural as they can be made. The part of the gardens around the rockwork should be picturesque, and, in any case, display a careless wildness resulting from the naturalization of beautiful hardy herbaceous plants, and the absence of formal walks, beds, etc. No tree should occur in or very near the rock-garden; hence a site should not be selected where it would be necessary to remove valuable or favourite specimens. The roots of trees would be almost sure to find their way into the masses of good soil provided for the choicer alpines, and thoroughly exhaust them. Besides, as the choicest alpine flowers are usually found on treeless and even bushless wastes, it is certainly wrong to place them under trees or in shaded positions, as has generally hitherto been their fate. It need hardly be added that it is an unwise practice to plant pines on rockwork, as has been lately done in Hyde Park and many other places. In large rock-gardens rhododendrons may be planted, if desired, without letting them occupy the surface suitable for alpine vegetation. It will, however, generally be in good taste to have some graceful, tapering young pines planted near, as this type of vegetation is usually to be seen on mountains, apart altogether from their great beauty and the aid which they so well afford in making the surroundings of the rock-garden what they ought to be. In small places, and in those where from unavoidable circumstances the rock-garden is made near a group of trees, the roots of which might rob it, it would be found a good plan to cut them off by a narrow drain, descending as deep as, or somewhat deeper than, the roots of the trees; this should be filled with rough concrete, and it will form an effectual barrier.

MATERIALS.

As regards the kinds of stone to be used, if one could

choose, sandstone or millstone grit would perhaps be the best; but it is seldom that a choice can be made, and happily almost any kind of stone will do, from Kentish rag to limestone; soft, slaty, and other kinds liable to crumble away, should be avoided, as also should magnesian limestone. It can hardly be necessary to add that the stone of the neighbourhood, if not very unsuitable, should be adopted for economy's sake, if for no other reason. Wherever the natural rock crops out, it is sheer waste to create artificial rockwork instead of embellishing that which naturally occurs. In the Central Park at New York there are scores of noble and picturesque breaks of rock, which have not been adorned with a single alpine flower or rock bush. Something of the same kind might be said of many of our country seats. In many cases of this kind nothing would have to be done but to clear the ground, and add here and there a few loads of suitable soil, with broken stones, etc., to prevent evaporation; the natural crevices and crests being planted where possible. Cliffs or banks of chalk, as well as all kinds of rock, should be taken advantage of in this way; many plants, like the dwarf Campanulas, Rock Roses, etc., thrive vigorously in such places. No burrs, clinkers, vitrified matter, portions of old arches and pillars, broken-nosed statues, etc., should ever obtain a place in a garden devoted to alpine flowers. Stumps and pieces of old trees are quite as objectionable as any of the foregoing materials; they are only fitted to form supports for rough climbers, and it is rarely worth while incurring any expense in removing or arranging them. Begin without attempting too much. Let your earliest attempts at "the first great evidences of mountain beauty" be confined to a few square yards of earth, with no protuberance more than a yard or so high. Be satisfied that you succeed perfectly with that before you try anything more ambitious. Never let any part of the rock-garden appear as if it had been shot out of a cart. The rocks should all have their bases buried in the ground, and the seams should not be visible; whenever a vertical or oblique seam of any kind occurs, it should be crammed with earth, and the plants put in this will quickly hide the seams. Horizontal fissures should be avoided as much as possible; they are only likely to occur in vertical faces of rock, and these should be avoided except where distant effect is sought. No vacuum should exist beneath the surface of the soil or surface-stones. The *detritus*, etc., should be so disposed that a vacuum cannot exist. Myriads of alpine plants have been destroyed from want of observing this precaution, the open crevices and loose texture of the soil permitting the dry air to destroy them in a very short time.

Mound of earth, with exposed points of rock.

In all cases where elevations of any kind are to be formed, the true way is to obtain them by means of a mass of soil suitable to the plants, putting a rock in here and there as the work proceeds; frequently it would be desirable to make these mounds of earth without any strata or "crags." The wrong and the usual way is to get the desired elevation by piling up arid masses of rock.

HIDDEN WEALTH.

While many go to great expense in embellishing their grounds with huge masses of artificial rock, made of old bricks and cement, and while many more are satisfied with the old bricks themselves, accompanied by clinkers and a great variety of offensive rubbish, very few trouble themselves about the rock treasures that often lay beneath the sod. Considering the large sums that are spent in sham rocks, etc., and the vast superiority in every way of natural rock, masses of it are most valuable to those who care for the picturesque in garden or park scenery. The illustration on the next page gives a feeble notion of one of the rocks that a friend of mine has succeeded in unearthing. The place originally was somewhat liberally embellished with rock on the surface; but the owner was anxious for more; in fact, he is like those "boys" out West who hunt for gold mines for years at a time. What tool he does his "prospecting" with, we are not certain; but by some means he ascertained the presence of ten feet of sand by the side of one huge mass of treasure. Then, by digging out the earth, he has formed a beautiful gorge between two flanks of rock that would reduce the cement-rock artist to despair; and

Unearthed Rocks in a Sussex Garden.

by clearing away the earth from the flanks of that nose of rock that just projects above a grassy knoll, he has discovered beautiful wrinkles, crevices, and other charms in it. Thus by a little persevering searching and digging has been produced a scene as striking and interesting as many in an alpine country, and one which offers such a variety of aspects and positions that every kind of hardy plant may be grown on it in the best manner, and arranged on it with the happiest effect. This subject is of the highest importance to the many who have ground on a rocky base, and who would be glad if this most precious stonework were brought to light.

PATHWAYS, ETC.

No formal walk—that is to say, no walk with regularly-trimmed edges of any kind—should ever be allowed to pass through, or even come near, the rock-garden. This need not prevent the presence of properly-made walks through or near it, as, by allowing the edges of the walk to be a little irregular and stony, and by permitting dwarf Sedums, Saxifrages, Linaria alpina, the lawn Pearl-wort, etc., to crawl

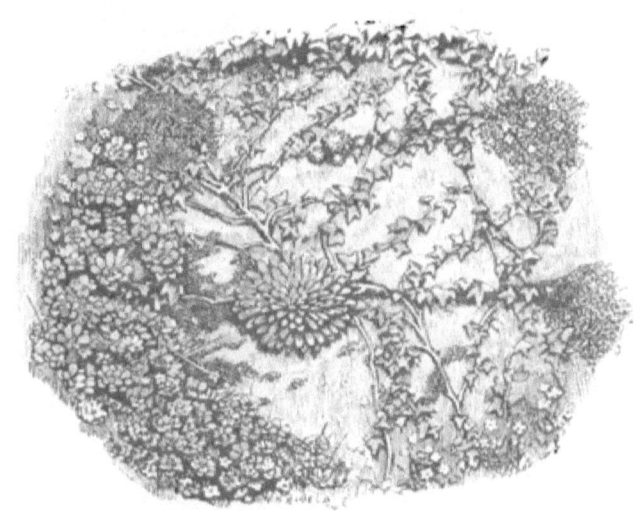

Vertical face of rock covered with narrow-leaved Ivy, and with various Alpine Plants in the chinks. (*From a Photograph.*)

into the walk at will, a perfectly unobjectionable effect will be produced. In every case where gravel walks pass through ferneries or rockeries, and are fringed by stonework, a variety of little plants should be placed at the sides, and allowed to crawl into the walk in their own wild way. There is no surface whatever of this kind that may not be thus embellished with interesting subjects. Violets and ferns, Myosotis dissitiflora, etc., will answer for the moister and shadier parts, and the Stonecrops, Saxifrages, Arenarias, and many others, will thrive in more arid parts and in the full sun. The whole of the surface of the alpine garden should be covered with plants, except the projecting points or crags; and even these should be covered, as far as possible, without completely concealing them. In moist districts, such alpines as Erinus alpinus and Arenaria balearica will grow wherever there is a resting-place for a seed on the face of the rocks; and even tall and vertical faces of rock may be embellished with a variety of plants; so that there is no reason whatever why any level surfaces of ground should be bare.

Passage in Rock-garden.

CONSTRUCTION.

In no case should regular steps be permitted in or near the rock-garden. Steps may be made quite irregular, and not only not offensive to the eye, but very beautiful, with violets and other small plants jutting from every crevice. No cement should be used in connection with the steps. The figure on the opposite page is from a photograph of the lower part of rude steps ascending abruptly from a deep and moist recess in a rock garden. It shows very imperfectly— no engraving could show it otherwise—the crowds of lovely plants that gather over it, except where worn bare by feet, thriving year by year as freely as they do on the most favoured spots in the Alps. In cases where the simplest type of rockwork only is attempted, and where there are no steps or rude walks in the rock-garden, the very fringes of the gravel walks may be gracefully enlivened by allowing such plants as the dwarfer Sedums to become established in them. The alpine Linaria is never more beautiful than when self-sown in a gravel walk. Rockwork which is so made that its miniature cliffs, etc., overhang, is useless for alpine vegetation; and all but such wall-loving subjects as Corydalis lutea quickly perish on it. The tendency to make it with overhanging peaks is everywhere seen in the cement rock-gardens now becoming rather common. Into the alpine garden this species of construction should never be admitted, except to get the effect of bold and distant cliffs, where this is desired and cannot be obtained in a more natural manner. When this system is admitted, the designer should be requested to obtain his picturesque effect otherwise than by making all his cliffs and precipices overhang. It is erroneous to suppose that heaps of stones or small rocks are absolutely necessary for the health of alpine plants. The great majority will thrive without their aid if the soil be suitable; and though all are benefited by them, if properly used as elsewhere described, it is important that it should be generally known how needless is the common system of inserting mountain plants among loose stones, burrs, etc. Half-burying rocks or stones in the earth round a rare species, which it is intended to save from excessive evaporation, and which has a deep body of soil to root into, is, however, quite a different and an excellent practice.

Rude stair from deep recess of Rock-garden, mossed over with Alpine Flowers.

SOIL.

The great majority of alpine plants thrive best in deep soil. In it they can root deeply, and when once they are so rooted they will not suffer from drought, from which they would quickly perish if planted in the usual way. Three feet deep is not too much for most species, and it is in nearly all cases a good plan to have plenty of broken sandstone or grit mixed with the soil. Any good free loam, with plenty of sand, broken grit, etc., will be found to suit the great majority of alpine and dwarf herbaceous plants, from Pinks to Gromwells. But peat is required by some, as, for example, various small and brilliant rock-plants like the Menziesias, Trillium, Cypripedium, Spigelia marilandica, and a number of other mountain and bog plants. Hence, though the general mass may be of the soil above described, it will be desirable to have a few masses of peat here and there. This is better than forming all the ground of good loam and then digging holes in it for the reception of small masses of peat. The soil of one or more portions might also be chalky or calcareous, for the sake of plants that are known to thrive best on such formations, as the pretty Polygala calcarea, the Bee Orchis, Rhododendron Chamæcistus, etc. Any other varieties of soil specially required by individual kinds can be given as they are planted. In the second part of this book the soil suitable for each plant there described is mentioned in its proper place.

WATER.

It is not well to endeavour to associate a small lakelet or pond with the rock-garden, as is frequently done. I do not remember to have met in alpine countries with any crowds of brilliant alpine flowers in the vicinity of small pools of grimy water; indeed, they usually crowd on fields high above the lake. If a picturesquely-arranged piece of water can be seen from the rock-garden, well and good; but water should not, as a rule, be closely associated with it. Hence, in places of limited extent it should not be thought of at all. If a pure rushing streamlet, with one or more cascades, can be introduced near the rock-garden with good effect, so much the better; but these things are better treated as incidental features.

Where a large rock-garden is being made, and where expense is no object, water should, if possible, be "laid on," as, without command of a strong pressure and a liberal hose, it is very difficult to water an extensive elevated rock-garden thoroughly, and very troublesome and expensive even to do it badly with watering-pots, etc. Several taps or outlets will be required in large rock-gardens.

We will now enter into particulars as to the various ways in which alpine plants may be grown, beginning with the best type of rock-garden—that in which (in addition to the low-lying, stony, and rocky banks and slopes, where numbers of hardy and vigorous species may be grown) there are

Right.

miniature peaks, cliffs, and ravines, with perhaps bog and water. The most usual and deplorable of the faults in making rockwork is that of so arranging the stones that they

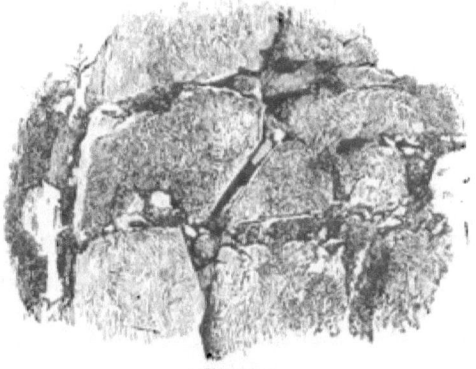

Wrong.

seem to have as little connection with the soil of the spot as if thrown out of a cart—indeed less so. Instead of allowing what may be termed the foundations, or apparent foundations, of the rock-garden to barely show their upper ridges above the earth, and thereby suggesting much more endurable ideas of "rock" than those arising from the contemplation of the bold and unnatural-looking masses usually seen, the stones are often placed on the ground with much the same idea that animates a bricklayer in setting bricks. The two cuts on page 15 will explain exactly what we mean; both are accurately engraved from photographs, both represent small portions of artificial rockwork—the ugliest of the two was much the most difficult and expensive to make. One well-selected stone allowed to peep from some gently rising isolated mound or open sunny spot, and arranged as shown in the accompanying little cut, would produce a better effect than several tons placed as in the lower figure on preceding page.

Half-buried Stone surrounded by Alpine Plants.

The surface of every part of the rock-garden should be so arranged that all rain will be directly absorbed by it; here, again, the objection to precipitous and overhanging faces holds good. If the elevations are obtained, as they should be, by gradually receding, irregular steps, rather than by abrupt crags, walls, etc., all the plants on the surface will be equally refreshed by rains. The illustrations on the next page will serve to show what is right and what is wrong in this respect. The upper surfaces of crags, mounds, etc., should in all cases be of earth, broken stones, grit, etc., as indeed should every spot where projecting stones or rock are not required for the sake of effect. All the soil-surfaces of the rock-garden should be protected from excessive evaporation by finely broken stones, pebbles, or grit scattered on the surface, or by means of small pieces of broken sandstone or millstone half buried in the ground.

If we merely want a certain surface of rock disposed in a picturesque way, such details as these may not be worthy of attention, but if we wish our rock-gardens to be faithful miniatures of those wild ones which are admitted to be the

Well-formed Sloping Ledges.

Artificial Rock on which plants do not thrive.

most exquisite of nature's gardens, then they are of much importance.

In dealing with the construction of the bolder masses of rockwork, we cannot have a better guide than Mr. James Backhouse, of York, to whom we are indebted for the following remarks :—

Lychnis and Silene in fissures.

"Comparatively few alpines prefer or succeed well in horizontal fissures. Those, however, which, like Lychnis Viscaria and Silene acaulis, form long tap roots, thrive well in such fissures, provided the earth in the fissure is continuous, and leads backward to a sufficient body of soil. Where the horizontal fissures are very narrow, owing to the main rocks being in contact in places, and leaving only irregular and interrupted fissures, such plants as the charming Lychnis Lagascæ, Lychnis pyrenaica, and others, bearing and preferring hot sunny exposures, do well. But many plants that would bear the heat and drought, *if* they could get their roots far enough back, would quickly die if placed in such fissures, from the want of soil and moisture near the front; therefore

Horizontal Fissure.

it is usually better, in building rockwork with these fissures, to keep the main rocks slightly apart by means of pieces of very hard stone (basalt, close-grained 'flag,' etc.), so as to leave room for a good intermediate layer of rich loam, stones, or grit, mingled with a little peat. The front view of such a structure would be as above—the dark spaces being firmly filled with the appropriate mixture of soil *before* the upper course of large rocks is placed.

As a rule, oblique and vertical fissures are both preferable to horizontal ones; but care should be taken with oblique fissures that the upper rock does not overhang. A plant

Wrong. Right.

placed at J will often die, when the same placed at H will live, because the rain falling on the sloping face of rock at I will drop off at J, and miss the fissure J altogether, while that falling on the sloping face of rock at K will all *run into* the fissure H. There are, however, some plants, like the rare Nothochlæna Marantæ and Androsace lanuginosa, which so much prefer positions dry in winter that a fissure like J would suit them better than one like H. Such, however, are rare exceptions to a general rule.

The best and worst general forms of steep rockwork we have tried are those indicated in the following figures. By making each rock slightly recede from the one below it, the rain runs consecutively into every fissure. Where the main fissures reverse this order, almost everything dies or languishes. Care should be taken to have the top made of mixed earth and stones—*not* of rock, unless use is intentionally sacrificed to scenic effect.

Vertical fissures (which suit many rare alpines best of all) should always, so far as possible, be made narrower at

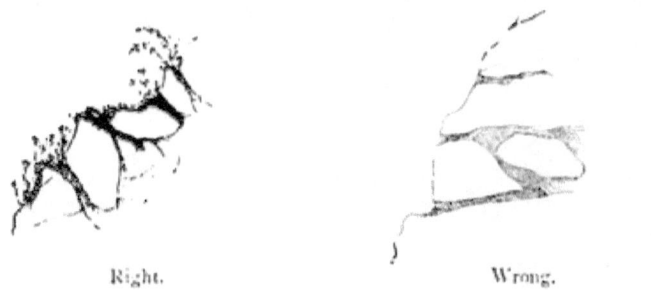

Right.   Wrong.

the bottom than at the top. If otherwise, the intervening earth, etc., leaves the sides of the rock as it 'settles,' instead of becoming tighter. In figure A, as the total mass of soil sinks, it becomes compressed against the sides of the rock; while in B, the soil *leaves* the sides of the fissures more and more as the mass sinks, and almost invariably forms distinct 'cracks' (separations between the soil and rock) sooner or later. The same principle applies to small stones and fissures. To prevent undue evaporation in the case of such fissures, stones, larger or smaller, may be laid on the *top* of the soil, care being taken not to cover too much of it, to the exclusion of rain.

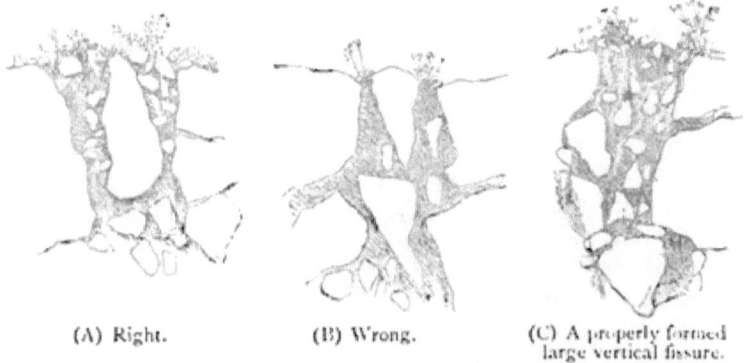

(A) Right.   (B) Wrong.   (C) A properly formed large vertical fissure.

Where a large fissure exists, the smaller pieces of stone *in* it are on this account best placed with the narrowest edge or point upwards (fig. C)—not downwards. It will easily be seen that the tendency of the mixed soil, both as a whole and in each of its subdivided parts, is to become more and more compressed by its own weight and by the action of rain."

In the construction and planting of every kind of rockwork, it should be distinctly remembered that every surface may and should be embellished with beautiful plants. Not alone on rocks or slopes, or favourable ledges, or chinks, or miniature valleys, should we see this kind of exquisite plant-life. Numbers of rare mountain species will thrive on the less trodden parts of footways; others, like the two-flowered Violet, seem to thrive best of all in the fissures between the rude steps of the rockwork; many dwarf succulents delight in gravel and the hardest soil, and numerous other plants will run wild in any wood or among low shrubs near the rock-garden.

Another very important principle to bear in mind in forming the rock-garden is, that, as a rule, much more vegetation than rocks should be seen. Where vast regions are inhabited by alpine plants, acres of crags with a stain of flower or fern here and there, are very attractive and imposing parts of the picture; but in gardens, where our creations in this way can only be Lilliputian, an entirely different method must be pursued; except in places where great cliffs are naturally exposed; and even in this case an abundant drapery of vegetation is desirable. A rockwork is rarely seen in which plants predominate as much as they ought. Frequently masses of stone, with an occasional tuft of vegetation, are met with under this name, every chink and joint between the stones being quite exposed. This should not be so; every minute chink should have its little line of verdure; and in this way we should not only have more plants, but hide the artificial nature of the structure. Where the ground is low and bank-like, there really is not the slightest necessity for placing stones all over the surface; an occasional one cropping up here and there from the mass of vegetation will produce the best effect. Alpine flowers are often seen in multitudes and in their loveliest aspect in some little elevated level spot, frequently without rocks being visible through it, and when they do occur merely peeping up here and there. They are lovely too in the desolate wastes of broken rock, where they cower down between the great stones in isolated, lonely-looking tufts; but it is only when Gentians and silvery Cudweeds, and minute white Buttercups, and strange large Violets, and Harebells that waste all their strength in flowers, and fairy Daffodils that droop their heads as gracefully as Snowdrops, are seen forming a

Ledge of Alpine Flowers (a Garden Sketch).

dense turf of living enamelled work, that alpine flowers are seen in all their beauty. Fortunately the flowery turf and stony mound are much more possible to us than the bare moraine blocks or arid cliffs.

In cultivating the very rarest and most minute alpine plants, the stony, or partially stony, surface is to be preferred. In their case we cannot allow the struggle for life to have its own relentless way, or we should often have to grieve at finding the Eritrichium from the high Alps of Europe overrun and exterminated by a dwarf American Phlox, and similar cases. Perfect exposure is also necessary to complete success with very minute plants, and the stones are very useful in preventing excessive evaporation from their roots. Few people have any conception of the great number of alpine plants that may be grown on the fully exposed level ground as readily as the common Chamomile; but there are, on the other hand, not a few that require some care to establish them, and there are usually new kinds to be added to the collection, which, even if vigorous ones, should be kept apart and under favourable conditions. Therefore, in every place where the culture of alpine plants is entered into with zest, there ought to be a select spot on which to grow the most

Alpine Plants growing on the level ground.

delicate, most rare, and most diminutive kinds. It should be fully exposed, and while sufficiently elevated to secure perfect drainage and all the effect desirable, should not be riven into miniature peaks or crags or cliffs.

The greatest watchfulness should be exercised over the plants on all such structures as this. They will not perish from cold or heat or wet, if properly planted, but many of them are so minute that they are not capable of affording a full meal to a browsing slug, and accordingly often disappear during a moist night. Now, as our gardens abound with slimy creatures that play havoc with many subjects colossal compared with our alpine friends, it is clear that one of the main points is to guard against slugs and snails, and as far as possible against worms. Mr. Backhouse has very cleverly fenced off the choicest parts of his rockwork from them by a very irregular little canal, so arranged and cemented that, while not an eyesore, it is perfectly watertight, and no slug can cross it. It thus becomes a much easier task to guard the plants from these enemies than when they are allowed to crawl in from all points of the compass. But even with this precaution, it is necessary to search continually for snails and slugs; and in wet weather the choicest plants should be examined in the evening, or very early in the morning; with a lantern, if at night. Sir Charles Isham, who is an enthusiastic cultivator of rock-plants, says that he not only protects toads, but does not forget to lay stones so as to form little retreats for them underneath. They prefer a stone just sufficiently raised to crawl under, and do a deal of good by destroying slugs, etc. He also protects frogs and all carnivorous insects. Ceaseless hand-picking, however, is the best remedy for slugs, and where this is not done, there is little hope of succeeding with many subjects, at least where slugs are as abundant as we usually find them in gardens.

CASCADES, ROCKY BRIDGES, ROCKY MARGINS, ISLETS.

As water is often introduced in connection with rockwork, and high cascades may be frequently attempted, and as the supply often flows from a woody knoll, it is well to take

Waterfall fringed with Yuccas, dwarf Pines, climbing and trailing plants.

advantage of this position for the arrangement of Yuccas, large grasses, herbaceous plants of noble port, and the like, that cannot well be arranged among the dwarf inhabitants of the rock-garden proper. Among the many plants suited for this position, the new Clematises raised by Jackman and others are the most magnificent. Planted high up on the rocks in a deep bed or vein of rich light soil, they will fall over the faces of the sunny crags, robing them as with imperial purple.

In connection with this subject, it may not be out of place

here to give a short description of Niagara, which is not only grand, as everybody knows, but also beautiful as a gigantic rock-garden. In fact, the noblest of Nature's gardens I have yet seen is that of the surroundings and neighbourhood of the Falls of Niagara; and very suggestive it is to those interested in forming artificial or improving natural cascades and the like. Grand as are the colossal falls, the rapids and the course of the river for a considerable distance above and below possess more interest and beauty.

Young Plants of Clematis falling over the face of Artificial Rock.

As the river courses far below the falls, confined between vast walls of rock—the clear water of a peculiar light-greenish hue, and white here and there with circlets of yet unsoothed foam—the effect is startlingly beautiful quite apart from the falls. The high cliffs are crested with woods; the ruins of the great rock-walls, forming wide irregular banks between them and the water, are also beautifully clothed with wood to the river's edge, often so far below that you sometimes look from the upper brink down on the top of tall pines that seem diminished in size. The wild vines scramble among the trees; many shrubs and flowers seam the high rocks; in moist spots here and there a sharp eye may detect many flowered tufts of the beautiful fringed Gentian, strange to European eyes, and beyond all, and at the upper end of the wood-embowered deep river-bed, a portion of the crowning glory of the scene—the falls—a vast cliff of illuminated foam, with a zone towards its upper edge as of green molten glass. Above the falls the scene is quite different, a wide and peaceful river carrying the surplus waters of an inland sea, till it gradually finds itself in the coils of the rapids, and is soon lashed into such a turmoil as we might expect if a dozen unpolluted Shannons or Seines were running a race together. A river no more, but a sea unreined. By walking about a mile above the falls on the Canadian shore this effect is finely seen, the breadth of the river helping to carry out the illusion. As the great waste of waters descends from its dark grey and smooth bed and falls whitening into foam, it seems as if tide after tide were gale-heaped one on

Bird's-eye view of Islands above the Falls of Niagara.

another on a sea strand. The islands just above the falls enable one to stand in the midst of these rapids where they rush by lashed into passionate haste; now boiling over some hidden swellings in the rocky bed, or dashing over greater but yet hidden obstructions with such force that the crest of the uplifted mass is dashed about as freely as a white charger's mane; now darkly falling into a cavity several yards below the level of the surrounding water, and, when unobstructed, surging by in countless eddies to the mist-crested falls below, and so rapidly that the drift-wood dashes on swift as swallow on the wing. Undisturbed in their peaceful shadiness, garlanded with wild vine and wild flowers, the islands stand in the midst of all this fierce commotion of waters—below, the vast ever-mining falls; above, a complication of torrents that seem fitted to wear away iron shores, yet there they stand, safe as if the spirit of beauty had in mercy exempted them from decay. Several islets are so small that it is really remarkable how they support vegetation; one looking no bigger than a washing-tub, not only holds its own in the very thick of the currents just above the falls, but actually bears a small forest, including one stricken and half cast-down pine. It looks a home for Gulliver in Brobdingnagian scenery. Most fortunate is it that these beautifully verdant islands and islets occur just above the falls, adding immeasurably to the effect of the scene. Magnificent it would have been without them, but their presence makes Nature seem as fair as terrible in her strength.

Where water occurs near the rock-garden, one or more little bridges are not unfrequently seen; but some such arrangement as that suggested in the accompanying woodcut

Stepping-stone Bridge, with Water Lilies and other Aquatic Plants.

would be more satisfactory and tasteful. It is, however, introduced here chiefly for the purpose of showing how well it enables one to enjoy various beautiful aquatic plants, from the fringed and crimson-tipped Bog-bean and graceful *Carex pendula* at the sides to the golden Villarsia and Water Lilies sailing among the stones. Arranged thus, a number of interesting plants not usually met with seem to crowd around for acquaintanceship. This mode of garden bridge-making, while infinitely more beautiful than the ordinary one, is less expensive. Care is, however, required to arrange it so that it may satisfy the eye, offer free passage to the water, and an easy means of crossing it at all times.

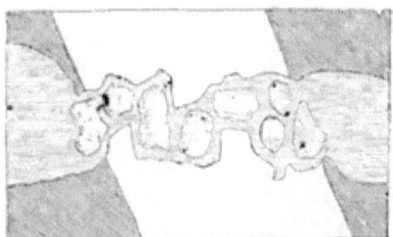

Plan of preceding figure.

Rockworks made on the margin of artificial water are very often objectionable—rigid, abrupt, unworn, and absurdly

Rock-garden near water, suited for bold and luxuriant types of vegetation.

unnatural. In no position is an awkwardness more likely to be detected; in none should more care be taken not to offend good taste. Charming effects may be produced on properly made rockwork near water, by planting it with a combination of choice moisture-loving rock-plants—Yuccas, Pampas Grass, and like subjects; but even the grace and beauty of the finest of these will not relieve the hideousness of the masses of brick-rubbish and stone that are frequently placed by the margins of water.

The next figure, showing the fringe of a little island in one of the lakes of Northern Italy, may serve to show how irregularly and prettily the waves carve the rocky shore. Frequently in such places diminutive islands from a few feet to a few yards across are seen, and, when tufted with Globe-

A glimpse at margin of Island in Lake Maggiore.

flowers, Ivy, Brambles, etc., are very charming. A few well-formed artificial islets may be introduced with good effect near a rocky margin.

Rocky Water-margin at Oak Lodge, Kensington.

THE ROCK-GARDEN FERNERY.

It is the fashion to make the hardy fernery in some obscure and sunless spot, in which it would be impossible to grow alpine plants, but there is no reason whatever why it should not be made in more open positions, and in connection with the rock-garden. No plants adhere more firmly to hard vertical surfaces, or better sustain themselves in perfect health without any soil, than ferns. In a wild state you find the Maidenhair Fern and many other species so rooted into mere little fissures in the hardest rocks that no effort can get out a particle of root. Some of our own small

British wild ferns are found on the face of dry brick walls when they are not to be found growing spontaneously on the ground, in the same neighbourhood.

The general idea is that ferns want shade, humidity, and sandy vegetable earth; but, though these suit a great number of ferns, others luxuriate under conditions the very opposite. M. C. Naudin, of the Institute, now settled down to carry out his experiments on the shores of the Mediterranean, informs me that the pretty little sweet-scented fern *Cheilanthes odora* is never found, even in that warm and sunny region, except on the south side of bare rocks and walls, where it is exposed to the full rays of the sun. It is sought for in vain on northern exposures, is rarely found to the east and west, and, when found there, is badly developed. Walls facing due south are covered with this little gem among ferns, and not a vestige of the species occurs on the opposite side. In the middle of winter it is in full vigour, by the end of spring the fronds begin to dry, and through the torrid summer, when the stones of the walls are burning hot, its roots, fixed between the hot stones, are the only parts with life. In humid valleys and recesses it is not found. Other ferns manifest analogous tendencies. This is merely by way of proof that some of the choicest ferns may not only be grown well in the most sunny and arid positions, but better on them than elsewhere.

I am informed by Mr. Atkins, of Painswick, who was the first to bring the charming little *Nothochlæna Marantæ* alive into this country, that he has had it in perfect health on a sunny rock for the last fourteen years, and without the least protection. It is reasonable to assume that many ferns, which in a wild state are found in half-shady spots, would, in our colder clime, flourish best if permitted to enjoy all the sun of our cloudy skies, while ferns that inhabit sunny rocks in countries not much warmer than our own should always have the warmest positions we can give them on the rockwork. And in the case of the species that require shade, it is quite possible to grow them in recesses in the rock-garden and in deep passages or miniature ravines leading through it, even if a portion be not specially designed as a fernery. Some small species and varieties may be used in any aspect as a graceful setting to flowering plants. The general subject of hardy fern culture is so well understood that there is no necessity for adverting to it here. Among the select lists, that of the ferns that thrive best in open exposed places may meet the wants of some, but where the fernery is specially designed as a part of the rock-garden, there is no necessity for any selection, as all hardy kinds may then be grown.

Even the rare Killarney fern, usually kept in houses, may be grown successfully in a cave in the rock-garden. The

Entrance to Cave for Killarney Fern.

illustration shows the entrance to Mr. Backhouse's cave for growing this plant. It is in a deep recess, perfectly sheltered and surrounded by high rocks and banks clothed with vegetation. Here in the darkness grows the Killarney fern, tufts of Hartstongue guarding the entrance. It is very likely that various kinds of New Zealand Trichomanes and Filmy ferns will prove as hardy as the Killarney fern, and, if so, this is likely to be one of the most attractive and interesting of all phases of out-door gardening.

ROCKWORK FORMED OF CONCRETE, CEMENT, ETC.

In connection with alpine gardens, the masses of rockwork occasionally made of brick-rubbish, concrete, and cement, demand some notice.

There can be no doubt that as picturesque effects may be produced in this way as in any other, and that this variety of artificial rockwork may be admirably associated with shrubs and trees, and vigorous climbing and trailing plants, but it is utterly unsuitable for true alpine vegetation. When properly constructed, care is taken to make the interior of the cemented masses with deep beds of earth, leaving holes here and there in the face of the structure from which plants can peep forth, while the top is left open, and may be planted with shrubs or trees. The new hybrid Clematises, with their noble flowers, will, if planted in these rich cases of earth and allowed to fall over the faces of the rocks, make an unrivalled display, and the position is also most suitable of all kinds of climbers, trailers, and shrubs; but the stony mound, free in every pore, or the rockwork constructed of separate pieces of stone, is infinitely the best

Masses of Artificial Rock.

for the small flora of the rocks. I have never seen on the large masses of cemented rock half the amount of beauty afforded in a few weeks after planting by the "alpine" bed shown further on. The plants that thrive luxuriantly on walls and old ruins, and send their roots far into the crevices of such, cannot obtain the slightest footing on these large masses coated with cement; and little plants stuck in the "pockets" which the constructors leave here and there on the face of the edifice rarely thrive or look happy. They should never be placed in such positions, and the rockwork made of natural stone should be preferred at almost any sacrifice. Where, however, natural stone cannot be obtained, the cemented work may be used with an excellent result to form the "peaks" and "cliffs" of the rock-garden, in the construction of cascades, etc., and in positions where only the distant and picturesque effect of rocks in garden scenery is sought. In places where it already exists, much improvement may be effected by the creation of patches of true alpine garden in open spots near the cemented rocks, covering the last gracefully with low shrubs and hardy shrubby climbers.

View of Artificial Rock at Oak Lodge.

The most successful of these cemented structures (as regards arrangement) that I have seen is the rockwork at Oak Lodge, Addison Road, Kensington; from which the next two illustrations, as well as that on page 28, have been sketched. The grounds here were laid out by Mr. Marnock, part of whose task consisted in converting a small formal pond and an ugly formal bank into something that would form a more pleasing feature in the landscape. The problem, however, was well solved. The ugly bank became a varied mass of picturesque rock, seamed with graceful ferns and trailing shrubs; the water fell into what seemed a natural hollow in the earth, and around it sprang up tufts of Iris and Yucca. Rich masses of specimen Rhododendrons crest the rocks. The rocks, in fact, form a sort of retaining wall for the masses of earth which accommodate these plants. Some old pear-trees and a pair of grand old Wych elms were carefully preserved, and greatly add to the effect of the scene. The mass of artificial rock is cleverly and artistically constructed, but, like all masses

Scene in the Gardens at Oak Lodge.

of the same species of rock, it is not suitable for alpine and rock-plants, etc. It would be easy, however, by depositing at its base a few cart-loads of soil interspersed with fragments of rock, properly arranged in the manner described in a previous page, to enjoy the pleasure of successfully cultivating an extensive variety of these plants. The rock-gardens at Kensington and many others throughout the country were formed by Mr. Pulham, who possesses good taste and skill in the work in a very high degree.

MINIATURE ROCK-GARDENS.

Hitherto we have chiefly considered the rock-garden on a somewhat extensive scale. As those who can afford this are less likely to want instruction than the much greater numbers who cannot, I propose now to treat of several successful modes of growing alpine flowers which may be carried out in the smallest gardens at a trifling expense. A well-arranged and well-planted alpine garden is somewhat costly, even where materials are easily obtained, and, moreover, requires much labour, skill, patience, and knowledge of plants to keep it in a perennially interesting condition. Local conditions, want of suitable materials, want of knowledge, and consequent want of interest in the plants, must, in many cases, prevent this most interesting phase of gardening from being enjoyed. I am therefore the more desirous to help the smaller and humbler attempts of those who cannot afford more than a very small patch of alpine garden, as well as to assist beginners of every class.

One of the simplest of all ways of cultivating alpine plants is in small rocky beds, arranged on the turf of some parts of the garden cut off by trees or shrubs from the ordinary flower-beds, without any of the pretensions of the ordinary rockwork; one of these will give much greater satisfaction than many a pretentious "rockwork," and by the exercise of a very little judgment is readily constructed so as not to offend the nicest taste. I once induced the owner of a garden in the northern suburbs of London to procure a small collection of alpines and try them in this way, and the result was so charming that a few words as to how it was attained may be useful.

A little bed was dug out in the clay soil to the depth of two feet, and a drain run from it to an outlet near at hand; the bed was filled with fine sandy peat and a little loam and leaf-mould, and, when nearly full, worn stones of different sizes were placed around the margin, so as to raise the bed on an average one foot or so above the turf. More soil was then put in, and a few rough slabs, arranged so as to crop out from the soil in the centre, completed the preparation for the neater Sedums and Sempervivums, such Saxifrages as *S. cæsia* and *S. Rocheliana*, such Dianthuses as *D. alpinus* and *D. petræus*, Mountain Forget-me-nots, Gentians, little spring bulbs, *Hepatica angulosa*, etc. They were planted, the finer and rarer things getting the best positions, and, when finished, the bed looked a nest of small rocks and alpine flowers.

In about eight weeks the plants had become established, and the bed looked quite gay from a dozen plants of *Calandrinia umbellata* that had been planted on the little prominences flowering profusely. This made the arrangement equal to one of bedding plants from the "effective" point of view. Another was made in the same manner, with more loam, however, and planted with subjects as different from those in the other bed as could be got;

Small rocky bed of Alpine Flowers.

confining them, however, to the choicest alpines, except on the outer side of the largest stones of the margin, where such plants as *Campanula carpatica bicolor* were planted with the best results.

The only attention these beds have required since planting has been to keep a free-growing species from overrunning a subject like *Gentiana verna*, to water the beds well in hot weather—to keep them in fact thoroughly moist—and to remove even the smallest weeds. With the exception of the exquisite *Gentiana bavarica*, every alpine plant grew well, and the beds presented fresh floral interest every week from the dawn of spring till late in autumn.

I have described the way by which this happy result has been brought about. An extended scheme of this sort would be admirable in some public garden, especially in those possessing large collections of alpine and herbaceous plants, from which many good selections could at once be made. Something of the sort might be made in any garden—nay, even in a London square, or in any other position fully exposed to the sun, and never under the shade and drip of trees.

Rockwork is, as a rule, made for the display of mountain plants, or those which by their dwarfness fall into the class commonly known as alpines. Some persons cover rockwork with climbing shrubs and dwarf bushes, but in every case, unless where a rock is introduced for its own effect in the landscape, the object is to grow plants. Now, as very few of the subjects above alluded to like shade, or even tolerate it, it follows that this is an ignorant and bad practice. Many persons who arrange such things doubtless fear the sun burning up their plants; yet the sun that beats down on the Alps and Pyrenees is fiercer than that which shines on the British garden. But, while the Alpine sun cheers the flowers into beauty, it also melts the snows above, and water and frost grind down the rocks into earth; and thus, enjoying both, the roots form perfectly healthy plants. Fully exposed plants do not perish from too much sun, but simply from want of water. Therefore it cannot be too widely known that full exposure to the sun is one of the first conditions of perfect rock-plant culture—abundance of free soil under the root, and such a disposition of the soil and rocks that the rain may permeate through and not fall off the rocks, being also indispensable.

The preceding plan can be carried out in the very smallest places. The next is quite as easily adopted on the fringe of any shrubbery. An open, slightly elevated, and, if possible, quite isolated spot should be chosen, and a small rock-garden so arranged as to appear as if naturally cropping out of the shrubbery. With a few cart-loads of stones and earth excellent effects may be produced in this way. The following illustration well explains my meaning: it represents an irregularly sloping border, with a few mossy bits of rock peeping from a swarming carpet of Sandworts, Mountain-pinks, Rock-cresses, Sedums and Saxifrages, Arabises and Aubrietias, with a little company of fern-fronds sheltered in the low fringe of shrub behind the mossy stones.

Having determined on the position of the bed, the next thing to do is to excavate the ground to a depth of two feet, or thereabouts, and to run a drain from it if very wet. If not, it is better let alone, as a good deal of the success depends upon the beds being continually moist; and in dry soils, instead of draining, it would be better to put in a substratum of spongy peat, so as to retain moisture for the stony matter that the cavity is to be filled with. As to soil, rock-plants are found in all sorts; but a good turfy loam, with plenty of silver or river sand added, will be found to suit a greater number of kinds than any other. The compost should be of a somewhat spongy character, and if not naturally so, it should be so made by the addition of well decomposed leaf-mould, cocoa-nut fibre, or, failing these, peat. If the trees of the shrubbery are of a nature likely to

send their roots into the mass of good compost prepared
for the rock-plants, it will be desirable to dig a narrow drain
as deep as their lowest roots, and fill it with concrete to
the surface : this will prevent the alpine plants from being
starved by their more vigorous neighbours.

With the soil should be incorporated the smallest and
least useful stones and débris among those collected for the
work, so that the plants to adorn the surface may send
down their roots through the mixture of earth and stone,
and revel in it. When this is well and firmly done, the
larger stones may be placed—half in the earth as a rule, and

Rock garden on margin of Shrubbery.

on their broadest side, so that the mass, when completed,
may be perfectly firm. Have nothing to do with tree roots
or stumps in work of this kind ; they crumble away, and are
at best a nuisance and a disfigurement to a garden. The
intervening spaces may then be filled up, half with the compost and half with the stony matter, and the smaller blocks
placed in position—the whole being made as tastefully

diversified as may seem desirable, taking the size of the structure into consideration. When finished, it should look like a bit of rocky ground, stones of different shapes protruding—here a straight-sided one, under the lee of which a shade-loving plant may flourish; there two in juxtaposition, between which a cliff alpine may find a place. Two or three feet will as a rule be high enough for the highest points of rocky fringes of this sort, though the plan admits of considerable variation, and it may be tastefully made twice or thrice as high. In some of our public and private gardens want of means is given as an excuse for the presence of the hideous masses of rockwork that disfigure them. The plan now recommended is as much less expensive than these as it is less offensive!

We will next discuss a most interesting way of growing alpines. Most of us have had opportunities of seeing how the most uninviting surfaces often yield a resting-place and nutriment to various forms of plant-life. The closest pavements, the stone roofs of old buildings, the stems and branches of trees, the faces of inaccessible rocks, and ruins, are all frequently embellished in the most charming way with ferns and wild flowers.

RUIN AND WALL GARDENS.

" Here stood a shattered archway gay with flowers,
And here had fallen a great part of a tower—

> Whole, like a crag that tumbles from a cliff,
> And like a crag was gay with wilding flowers;
> And high above a piece of turret stairs,
> Worn by the feet that now were silent, wound
> Bare to the sun."—

These lines from Tennyson's 'Idylls' are a true picture of the plant-life on many old ruins; and on many comparatively new structures we also see flowers and ferns quite at home. Many plants that in gardens have carefully prepared soil grow naturally on the barest and most arid surfaces. This fact must not be supposed to be contradictory of previous statements, as to the necessity of giving alpine plants a suitable material to root into; it is the open loose texture of the ordinary rockwork, or its solidly cemented masses, into which the plants cannot root, that does the mischief.

It is not without considerable observation of the capabilities of walls—even walls in good repair—to grow numerous rare and pretty plants, and, moreover, keep them in perpetual health without requiring any trouble, that I recommend everybody who takes an interest in the matter to have the fullest confidence in growing them easily in this way. Most of those who are blessed with gardens have usually a little wall surface at their disposal; and all such should know that some plants will grow thereon better than in the best soil. A mossy old wall, or an old ruin, would afford a position for many dwarf rock-plants which no specially prepared situation could rival; but even on straight and well-preserved walls we can establish some little beauties, which year after year will abundantly repay the tasteful cultivator for the slight trouble of planting or sowing them. Those who have observed how dwarf plants grow on the tops of mountains, or on elevated stony ground, must have seen in what unpromising positions many flourish in perfect health—fine tufts sometimes springing from an almost imperceptible chink in an arid rock or boulder. They are often stunted and diminutive in such places, but always more floriferous and long-lived than when grown vigorously upon the ground; in fact, their beauty is often intensified by starvation and aridity. Now, numbers of alpine plants perish if planted in the ordinary soil of our gardens, and many do so where much pains is taken to attend to their wants. This results from over-moisture at the root in winter, the plant being rendered more susceptible of injury by our moist green winters inducing it to make a lingering growth. But it is interesting and useful to know that, by placing many of these delicate plants where their roots can secure a comparatively dry and well-drained medium, they remain in perfect health. My attention was first called to the great adaptability of walls, ruins, etc. for growing many choice rock-plants, while visiting Dublin a few years ago. Near Lucan, I observed the upper portion of the old brick wall of a garden—indeed, all of it that was out of convenient reach—covered with a dwarf, green,

"The garland forest, which the gray walls wear."—Pyrus.

moss-like plant, and before coming close to it, I asked the gardener what it was that made the wall so green. "It is," he replied, "a plant like a moss, but every spring it is covered with the most 'beauteeful' flowers." And "sure enough" that is its character, for it proved to be the pretty little *Erinus alpinus*, which would have had little or no chance of existing on the level ground in the same place, and which had, at some time or other, escaped to the wall, and there found a home as congenial as its native one. This will suggest at once that many plants from latitudes a little farther south than our own, and from alpine regions, may find on walls, rocks, and ruins, that dwarf, ripe, sturdy growth, stony firmness of root medium, and dryness in winter, which go to form the very conditions that make them at home in a climate entirely different from their own. There

Rock-plants established on an old fort wall.

are many alpine plants now cultivated with difficulty in frames, even in places where there is a fine collection and much

The great Pyrenean Saxifrage (*S. Longifolia*), one foot in diam.
(*From a Photograph.*)

knowledge of these subjects, that the most unpractised may grow in such positions as I describe.

The reader will do well not to ask if what I advise is practised in gardens growing collections of alpines, but to put the matter to the test of experiment. The idea of growing such splendid alpine plants as the true *Saxifraga longifolia* of the Pyrenees on the straight surface of a wall, has never even entered into the heads of the managers of our largest gardens, and probably some of them would laugh at it; but I affirm that it is in the power of any person to succeed with them, and the trial can be made at a merely nominal cost.

I have no doubt whatever that at least 400 species of cultivated rock and alpine plants would thrive well on old walls and ruins if sown thereon. Nor must it be supposed that a moist district is necessary, for the Pansy shown in the accompanying cut grew by chance on a wall at Kew—the brick wall behind the narrow border for herbaceous plants. It sprung forth at a foot or so below a coping, which prevented it from getting much or any rain, and one would scarcely have expected a Pansy to have existed in such a position. It not only did so, but flowered well and continuously. No doubt the seed fell in the chink by chance.

Pansy on dry brick-wall.

The best way to establish plants on walls is by seed. The Cheddar Pink, for example, grows on walls at Oxford much better than I have ever known it do on rockwork or on the level ground, in which last position indeed it soon dies. A few seeds of this plant, sown in a mossy or earthy chink, or even covered with a dust of fine soil, would soon take root and grow into neat little specimens, living, moreover, for years in that dwarf and perfectly healthful state so agreeable to the eye. So it is with most of the plants enumerated; the seedling roots vigorously into the chinks, and gets a hold which it rarely relaxes. But of some plants seeds are not to be had, and therefore it will be often necessary to use plants. In all cases young plants should be selected, and, as they will have been used to growing in fertile ground, or good soil in pots, and have all their little feeding roots compactly gathered up near the surface, they must be placed in a chink with a little moist soil, which will enable them to exist until they have struck root into the interstices of the wall. In this way I have seen several interesting species of ferns established, and also the silvery Saxifrages, and can assure the reader that the appearance of the starry rosettes of these little rock-plants (the kinds with incrusted leaves, like *S. longifolia*, and *S. lingulata*) growing flat against the wall will prove strikingly beautiful. All the best kinds for our purpose, those that can be readily obtained and established without trouble, are marked with an asterisk in the list of selections which will be found farther on, and should be chosen by the doubters and beginners in this culture.

While many have old ruins and walls on which to grow alpine plants, others will have no means of enjoying them this way; but all may succeed perfectly with the plan suggested in the next figure. By building a rough stone wall, and packing the intervals as firmly as possible with loam and sandy peat, and putting, perhaps, a little mortar on the outside of the largest interstices, a host of brilliant gems may be grown with almost as little attention as we bestow

A rude stone wall covered with Alpine Plants.

on the common Ivy. Thoroughly consolidated, the materials of the wall would afford precisely the kind of nutriment required by the plants. To many species the wall would prove a more congenial home than any but the best constructed rock-garden. In many parts of the country the rains would keep the walls in a sufficiently moist condition, the top being always left somewhat concave; in dry districts a perforated copper pipe laid along the top will diffuse the requisite moisture. In very moist places, natives of wet rocks, and trailing plants like the *Linnæa*, might be interspersed here and there among the other alpines; in dry ones it would be desirable to plant chiefly the Saxifrages, Sedums, small Campanulas, Linarias, and subjects that, even in hotter countries than ours, find a home on the sunniest and barest crags. The chief care in the management of this wall of alpine flowers would be in preventing weeds or coarse plants from taking root and overrunning the choice gems. When these intruders are once observed, they can be easily prevented from making any further progress by continually cutting off their shoots as they appear; it would never be necessary to disturb the wall even in the case of a thriving Convolvulus. The wall of alpine plants may be placed in any convenient position in or near the garden: there is no reason why a portion of the walls usually devoted to climbers should not be prepared as I describe. The boundary walls of multitudes of small gardens would look better if graced by alpine flowers than bare as they usually are. However, when once it is generally known that the very walls may be jewelled with this exquisite plant-life, it need not be pointed out where opportunities may be found for developing it.

ALPINE FLOWERS IN THE MIXED BORDER.

The old-fashioned mixed border offers a capital means of growing, without trouble, numbers of first-class alpine plants.

This much abused, much misunderstood, sometimes overpraised method of arranging plants is now rarely or never seen with us in what are called "good gardens." When seen, it is usually a poor sight, and worthy of the ridicule bestowed by some persons on what they have never seen in perfection, and know little about. They misunderstand this old system, and abuse it. However, its ancient admirers were not backward in the first respect, as they filled it with tall, weedy, and strong Asters, Solidagos, and the like, possessing no merit, and therefore soon brought the system into contempt. It is undervalued by nearly everybody; curators of botanic gardens—the very men who ought to know and appreciate its merit—have sneered at it; great "bedding-out people" have given it no mercy, when it was nearly or quite finished without their aid; and finally, the very people of whose gardens it was the life—the owners of cottage gardens—have too often neglected it in favour of a few kinds of tender bedding plants.

Even yet, however, you may see a trace of it about country cottages, and nothing can be prettier than to find one surrounded by a nice variety of hardy plants, from Roses and Honeysuckle to double Saxifrage and Lily-of-the valley; but, unhappily, these poor cottagers are also beginning to run after strange gods, as would appear from the following extract from a letter addressed by a Nottingham clergyman to the "Field":—

"It is, I confess, with deep regret that in the last few years I have seen the 'posy gardens' of several cottages in my parish destroyed—the Moss-roses, Clove Carnations, aye, and the Ladslove and the Lemon Thyme, rooted out, and their place supplied by a ridiculous grass plat, with a hole in the centre, empty for eight months in the year, and containing for the other four months Scarlet Geraniums and Verbenas purchased at sixpence each from some neighbouring nursery, and forming a wretched parody upon the 'masses of colour' which weary my eyes and try my temper when I am conducted by lady friends through their blazing parterres, which, notwithstanding their perpetual sameness, I am expected to admire."

Such is the happy result we have arrived at by "improving" the flower garden. Persons with houses and frames and other garden conveniences can manage very well; but what a sorry thing it is to think that those with only means to grow hardy flowers have rooted them out, and are obliged to buy or to beg a few plants every spring! For them the exquisite flora of the Alps has no attractions. To them the vast families of plants that garnish with unsurpassed beauty the woods and wilds of northern and temperate climes offer not a sole specimen which they consider worthy of cultivation. But where is the interest or true beauty of their gardens? It does not exist; and thus the delightful art of gardening has become with them a thing more contemptible than the production of wall-paper patterns. Instead of gathering round their homes much of the choicest interest of the vegetable kingdom—a thing which anyone can do without a particle of expense for artificial heat—they make a series of blotches, and boast that there is scarcely a leaf to be seen. "I have," says the gentleman above quoted, "amidst hundreds of plants in my own garden, which recall absent friends and far-off scenes—I have flourishing in my flower-beds Acanthus from the walls of the Coliseum, Cyclamen from the tomb of Virgil, and Anemone from the cliffs of Sorrento." Where are the associations of the common "bedding" gardener? where even the fragrance or the beauty of his flowers? They are mostly devoid of any such thing, simply affording telling colour of some kind—it matters not whether by leaves or flowers. We must change all this, without destroying any good feature of "bedding out." We must again have our mixed borders, not the old mixed borders, but improvements on them.

There are several other ways of arranging hardy plants in a more beautiful, natural, and pleasing manner, but the mixed border forms a sort of reception-room for all comers and at all times. On its front margin you may place your newest Sedum or silvery Saxifrage; at the back or in the centre your latest Delphinium, Phlox, or Gladiolus; and therefore it is, on the whole, the most useful arrangement, though it should as a rule be placed in a rather isolated part of the garden, where the extent of the place permits of that. Not that a mixed border is not sufficiently presentable for any position; but, as many other arrangements require the more open and important parts of the garden, this had better be kept in a quiet, retired place, where indeed its interest may be best enjoyed. If no better situation be offered than the kitchen-garden, make a mixed border there by all means. The little nursery department, if there be one, will also suit; but best of all, in a large place, would be a quiet strip in the pleasure-ground or flower-garden, separated, if the garden be in the natural style, by a thin shrubbery, from the general scene of the flower-garden. It is vain to lay down any precise rules as to the position or arrangement of this or anything else; for, even if we succeeded in having them adopted, what a sad end would it not lead to—every place being like its neighbour! That, above all others, is a thing to be avoided. In old times, the borders on each side of the main walk of the kitchen-garden were principally appropriated to herbaceous plants; and, if well done, this is a good practice, especially if the place be small. A border arranged in this way in a small villa garden will prove a very attractive feature, especially if cut off from the vegetable and fruit quarters by a trellis-work completely covered with good strong-growing varieties of Roses on their own roots.

The mixed border is capable of infinite variation as to plan as well as to variety of subjects. The most interesting variety is that composed of choice hardy herbaceous plants,

Alpine Plants growing in a level border.
(A Garden Sketch, June, 1871.)

bulbs, and alpine plants. Another of a very attractive description may be made by the use of bedding plants only, from Dahlias and Gladioli to the smallest kinds, but we will now confine ourselves to the old-fashioned sort made with hardy plants alone. There is a symmetrical system, which must be entirely avoided—that of placing quantities of one thing, good or bad, as the case may be, at regular intervals from each other. The very reverse of that is the true system for the best and most interesting kind of mixed border. In a well-arranged one, no six feet of its length should resemble any other six feet of the same border. Certainly it may be desirable to have several specimens of a favourite plant; but any approach to planting the same thing in numerous places along the same line should be avoided. I should not, for instance, place one of the neat Saxifrages along in front of the border at regular intervals, fine and well suited as it might be for that purpose; but, on the contrary, attempt to produce in all parts totally distinct types of vegetation.

A Mixed Border of Hardy Flowers, with Alpine Plants on its margin.

PART I.   *ALPINE FLOWERS IN MIXED BORDERS.*   49

PLAN OF PORTION OF MIXED BORDER WITH TUFTS OF ALPINE FLOWERS ON ITS MARGIN.

Trellis of galvanized wire covered with Roses growing on their own roots.

Hibiscus militaris.	Lilium tigrinum Fortunei.				Tritoma grandis.		
Baptisia exaltata.		Tall Delphinium.	Phlox, tall kind.		Anemone Honorine Jobert.		
Echinops ruthenicus.			Eryngium amethystinum.				
Aster turbinellus.		Rudbeckia Newmanni.					
	Pyrethrum roseum, var.			Stenactis speciosa.		Iris Victorine.	
Symphytum caucasicum.		Trollius napellifolius.		Symphytum bohemicum.		Dielytra spectabilis.	
Anemone sylvestris.		Funkia grandiflora.		Orobus vernus.	Gentiana asclepiadea.	Anemone fulgens.	
Aster versicolor.							
Saponaria ocymoides.	Anthyllis montana.	Sempervivum calcareum.	Silene alpestris.	Hepatica triloba.	Erica carnea.	Antirrhinum distichon.	Gentiana acaulis.	Iberis corifolia.	Phlox verna.

Walk.

4

The plan on page 49 shows a small portion of what I conceive to be a tastefully arranged mixed border, and, at the same time, the proper position for the alpine plants in the front line. Each of the dwarf plants in front should be allowed to grow into a strong spreading tuft.

The borders should be deeply prepared, and of a fine free texture—in short, of good, rich, sandy loam. That is the chief point in the culture. It is a great mistake to *dig* among choice hardy plants, and therefore no amount of pains should be spared in the preparation of the ground at first. If thoroughly well made then, there will be no need of any digging of the soil for a long time.

Many alpine plants, when grown in borders, are much benefited by being surrounded by a few half-buried rugged stones or pieces of rock. These are useful in preventing excessive evaporation, in guarding the plant, when small and

Alpine Plant on border surrounded by half-buried stones.

young, from being trampled upon or overrun by coarse weeds or plants, and in keeping the ground firmer. Besides, many mountain plants look much more at home when arranged somewhat as shown in the accompanying illustration than in any other way on borders.

A few barrowfuls of stones—the large flints of which edgings are often made will do well, if better cannot be obtained—will suffice for many plants; and this simple plan will be found to suit many who cannot afford the luxury of a properly formed rockwork. Lists of alpine plants suitable for the mixed border will be found in the selections at the end of the book.

ALPINE PLANTS IN SHRUBBERY BORDERS.

Lastly, I will speak of the capabilities of common shrubbery borders, etc., for growing a very considerable number of alpine plants. No practice is more general, or more in accordance with ancient custom, than that of digging shrubbery borders, and there is none in the whole course of gardening less profitable or worse in its effects. When winter has once come, almost every gardener, although animated with the best intentions, simply prepares to make war upon the roots of everything in his shrubbery border. The generally accepted practice is to trim, and often to mutilate the shrubs, and to dig all over the surface that must be full of feeding roots. Delicate half-rooted shrubs are often disturbed; herbaceous plants, if at all fragile and not easily recognised, are destroyed; bulbs are often displaced and injured, and a bare depopulated aspect is given to the margins, while the only "improvement" that is effected by the process is the annual darkening of the surface by the upturned earth.

Illustrations of my meaning occur by miles in our London parks in winter. Walk through any of them at that season, and observe the borders round masses of shrubs, choice and otherwise. Instead of finding the earth covered, or nearly covered, with vegetation close to the margin, and each individual plant developed into a representative specimen of its kind, we find a spread of recently-dug ground, with the plants upon it, exhibiting the air of having recently suffered from a whirlwind, or some other visitation that necessitated the removal of mutilated branches. Rough-pruners precede the diggers, and bravely trim in the shrubs for them, so that nothing may be in the way; and then come the spadesmen, who sweep along from margin to margin, plunging deeply round and about plants, shrubs, or trees. The first shower that occurs after this digging exposes a whole network of torn-up roots. There is no relief to the spectacle; the same thing occurs everywhere—in a London botanic garden as well as in our large West-end parks; and year after year is the process repeated. While such is the case, it will be impossible to have an agreeable or interesting margin to a shrubbery; albeit the importance of the edge, as compared to the hidden parts, is pretty much as that of the face to the back of a mirror.

Of course all the labour required to produce this unhappy result is worse than thrown away, as the shrubberies would do better if left alone, and merely surface-cleaned now and then. By utilising the power thus wasted, we might highly beautify other parts of our grounds which are now in a very objectionable condition.

If we resolve that no annual manuring or digging is to be permitted, nobody will grudge a thorough preparation at first. The planting should be so arranged as to defeat the digger. To graduate the vegetation from the taller subjects behind to the very margin of the grass is of much importance, and this can only be done thoroughly by the greater use of permanent evergreen and very dwarf subjects. Happily, there are quite enough of these to be had suitable for every soil. On light, moist, peaty, or sandy soils, where such subjects as the sweet-scented *Daphne Cneorum* would spread forth its dwarf cushions, a better result would ensue than, say on a stiff clay; but for every position suitable plants might be found. Look, for example, at what we could do with the dwarf green Iberises, Helianthemums, Aubrietias, Arabises, Alyssums, dwarf shrubs, and little conifers like the creeping Cedar (*Juniperus squamata*), and the Tamarix-leaved Juniper! All these are green, and would spread out into dense wide cushions, covering the margin, rising but little above the grass, and helping to cut off the formal line which usually divides margin and border. Behind them we might use very dwarf shrubs, deciduous or evergreen, in endless variety; and of course the margin should be varied also.

In one spot we might have a wide-spreading tuft of the prostrate Savin pushing its graceful evergreen branchlets out over the grass; in another the little dwarf Cotoneasters might be allowed to form the front rank, relieved in their turn by pegged-down roses; and so on without end. Herbaceous plants, that die down in winter and leave the ground bare afterwards, should not be assigned any important position near the front. Evergreen alpine plants and shrubs, as before remarked, are perfectly suitable here; but the true herbaceous type, and the larger bulbs, like Lilies, should be placed between spreading shrubs rather than be allowed to monopolise the ground. By so placing them, we should not only secure a far more satisfactory general effect, but highly improve the aspect of the herbaceous plants themselves. The head of a white Lily, seen peeping up between shrubs of fresh and glistening green, is infinitely more attractive than when forming one of a large batch of its own or allied kinds, or associated with a mass of herbaceous plants. Of course, to carry out such planting properly, a little more time at first and a great deal more taste than are now employed would be required; but what a difference in the result! In the kind of border I advocate, nearly all the trouble would be over with the first planting, and labour and skill could be successively devoted to other parts of the place. All that the covered borders would require would be an occasional weeding or thinning, etc., and perhaps, in the case of the more select spots, a little top-dressing with fine soil. Here and there, between and amongst the plants, Forget-me-nots and Violets, Snowdrops and Primroses, might be scattered about, so as to lend the borders a floral interest, even at the dullest seasons; and thus we should be relieved from the periodical annoyance of digging and its consequences, and see our borders alive with exquisite plants. A list of species suitable for this purpose will be found among the selections.

And now, having spoken of growing alpine flowers in various ways, I will say a few words in favour of such of them as happen to be among the plants usually termed "florists' flowers." What is a "florists'" flower? Well, merely one that has been a great favourite with gardeners, and, being much raised from seed by them, has sported into such a number of distinct varieties in their hands that it forms a sort of little isolated family in a corner apart from botanical classification, so to speak. The term is, in short, a bad one to designate flowers that have been much grown by man, or rather which, exhibiting considerable variation under his care, have been preserved by him in their most striking and admired forms. They are in many cases double flowers that belong to these florists' groups—the Hollyhock and Dahlia, to wit—though not a few are single, like the Gladiolus and Auricula. Florists' flowers that have sprung from high mountain or rock-plants, like the Auricula or the Carnation, are perhaps more worthy of attention than any others, in consequence of their rich and elegant markings, perfect hardiness, neatness of habit, shape of bloom, and adaptability to the wants of cultivators in all parts of the country. They ought to be in every garden—not of necessity to be cultivated as "florists' flowers," but treated as ordinary hardy plants. The true florist tends his flowers almost as carefully as if they were so many tender exotics, and is precise as to their position, soil, and every other condition; but these are such very hardy subjects that they may be well enjoyed without any attention beyond planting them in a suitable position in the first instance.

We may assign some cause why many interesting plants and classes of plants have gone out of cultivation; but there is one thing that can hardly be accounted for, and that is, why the fragrant, beautiful, and neat classes of hardy florists' flowers—from elegantly laced Picotees to richly stained Polyanthuses—should have almost disappeared from our gardens, and be now in want of the least advocacy from me. In them we have flowers of unimpeachable merit, equally worthy of cultivation in garden of peer or cottager. They are as hardy as our native plants, require no steaming in houses at any time of their lives, are generally pleasing in habit, whether in or out of flower, sometimes useful for the spring garden, and in nearly all cases among the very best plants which we can grow for cutting from; and yet, with all these undoubted merits, where are they? Generally speaking, fallen into "the abyss of things that were." They have, of course, been driven from the field by the bedding system; but so surely as the perception of what is really beautiful still lives amongst us, so surely will they come into our gardens again, and be grown more than ever.

THE NATURAL ROCK-GARDEN.

Perhaps the most fortunate of all lovers of alpine flowers are those who have opportunities of growing them where there is a natural rock-garden—a not uncommon case in many parts of these islands. Where the rock crops up naturally in any way approaching that shown on the opposite page, a very trifling expense, a little taste, and some knowledge of suitable plants is all that is required to produce a magnificent result. Numbers of exotic herbs are sufficiently vigorous to take care of themselves among the weeds that grow in such places, while a select open spot may be easily cleared for the rarer and more delicate alpine plants. Even if only a few points of rock show, excavating or procuring smaller masses, and arranging them so that they seem to peep naturally from the earth, cannot be a matter

A Natural Rock-garden.

of difficulty. The nobler herbaceous plants, from the stately Pampas Grass to the brilliant Tritoma, might here be associated with the Brake-fern and the Struthiopteris; the light tracery of the various splendid everlasting peas, white and crimson, might twine in undisciplined loveliness amongst the huge leaves of such plants as *Rheum Emodi* and *Acanthus latifolius;* the superb new purple Clematises, with countless blooms like saucers of purple, and many trailing mountain herbs, might drape over the rocks not too thickly studded with ferns or flowers; the Cyclamens and Lilies, and many brilliant hardy bulbs from the sunny hills of Italy and Greece might here bloom in company with the Linnæa of North Europe and Scotland, and the many interesting plants that haunt the bogs and mossy woods of northern and arctic regions; and with all these and many more might be carried out Lord Bacon's conception of a "Naturall wildnesse. *Trees* I would have none in it; But some *Thickets*, made only of *Sweet-Briar*, and *Honny-suckle*, and some *Wilde-Vine* amongst; and the Ground set with *Violets, Strawberries*, and *Prime-Roses*. For these are Sweet, and prosper in the Shade. And these to be in the *Heath*, here and there, not in any Order. I like also little *Heaps*, in the *Nature* of Mole-hils, (such as are in *Wilde Heaths*) to be set, some with Wild Thyme; Some with Pincks; Some with Germander, that gives a good Flower to the Eye; Some with Periwinckle; Some with Violets; Some with Strawberries; Some with Cowslips; Some with Daisies; Some with Red-Roses; Some with Lilium Convallium; Some with Sweet-Williams Red; Some with Beares-Foot; And the like Low Flowers, being withal Sweet, and Sightly."

Where natural rock appears in only one spot, and we desire to make the most of it, it is better to clear away any wood or coarse undershrub that may surround it, so as to permit the full development of alpine and rock-plants; but should it crop up in more than one or in several positions in woods, it would be better to leave at least one such spot as much shaded with trees as possible, so that wood and copse-plants and shade-loving ferns might be there fully developed. Such a spot would form a very agreeable retreat in hot days. A few groups of the noble-leaved Berberises in the way of *B. nepalensis* would thrive admirably in peat near such a position; in an open, sunny but sheltered nook, a wild arrangement of Cannas and other subtropical plants would form a fine feature, while various low wood-shrubs, like the American *Rubus nutkanus* and *R. spectabilis*, would be seen to greater advantage running wild near such positions than in any others. And so also would a number of interesting hardy grasses, herbs, and shrubs, and dwarf wood-plants like the Pyrolas.

ALPINE PLANTS ON WINDOW SILLS.

Hitherto, all the arrangements treated of, whether large or small, ambitious or humble, require some kind of garden in which to carry them out. We will next consider the case of the owners of those limited sites for gardens—window sills. On these numbers of diminutive and interesting alpine plants may be easily grown. My first proposal is to pick out some of the prettiest and most diverse of the Stonecrops, Sempervivums, silvery Saxifrages, etc., to plant them in a goodly-sized box, and use a few rough stones by way of miniature rocks. I would place the box or boxes in the full sun, and give them plenty of water from a rose in warm spring and summer weather, and, indeed, at all times when they are dry, which is not likely to occur often during the dull months of winter. Among and between the alpine plants I would in autumn place here and there a diminutive spring-flowering bulb—say *Bulbocodium vernum*, *Scilla sibirica*, and *S. bifolia*, small Daffodils, Snowdrops, Snowflakes; and, if the box be large, a few of the delicately coloured Crocuses. The boxes should never be taken indoors, except for re-arrangement. When the snow comes the plants are comfortable, as it is their natural protection in a wild state; frost or rain hurts them not, and even London smoke is not able to destroy their little lives if they are tolerably attended to. The boxes most suitable for this purpose are wooden ones, with zinc troughs, decorated externally with chippings of oak and apple trees, fir cones, etc.; or what may be called architectural boxes, of wood also, but painted stone-colour externally, and designed so as to suit buildings, which the rustic ones do not. Both these kinds of boxes are made in good form by various firms in London and elsewhere. No matter what pattern is adopted, it is desirable to allow some plants of a trailing habit to fall over its outer edge. If the outer margins of the boxes were well covered, it would matter little what form were used. The common Stonecrop, *Sedum Sieboldii, Thymus lanuginosus,* the woolly-leaved Cerastiums, and many other hardy plants, will do this effectively.

A yet more satisfactory window rock-garden can be made outside of a window to which light has free access, by forming a miniature alpine garden on the sill. It is done by simply putting a few irregular stones along the front margin, and packing a few small bits of turfy peat or loam inside them to prevent the fine soil, afterwards to be added, from being washed out. Then fill in the hollow with sandy loam, mixed, if convenient, with morsels of broken sandstone. A few mossy or ancient-looking stones should be half-buried

The Window Rock-garden (interior view).

on the upper surface, and then the whole should be planted, the best time to do this being in April. It is not merely possible to keep alpine succulents in this way: it is easy to grow a multitude of the most interesting and beautiful kinds! I never in garden or wild saw these plants in better health, or looking more at home, than on the outside of a low sunny window in Mr. Peter Barr's house at Tooting. The accompanying figure shows a view of this from the interior; it was no less pretty seen from without. It is, however, impossible to show in an engraving the exquisite effect of the Lilliputian succulents. The attention required is very trifling—some little taste in forming and planting, a judicious selection of plants, and thorough waterings through the dry season. I need hardly add that small and brilliant-flowered spring bulbs might be employed to light up this tiny garden in spring as well as that previously mentioned. It would also be desirable to plant subjects of a drooping character on the outer margin. The alpine succulents are all thoroughly hardy, and would remain in good condition during the winter, but a little changing and replanting every spring would be very desirable.

ALPINE PLANTS IN POTS.

Hitherto alpine plants have generally been grown in pots, and it might perhaps be supposed from this fact that something like perfection was arrived at in their culture. It is not so. I do not advocate their culture in pots at all where an opportunity of making even the smallest type of rockwork exists; but there are many cases in which they cannot be well grown in any other way. It is desirable to keep some kinds in pots till sufficiently plentiful, and it is also very desirable to grow a number of distinct and handsome kinds in this way for the purpose of exhibiting them at flower shows.

We are pre-eminently great at exhibiting; our pot-plants are far before those of other countries; specimens are to be seen at every show which are models not only as regards beauty, but as showing a remarkable development of plant from a very small portion of confined earth, exposed to many vicissitudes; yet in one respect we have made no progress whatever, and that is, in the pot-culture of alpine and herbaceous plants for exhibition purposes.

Prizes are frequently offered and usually awarded at our flower shows for these plants, but the exhibitors rarely deserve a prize at all, for their plants are generally badly selected, badly grown, and such as ought never to appear on a stage at all. In almost every other class the first thing the exhibitor does is to select appropriate kinds—distinct and beautiful, and then he makes some preparation beforehand for exhibiting them; but in the case of hardy plants, anybody who happens to have a rough lot of miscellaneous plants exhibits them, and thus it is that I have seen such beauties as the following more than once shown: a common Thrift with the dead flower-stems on it, and drooping over the green leaves; a plant of *Arabis albida* out of flower; the Pellitory-of-the-wall, which has as little beauty in flower as out of it: not to speak of a host of worthless things not in themselves ugly, but far inferior to others in the same families. What would become of our shows if the same tactics were carried out in other classes? Even the most successful exhibitors are apt to look about them a day before the show for the best flowering cuttings of such plants as *Iberis correæfolia*, and, sticking four or five of these into a pot, present them as "specimens." Now, what is so easily grown into the neatest of specimens as an Iberis? By merely plunging in the ground a few six-inch pots filled with rich soil, and putting in them a few young cutting plants, they would, if "left to nature," be good specimens in a short time, while with a little pinching, and feeding, and pegging-down, they would soon be fit to grace any exhibition. So it is with many other plants of like habit and size—the dwarf shrubby *Lithospermum prostratum*, for example; a little time and the simplest skill will do all that is required. Such subjects as the foregoing, with tiny shrubs like *Andromeda tetragona* and *A. fastigiata*, the Menziesias and *Gaultheria procumbens*, the choicer Helianthemums and dwarf Phloxes, and many others enumerated in the selections of exhibition plants, might be found pretty enough to satisfy even the most fastidious growers of New Holland plants.

The very grass is not more easily grown than plants like Iberises and Aubrietias, yet to ensure their being worthy of a place, they ought to be at least a year in pots so as to secure well-furnished plants. Such vigorous subjects, to merit the character of being well grown, should fall luxuriantly over the edge of the pots, and in all cases as much as possible of the crockeryware should be hidden. The dwarf and spreading habit of many of this class of plants would render this a matter of no difficulty. In some cases it would be desirable to put a number of cuttings or young rooted plants into six-inch pots, so as to form specimens quickly. Pots of six inches diameter are well adapted for growing many subjects of this intermediate type; and with good culture, and a little liquid manure, it would be quite possible to get a large development of plant in such a comparatively small pot, but if very large specimens were desired, a size larger might be resorted to.

To descend from the type that seems to present to the cultivator the greatest number of neat and attractive flowering plants, we will next deal with the dwarf race of hardy succulents, and the numerous minute alpine plants that associate with them in size—a class rich in merit and strong in numbers. These should, as a rule, be grown and shown in pans: they are often so pretty and singular in aspect, as in the case of the little silvery Saxifrages, that they will be very attractive when out of flower, while the flowers are none the less beautiful because the leaves happen to be ornamental in an unusual way. Many of a similar size, as *Erpetion reniforme* and *Mazus Pumilio*, must be shown in good flower. All these little plants are of the readiest culture in pans, with good drainage and light soil. Of course the speediest way to form good specimens of the most diminutive kinds is to dot young plants over the surface of the pot or pan at once.

Some few alpine plants are somewhat delicate or difficult to grow; and amongst the most beautiful and interesting of these are the Gentians, and certain of the Primulas. There are many beginners who will be ambitious to succeed in cultivating them, but, in a general way, it would be better to avoid, at first, all such difficult subjects, since a failure with them is apt to be disheartening. I believe that a more liberal culture than is generally pursued is what is wanted for these more difficult kinds, and for such as are usually considered impossible to cultivate. The plants are often obtained in a delicate and small state; then they are, perhaps, kept in some out-of-the-way frame, or put where they receive but chance attention; or, perhaps, they die off from some vicissitude, or fall victims to slugs, which seem to relish their flavour, considering how thoroughly they eat off some kinds; or, if a little unhealthy about the roots, are injured by earth-worms, whose casts serve to clog up the drainage, and thus render the pot uninhabitable. With strong and healthy young plants to begin with, good, and more liberal culture, and plunging in the open air in beds of coal-ashes through the greater part of the year, the majority of those supposed to be unmanageable would soon flourish beautifully. I have taken species of Primula, usually seen in a very weakly and poor state, divided them, keeping safe all the young roots, put one sucker in the centre, and five or six round the sides of a 32-sized pot, and in a year made "perfect specimens" of them, with. of course, a greater profusion of bloom than if I had depended on one plant only. Annual or biennial division is an excellent plan to pursue with many of these plants, which in a wild state run each year a little farther into the deposit of decaying herbage which surrounds them, or, it may be, into the sand and grit which are continually being carried down by natural agencies. In our long summer, some of the Primulas will make a tall growth and protrude rootlets on the stem—a state for which dividing and replanting them firmly, nearly as deep down as the collar, is an excellent remedy.

There are many plants which demand to be permanently established, and with which an entirely different course must be pursued, *Spigelia marilandica*, *Gentiana verna*, *G. bavarica*, and *Cypripedium spectabile*, for example. The Gentians are very rarely well grown, and yet I am convinced that few will fail to grow them if they procure in the first instance strong established plants ; pot them carefully and firmly in good sandy loam, well drained, using bits of grit or gravel in the soil ; plunge the pots in sand or coal-ashes to the rim, in a position fully exposed to the sun ; and give them abundance of water during the spring and summer months, taking, of course, all necessary precautions against worms, slugs, and weeds. And such will be found to be the case with many other rare and fine alpine plants. The best position in which to grow the plants would be in some open spot near the working sheds, where they could be plunged in coal-ashes, and be under the cultivator's eye. And, as they should show the public what the beauty of hardy plants really is, so should they be grown entirely in the open air in spring and summer. To save the pots and pans from cracking with frost, it would in many cases be desirable to plunge them in shallow cold frames, or cradles, with a northern exposure in winter; but, in the case of the kinds that die down in winter, a few inches of some light covering thrown over the pots, when the tops of the plants have perished, would form a sufficient protection.

Alpine and herbaceous plants in pots, and kept in the open air all the winter, are best plunged in a porous material on a porous bottom, and on the north side of a hedge or wall, where they would be less exposed to changes of temperature, and less liable to be excited into growth at that season.

The most suitable kind of pots for alpine plants that I have yet seen are those used by Mr. G. Maw, in his gardens at Benthall Hall, near Brosely. These pots are of a peculiar size—eight inches broad by four inches deep. They seem peculiarly well suited to the wants of alpine plants, securing, as they do, a good body of soil, not so liable to rapid changes as that in a small vessel; while in stature, being only four inches high, they are exactly what is wanted for these dwarf plants. The common garden pan suits some alpine plants well; but is not so well suited to the stature of alpine plants, or the wants of their roots, as a pot of this pattern.

For growing the Androsaces and some rare Saxifrages a modification of the common pot may be employed with a good result. This is effected by cutting a piece out of the

Pot for Androsaces, etc.

Alpine Plant growing between stones in a Pot.

side of the pot, one and a half or two inches deep. The head of the plant potted in this way is placed outside of the pot, leaning over the edge of the oblong opening, its roots within in the ordinary way, among sand, grit, stones, etc. (See fig.). Thus water cannot lie about the necks of the plants to their destruction. This method, which I first observed in M. Boissier's garden, near Lausanne, in 1868, is undoubtedly an advantageous one for delicate tufted plants which are liable to perish from this cause. The pots used there were taller proportionately than those we commonly use, so that there was plenty of room for the roots after the rather deep cutting had been made in the side of the pot.

A yet more desirable mode than the preceding one is

that of elevating the collar of the plant somewhat above the level of the earth in the ordinary pot by means of half-buried stones, as shown in the cut on page 63.

In this way we not only raise the collar of the plant so that it is less liable to suffer from moisture, but, by preventing evaporation, preserve conditions much more congenial to alpine plants, and keep the roots firm in the ground; besides, the small plants look more at home springing from and spreading over their little rocks. It should, however, be distinctly understood that no such attention is required by the great majority of alpine plants.

No matter in what way these plants may be grown in gardens, it is desirable to keep the duplicates and young stock in small pots plunged in sand or fine coal-ashes, so that they may be carefully removed to the rockwork, or sent

Bed of small Alpine Plants in pots plunged in sand.

away at any time. The best way of doing this is shown in the accompanying illustration, which represents a four-foot bed in which young alpine plants are plunged in sand, the bed being edged with half-buried bricks. In bottoms of beds of this kind there should be half-a-dozen inches of coal-ashes, so as to prevent worms getting into the pots, in which they always prove very injurious. Sand, or grit, or fine gravel, from its cleanliness and the ease with which the plants may be plunged in it, is to be preferred, but finely sifted coal-ashes will do if sand cannot be spared for this purpose.

Such beds should always be in the full sun, near to a

good supply of water, and, if several or many are made, should be separated by gravelled alleys of about two feet wide. The watering is very important. In a large nursery it should be laid on and given with a fine hose. This certainly is the most convenient and economical way. Over some of the beds in Mr. Backhouse's nursery at York may be seen an ingenious way of giving a constant supply of water to Primulas, Gentians, and other plants. Two perforated half-inch copper pipes are laid just above the plants in the beds as shown in the cut. From the perforations in every two feet or so of the pipe, drops continually trickle down in summer, satura-

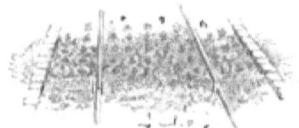

Bed kept saturated by perforated pipes.

ting the beds of sand, and of course the porous pots and their contents. In winter or very wet weather the water can be readily turned off. I do not believe there is any necessity for this system, provided the water is laid on and applied copiously with a fine hose.

ALPINE PLANTS FROM SEED.

A large number of alpine plants may be raised from seed, and in every place where there is a collection, it is desirable to sow the seeds of as many rare and new kinds as are worth raising in this way. A good deal will depend on the appliances of the garden as to the precise way in which they are to be raised; but whether there be greenhouses on the premises or not even a glass hand-light, alpine plants and choice perennials may be raised there in abundance. Supposing we are supplied with a good selection of seeds in early spring, and have room to spare in frames and pits, some time might be gained by sowing in pans or pots, and by placing them in those frames, or by making a very gentle hotbed in a frame or pit, covering it with four inches or so of very light earth, and sowing the seeds on that. If this mode be adopted, they may be sown in March; and, thus treated, many will flower the first year. In gardens without any glass they may be raised in the open air. The best time to sow is in April, choosing mild open weather, when the ground is more likely to be in the comparatively dry and friable condition so desirable for seed-sowing. But it should be borne in mind that they may be sown at any convenient time from April till August, as it is not till the year after they are sown that they display their full beauty or perhaps flower at all; and, therefore, should a packet or more of choice seed come to hand during the summer months, it is always better to sow it at once than to keep it till the following spring, as thereby nearly a whole season is lost. Those who already possess a collection of good hardy flowers may find a choice perennial—say, for instance, an evergreen *Iberis*, a *Campanula*, or a *Delphinium*—ripening a crop of seed in May, June, or July. Well, suppose we want to propagate and make the most of it, the true way is to sow it at once instead of keeping it over the winter, as is usually done. By winter, the seedlings will be strong enough to take care of themselves, and be ready to plant out for flowering wherever it may be desired to place them.

As to the immediate subject of raising them in spring, we will suppose the seeds provided, and the month of April to have arrived. If not already done, a border or bed should be prepared for them in an open but sheltered and warm position, and where the soil is naturally light and fine, or made so by artificial means. It would be as well to prepare and devote two or three, or more, little beds to this purpose of raising hardy flowers. They would form a most useful nursery-like kind of reserve ground, from which plants could be taken at any time to fill up vacancies, to exchange with those having collections, and to give away to friends; for assuredly it is one of the greatest pleasures of gardening to be able to give away a young specimen to a friend who happens to see and admire one of our "good things" in flower; and by raising them from seed we can always do this with ease. I have said that the seed-bed should be in a warm position, but let it, if possible, be in or near what is often called the reserve-garden in large places, or, in smaller, in the kitchen-garden—anywhere but in the portion of the gardens devoted to ornament. If the ground happen not to be naturally fine, light, and open, make it so by adding plenty of sand and leaf-mould, and then surface it with a few inches of fine soil from the compost-yard or potting-shed. The sifted refuse of the potting-bench will do well.

Then level the beds, and form little shallow drills in them for the reception of the seed. Let the beds be about four feet wide, with a little footway or alley between each about fifteen inches wide, and let them run from the back to the front of the border, not along it. Make the little drills across the beds, and, instead of making these drills with a hoe or anything of the kind, simply take a rake handle, a measuring rod, or anything perfectly straight that happens to be at hand, and, laying it across the little bed, press it gently down till it leaves a smooth impression about one inch deep. Do this at intervals of about six inches, and then your little nursery bed is ready for the seed. From these smooth and level drills the seeds will spring up evenly and regularly.

Before opening the seed packets, it is necessary to have clearly written wooden labels at hand on which to write the name of each species, so that there may be no confusion when the plants come up. These labels should be about eight or nine inches long, and an inch wide, and the name should be written as near the upper end as possible, so that it may not be soon obliterated by contact with the moist earth. Now, this labelling process is usually performed in all such cases at the time of sowing the seeds, but a very much speedier and better way is to lay out all the seeds on a table some wet day when out-of-door work cannot be done, and there and then arrange them in the order of sowing. Write a label for each kind, tie the packet of seeds up with a piece of bast, and then, when a fine day arrives for sowing them, it can be done in a very short time. In sowing, put in at the end of the first little drill the label of the kind to be sown first, then sow the seed, inserting the label for the following kind at the spot to which the seed of the first has reached, and so on. Thus there can be no doubt as to the name of a species when the same plan is pursued throughout. Near at hand, during the sowing, should be placed a barrow of finely-sifted earth; with this the seeds should be covered more or less heavily according to size, and then well watered from a very fine rose. Minute seed like that of Campanula will require but a mere dust of the sifted earth to cover it.

Once sown, the rest may be left to nature, save and except the keeping down of weeds, the seeds of which abound in the earth in all places, and will be pretty sure to come up among the young plants. But these being in drills, we can easily tell the plant from the weed, and nothing is required but a little persevering weeding. In these little beds the finest perennials will come up beautifully, and may be left exactly where sown till the time arrives for transplanting them to the rockery, spring garden, or mixed border. This is a better way than sowing in pots, where they are liable to much vicissitude, and from which they require to be "potted off." Of course in the case of a very rare or admired kind, the seedlings might be thinned a little and the thinnings dibbled into a nursery bed, but, by sowing rather thinly, the plants will be quite at home where first sown till the time arrives for planting them out finally.

I am convinced that in finely pulverised earth, with, if convenient, an inch or so of cocoa-fibre and sand between the drills to prevent the ground getting hard and dry, much better results will be obtained than by sowing in pots. In the open air they come up much more vigorously, and never suffer from transplantation or change of temperature afterwards. Nevertheless, as few will venture the very finest and rarest kinds of seed in the open air, how to treat them in frames is of some importance, and the following observations on this matter are by Mr. Niven, of the Hull Botanic Garden, one of the most successful cultivators of alpine plants, and who possesses, chiefly in pots, one of the most complete collections ever made. They were communicated to the "Gardener's Chronicle."

"Much disappointment is often experienced in raising the seeds of perennial plants, and blame is attributed to the vendor of the seeds, that ought in reality to be awarded nearer home. Presuming that the selection of the seeds is made, and that the seeds themselves are in the hands of the purchaser, the operation of sowing should take place as early as may be practicable in March. First of all, the requisite number of five or six-inch pots should be obtained, so that each seed packet can have a separate pot for itself. Some nice light soil, mixed with a fair amount of sand and leaf-mould (if obtainable), should be prepared, and passed through a coarse sieve, keeping a sharp eye after worms, and at once removing them; the rough part which remains in the sieve should be placed above the drainage in the bottom of the pots to the extent of two-thirds of the depth, filling the remaining third with the fine soil; the whole should then be well pressed down, so that the surface for the reception of the seeds may be half an inch below the brim of the pot, and tolerably even. Each packet of seed should then be sown, and covered with a sprinkling of fine soil, which should be pressed down by means of a flat piece of wood, or, what will be perhaps more readily available, by the bottom of a flower-pot.

"The best guide as to the thickness of covering required is to arrange so that no seeds shall be seen on the surface after the operation. If the seeds are minute, a very small quantity will be required to attain this end; if they are large, more will be requisite. This completed, and each pot duly labelled with the name of the plant and height of growth, the pots should then be placed in a cold frame tolerably near the glass, taking care that each pot is set level or as nearly so as practicable.

"In preparing the frame for their reception, it is desirable to have a good thickness of lime-rubbish in the bottom, say from nine to twelve inches deep, as a protection against worms.

"Many seeds come up a long time after others; in fact, seed-pots are often thrown away in the supposition that the seeds are dead, when they are perfectly sound; and some will come up a year or so after being sown. All that is necessary with the seeds that do not come up during the spring is to give them occasional watering, and to guard against the growth of the Linchen-like Marchantia. This is frequently a great pest in damp localities, and is only to be kept in check by carefully removing it on its first appearance, for if allowed to make too much headway, any attempt at removal carries away the surface soil, and with it the seeds. In the month of October each pot should be surfaced with a sprinkling of fine soil, well pressed down; in fact, the process before described after sowing should be repeated. The pots may remain in the frame till the spring, nor should they be despaired of altogether till May or June, or in some instances later.

"To those who may not have the advantage of a cold frame to carry out the foregoing instructions, I would still recommend the use of flower-pots rather than sowing in the open ground; but under these circumstances I would say—sow one month later; place the pots in a warm sunny corner, and arrange some simple contrivance so that you can shade with mats during hot sunshine, and also cover up at night, in order to keep off heavy rains; the same care in watering should be observed, and the same watchful eye after snails, woodlice, and other depredators, should be maintained.

"So much for the seeds in their seed-pots. Now a word or two as to the treatment of the plants afterwards. My practice is to pot off, as soon as they are sufficiently strong to handle, as many as are required, in three or four-inch pots, say three in each pot. In these they will grow well during the summer, and become thoroughly rooted, ready for consigning to their final habitat, be it rockery, border, or shrubbery, in the early part of spring, after the borders have been roughly raked over; thus giving them ample time to establish themselves before autumn arrives, and their enemy, the spade, is likely to come in their way. Failing a supply of pots sufficient for all, some of the stronger-growing ones may be planted in a sheltered bed of light soil, care being taken to shade them for a few days after being planted; or a few old boxes, five or six inches deep, may be used with even greater advantage for the same purpose, as they may readily be moved from the shady side of a wall to a more sunny locality after they have sufficiently recovered the process of transplanting; and, finally, they may receive the shelter of a cold frame as soon as winter sets in. This recommendation must not be considered as indicative of their inability to stand the cold weather, but as a preventive of the mechanical action of frost, which, in some soils especially, is apt to loosen their root-hold, and force the young plants, roots and all, to the surface.

"In the case of the smaller-growing alpines, such as the Drabas, Arabises, etc., I generally find that they stand the first winter best in pots of the smallest size, and in this form they may be the more readily inserted in interstices of a rockery, where they will permanently establish themselves."

WATERING.—Than this question there is nothing of more importance in connection with our subject. The popular and erroneous notion that alpine plants want shade arises from the fact that those placed in the shade do not perish so soon from drought as those in the sun. The reason that alpine plants perish so soon on bare flower-borders, the surface of which may be saturated with rain one day and be as dry as snuff the next, at least to the depths to which the roots of a small or young alpine plant would penetrate, is therefore very easily accounted for. Matted through a soft carpet of short grass in their native hills, or rooted deeply between stones and chinks, they can stand many degrees more heat than they ever encounter in this country. As a rule, it is impossible to water them too freely if the drainage be good, which of course it will be in a well-formed rockgarden. To have the water laid on and applied thoroughly and regularly with a fine hose is the best plan in districts not naturally very moist, and where there is a large rockgarden; small rockworks may be supplied in the ordinary way from pots or barrels, and in some parts of the country the natural moisture will suffice. Some lay small copper pipes through the masses and to the highest points of the rock, allowing the water to gently trickle from these, but, except in special cases, the plan is not so good as the hose. It may, however, be worth adopting for one spot in which *Gentiana bavarica* and other plants that like abundant moisture are planted. Whatever system be adopted, the rule should be: Never water unless you thoroughly saturate the soil, say with from one and a half to two inches deep of water over the whole surface. As a rule, ambitious, wall-like, erect masses of rockwork require half-a-dozen times as much water as those constructed on a proper principle with plenty of soil so arranged that it is saturated by the rains. Indeed, nothing but ceaseless watering could preserve plants, even for a short time, in a healthy state on the rockwork commonly made. As regards the time of watering, it is a matter of very little importance, though for convenience' sake it is better not done in the heat of the day. The really important point is to see that it is equably and thoroughly done.

There is a mischievous way of planting almost every kind of small plant, which is particularly injurious in the case of the hardy orchids (whose roots are easily injured), and of all rare hardy plants. I refer to the practice of making a hole for the plant, and after a little soil has been shaken over the roots, pressing heavily with the fingers over the roots and near the neck of the unfortunate subject. What is meant will be understood from fig. 2, if the reader assumes that there is a little soil between the fingers and the roots. Where the roots are not all broken off in this way, many of them are mutilated, or those near the collar of the plant are thrust deeper into the earth. Not unfrequently plants

1.   2.

perish from this cause. The right way, after preparing the ground, is, to make it firm and level, and then make a little cut or trench. The side of this trench should be firm and smooth, and the plant placed against it, the roots spread out, and the neck of the plant set just at the proper level, as in fig. 1. Then a good deal of the fine earth of the little trench is to be thrown against the roots, and as much *lateral* pressure applied as may be necessary to make the whole quite firm. Once the subject is carefully planted, as much surface-pressure as you like may be given. In this way not a fibre of the most fragile plant will be injured. This, of course, only applies to subjects not planted with bails, and is the best way to plant them.

THE BOG-GARDEN.

The bog-garden is a home for the numerous children of the wild that will not thrive on our harsh, bare, and dry garden borders, but must be cushioned on moss, and associated with their own relatives in moist peat soil. Many beautiful plants, like the Wind Gentian and Creeping Harebell, grow on our own bogs and marshes, much as these are now encroached upon. But even those acquainted with the beauty of the plants of our own bogs have, as a rule, but a feeble notion of the multitude of charming plants, natives of northern and temperate countries, whose home is the open marsh or boggy wood. In our own country, we have been so long encroaching upon the bogs and wastes that some of us come to regard them as exceptional tracts all over the world. But when one travels in new countries in northern climes, one soon learns what a vast extent of the world's surface was at one time covered with bogs. In North America day after day, even by the margins of the railroads, one sees the vivid blooms of the Cardinal-flower springing erect from the wet peaty hollows. Far under the shady woods stretch the black bog-pools, the ground between being so shaky that you move a few steps with difficulty. One wonders how the trees exist with their roots in such a bath. And where the forest vegetation disappears the American Pitcher-plant (*Sarracenia*), Golden Club (*Orontium*), Water Arum (*Calla Palustris*), and a host of other handsome and interesting bog-plants cover the ground for hundreds of acres, with perhaps an occasional slender bush of Laurel Magnolia (*Magnolia glauca*) among them. In some parts of Canada, where the painfully long and straight roads are often made through woody swamps and where the few scattered and poor habitations offer little to cheer the traveller, he will, if a lover of plants, find conservatories of beauty in the ditches and pools of black water beside the road, fringed with the sweet-scented Button-bush, with a profusion of royal and other stately ferns, and often filled with masses of the pretty Sagittarias.

Southwards and seawards, the bog-flowers become tropical in size and brilliancy, as in the splendid kinds of herbaceous *Hibiscus*, while far north, and west, and south along the mountains, the beautiful and showy Mocassin-flower (*Cypripedium spectabile*) grows the queen of the peat bog and of hardy orchids. Then in California, all along the Sierras, you see a number of most delicate little annual plants growing in small mountain bogs long after the plains have become quite parched, and annual vegetation has quite disappeared

THE ALPINE BOG-GARDEN.
Cypripedium. Trillium. Sarracenia. Helonias. Pinguicula.

from them. But who shall record the beauty and interest of the flowers of the wide-spreading marsh-lands of this globe of ours, from those of the vast wet woods of America, dark and brown, and hidden from the sunbeams, where the fair flowers only meet the eyes of water-snakes and frogs, to those of the breezy uplands of the high Alps, far above the woods, where the little bogs teem with Nature's most brilliant jewellery, joyous in a bright sun, and dancing in the breeze? No one worthily; for many mountain-swamp regions are as yet as little known to us as those of the Himalaya, with their giant Primroses and many strange and lovely flowers. One thing, however, we may gather from our small experiences—that many plants commonly termed "alpine," and found on high mountains, are true bog-plants. This must be clear to anyone who has seen our pretty Bird's-eye Primrose in the wet mountain-side bogs of Westmoreland, or the Bavarian Gentian in the spongy soil by alpine rivulets, or the Gentianella (*Gentiana acaulis*) in the snow ooze. We enjoy at our doors the plants of hottest tropical isles, but many wrongly think the rare bog-plants, like the minute alpine plants, cannot be grown well in gardens. Like the rock-garden, the bog-garden is rarely or never seen properly made and embellished with its most suitable ornaments. Indeed, bog-gardens of any kind are very rare, and only attempted by an individual here and there, who usually confines them to the accommodation of a few plants found in the neighbouring bogs. I will now proceed to point out how these may be made with a certainty of success.

In some places, naturally boggy spots may be found which may be readily converted into a home for some of the subjects to be named hereafter. But in most places an artificial bog is the only possible one. It should only be made in a picturesque part of the grounds. It may be associated with a rock-garden with good effect, or it may be in a moist hollow, or may touch upon the margins of a pond or lake. By the margins of streamlets, too, little bogs may be made with excellent taste. A tiny streamlet may be diverted from the main one to flow over the adjacent grass—irrigation on a small scale. No better bog than this can be devised, and none so easily made. Another very good kind could be made at the outlet of a small spring. It was in such little bogs that I found the Californian Pitcher-plant in dry parts of California, where there were no large bogs. In some of these positions the ground will often be so moist that little trouble beyond digging out a hole to give a different soil to some favourite plant will be needed. Where the bog has to be made in ordinary ground, and with none of the above aids, a hollow must be dug to a depth of at least two feet, and filled in with any kind of peat or vegetable soil that may be obtainable. If no peat is at hand, turfy loam with plenty of leaf-mould, etc., must do for the general body of the soil; but, as there are some plants for which peat is indispensable, a small portion of the bog-bed should be composed entirely of that soil. The bed should be slightly below the surface of the ground, so that no rain or moisture may be lost to it. There should be no puddling of the bottom, and there must be a constant supply of water. This can be supplied by means of a pipe in most places—a pipe allowed to flow forth over some firmly-tufted plant that would prevent the water from tearing up the soil.

As to planting the select artificial bog, all that is needed is to put as many of the undermentioned subjects in it as can be obtained, and to avoid planting in it any rapid-running sedge or other plant, or all satisfaction with the bog is at an end. Numbers of Carexes and like plants grow so rapidly and densely that they soon exterminate all the beautiful bog-plants. If any roots of sedges, etc., are brought in with the peat, every blade they send up should be cut off with the knife just below the surface; that is, if the weed cannot be pulled up on account of being too near some precious subject one does not like to disturb. All who wish to grow the tall sedges and other coarse bog-plants should do so by the pond-side, or in one or more moist or watery places set apart for the purpose. Given the necessary conditions as to soil and water, I can testify that the success of the bog-garden will depend on the continuous care bestowed in preventing rapidly-growing or coarse plants from exterminating others, or from taking such a hold in the soil that it becomes impossible to grow any delicate or minute plant in it. Couch and all weeds should be exterminated when very young and small.

The following are the bog and marsh plants at present most worthy of culture; but there are numbers not yet in cultivation, equally lovely.

A selection of choice Bog-Plants.

Anagailis tenella ; Butomus umbellatus ; Calla palustris ; Caltha in var.; Campanula hederacea ; Chrysobactron Hookeri ; Coptis trifolia ; Cornus canadensis ; Crinum capense ; Cypripedium spectabile ; Drosera in var. ; Epipactis palustris ; Galax aphylla ; Gentiana Pneumonanthe ; Helonias bullata ; Hydrocotyle bonariensis ; Iris graminea, Monnieri, ochroleuca, sibirica ; Leucojum æstivum, Hernandezii ; Linnæa borealis ; Parnassia palustris ; Lobelia syphilitica ; Lycopodium in var. ; Menyanthes trifoliata ; Myosotis dissitiflora, palustris ; Nierembergia rivularis ; Orchis latifolia and vars., laxiflora, maculata ; Orontium aquaticum ; Pinguicula in var.; Primula Munroi, sikkimensis, farinosa, Rhexia virginica ; Sagittaria in var. ; Sarracenia purpurea ; Saxifraga Hirculus ; Spigelia marilandica ; Swertia perennis ; Tofieldia in var. ; Tradescantia virginica ; Trillium ; Lastrea Thelypteris.

The above are most suitable for the select bog-bed kept for the most beautiful, rare, and delicate plants ; and among these, as has been stated, should be planted nothing which cannot be readily kept within bounds. To them lovers of British plants might like to add such native plants as Malaxis paludosa ; but it is better, as a rule, to select the finest, no matter whence they come. Some may doubt if the American Pitcher-plant, (*Sarracenia purpurea,*) would prove hardy in the open air in this country. It certainly is so, as one might expect from its high northern range in America. It will thrive in the wettest part of the bog-garden. In America I usually observed the Pitchers half buried in the water and sphagnum, the roots being in water. In British gardens it usually perishes from want of water.

HARDY AQUATIC PLANTS.

As ornamental water and aquatic plants are often intimately associated with rock and bog-gardens, something requires to be said of the most desirable water-plants.

A great deal of beauty may be added to the margins, and here and there to the surface, of ornamental water, by the use of a good collection of hardy aquatics arranged with some taste, but, so far as I have seen, this has not yet been fairly attempted by any designer of a garden or piece of water. Usually you see the same monotonous vegetation all round the margin if the soil be rich; in some cases, where the bottom is of gravel, there is little or no vegetation, but

The White Water Lily.

an unbroken ugly line of washed earth between wind and water. In others, water-plants accumulate till they are a nuisance and an eyesore—I do not mean the submerged plants like *Anacharis*, but such as the Water Lilies, when they get matted. Now a well-developed plant or group of plants of the queenly Water Lily, floating its large leaves and noble flowers, is a sight not surpassed by any other in our gardens; but when it increases and runs over the whole or a large part of a piece of water,—thickening together and being in consequence weakened,—and the fowl cannot make their way through it, then even the queen of British water-plants loses its charms. No garden water, however, should be without a few fine plants or groups of the Water Lily, and if the bottom did not allow of the free development of the plant, scrapings or rubbish might be accumulated in the spot where it was desired to exhibit the beauties of *Nymphæa*. Thus arranged, it would not spread too much. But it is not difficult to prevent the plant from spreading; indeed we have known isolated plants and groups of it remain of almost the same size for years, and where it increases too much, reduction to the desired limits is of very easy accomplishment, either by cutting off the leaves or getting at the roots in the bottom.

The Yellow Water Lily, *Nuphar lutea*, though not so beautiful as the preceding, is worthy of a place; and also the little *N. pumila*, a variety or sub-species found in the lakes of the North of Scotland. Then there is the fine and large *N. advena* (a native of N. America), which pushes its leaves boldly above the water, and is very vigorous in habit. It is very plentiful in the Manchester Botanic Garden, and

The Yellow Water Lily.

will be found to some extent in most gardens of the same kind. The American White Water Lily (*N. odorata*) is a noble species which would prove quite hardy in Britain. In collecting these things, the true and the only way is to get as many as possible from ordinary sources at first, and then exchange with others having collections, whether they be the curators of botanic gardens or private gentlemen fond of interesting plants. With a little perseverance, many good things may soon be collected in this way.

One of the prettiest effects I have ever observed was afforded by a sheet of *Villarsia nymphæoides* belting round the margin of a lake near a woody recess, and beyond it, more towards the deep water, a fine group of Water Lilies. The beauty of this Villarsia is too seldom seen in garden waters. It is a charming little water-plant, with its Nymphæa-like leaves and numerous golden-yellow flowers, which furnish a beautiful effect on fine days under a bright sun. It is not very commonly distributed as a native plant, though, where found, it is generally very plentiful, and not difficult to obtain in gardens where aquatic plants are grown. It is in all respects one of the most serviceable of hardy water-plants.

Not rare—growing, in fact, in nearly all districts of Britain—but exquisitely beautiful and singular, is the Buckbean or Marsh Trefoil (*Menyanthes trifoliata*), with flowers elegantly and singularly fringed on the inside with white filaments, and the round unopened buds blushing on the top with a rosy red like that of an apple-blossom. In early summer, when seen trailing in the soft ground near the margin of a stream, this plant has more charms for me than any other marsh-plant. It will grow in a bog or any moist place, or by the margin of any water. Though a rather common native plant, it is not half sufficiently grown in its garden waters. For grace and singularity combined, nothing can surpass *Equisetum Telmateia*, which, in deep soil, in shady and sheltered places near water, often grows several feet high, the long, close-set, slender branches depending from each whorl in a singularly graceful manner. It grows in many parts of England, but does not penetrate far into Scotland, and may be seen finely developed against the wall near the fernery in the Oxford Botanic Garden: I doubt not that many who see it there conclude it to be a foreigner, so distinct is it from our ordinary native vegetation.

The Great Water Dock.

For a bold and picturesque plant on the margin of water nothing equals the great Water Dock (*Rumex Hydrolapathum*), which is rather generally dispersed over the British Isles; it has leaves quite subtropical in aspect and size, becoming of a lurid red in the autumn. It forms a grand mass of foliage on rich muddy banks. The *Typhas* must not be omitted, but they should not be allowed to run everywhere. The narrow-leaved one (*T. angustifolia*) is more graceful than the common one (*T. latifolia*). *Carex pendula* is excellent for the margins of water, its elegant drooping spikes being quite distinct in their way. It is rather common in England, more so than *Carex Pseudo-cyperus*, which grows well in a foot or two of water or on the margin of a muddy pond. *Carex paniculata* forms a strong and thick stem, sometimes three or four feet high, somewhat like a tree-fern, and with luxuriant masses of drooping leaves, and on that account is transferred to moist places in gardens, and cultivated by some, though generally these large specimens are difficult to remove and soon perish. *Scirpus lacustris* (the Bulrush) is too distinct a plant to be omitted, as its stems, sometimes attaining a height of more than seven and even eight feet, look very imposing; and *Cyperus longus* is also a desirable plant, reminding one of the aspect of the *Papyrus* when in flower. It is found in some of the southern counties of England. *Poa aquatica* might also be used. *Cladium Mariscus* is also another distinct and rather scarce British aquatic which is worth a place.

If one chose to enumerate the plants that grow in British and European waters, a very long list might be made, but the enumeration and recommendation of those which possess no distinct character or no beauty of flower are precisely what I wish to avoid, believing that it is only by a judicious selection of the very best kinds that horticulture of this kind can give satisfaction; therefore, omitting a host of inconspicuous water-weeds, I will endeavour to indicate all others of real worth.

Those who have seen the flowering Rush (*Butomus umbellatus*) in flower, are not likely to omit it from a collection of water-plants, as it is conspicuous and distinct. It is a native of the greater part of Europe and Russian Asia, and is dispersed over the central and southern parts of England and Ireland. Plant it not far from the margin, as it likes rich muddy soil. The common *Sagittaria*, very frequent in England and Ireland, but not in Scotland, might be associated with this; but there is a very much finer double exotic kind to be had here and there, which is really a handsome plant, its flowers being white, and resembling, but larger than, those of the old white double Rocket. This I once saw in abundance in the pleasure gardens of the Rye House at Broxbourne, where it filled a sort of oblong basin or wide ditch, and looked quite attractive when in flower. It has the peculiarity of forming large egg-shaped tubers, or rather receptacles of farina, and I have found that in searching for these, ducks, or something of the kind, have destroyed the plants. This makes me suspect that it might prove a useful plant for feeding wild fowl, and that it might be worthy of trial in that way. No native water-plant that I am acquainted with has anything like such a store of farina as is laid up in the tubers of this plant. *Calla palustris* is a beautiful bog-plant, and I know nothing that produces a more pleasing effect over a bit of rich, soft, boggy ground. It will also grow by the side of water. *Calla æthiopica*, the well-known and beautiful "Lily of the Nile," is hardy enough in some places if planted rather deep, and in nearly all it may be stood out for the summer; but except in quiet waters, in the South of England and Ireland, I doubt if it would make any progress. However, as it is a plant so commonly cultivated, it may be tried without loss in favourable positions. The pine-like Water Soldier (*Stratiotes aloides*) is so distinct that it is worthy of a place: there is a pond quite full of this plant at Tooting, and it is common in the fens. It is allied to the Frogbit (*Hydrocharis Morsus-ranæ*), which, like the species of Water Ranunculi and some other fast-growing and fast-disappearing families, I must not here particularise; they cannot be "established" permanently in one spot like the other plants mentioned. The tufted Loose-strife (*Lysimachia thyrsiflora*) flourishes on wet banks and ditches, and in a foot or two of water. It is curiously beautiful when in flower; rather scarce as a British plant, but found in the North of England and in Scotland. *Pontederia cordata* is a stout, firm-rooting, and perfectly hardy water-herb, with erect and distinct habit, and blue flowers; not difficult to obtain from botanic garden or nursery. There is a small Sweet-flag (*Acorus gramineus*) which is worth a place, and has also a well variegated variety, while the common Acorus, or Sweet-flag, will be associated with the Water Iris (*I. Pseud-acorus*), the rather ornamental Water Plantain (*Alisma Plantago*), and the pretty *Alisma ranunculoides*, if it can be procured; it is not nearly so common as the Water Plantain. The pretty and interesting little Star Damasonium of the southern and eastern counties of England is very interesting, but, being an annual, is not to be recommended to any but those who desire to make a full collection, and who could and would provide a special spot for the more minute and delicate kinds. In such a spot, or even in the basin of a fountain, where they should be safely watched from being choked by larger weeds, the very tiny and pretty yellow Water Lily, *Nuphar Kalmiana*, the little white *Nymphæa odorata*, *Lobelia Dortmanna*, and not a few others, might be grown. The Water Lobelia does not seem to thrive away from the shallow parts of the northern lakes, getting choked by the numerous water weeds. *Aponogeton distachyon* is a native of the Cape of Good Hope, a singularly pretty plant, which is nearly hardy enough for our climate generally, and, from its sweetness and curious beauty, a most desirable plant to cultivate either in a basin or fountain in the greenhouse, or in a warm spot in the open air. It is largely grown in one or two places in the south, and it nearly covers the surface of the only bit of water in the Edinburgh Botanic Garden with its long green leaves, among which the sweet flowers float abundantly. The curator of the garden accounts for the plant doing so well by the fact that there are springs in the bottom of the water which, to some extent, elevate its temperature. In any sort of a greenhouse or conservatory aquarium, where it may have room to develop itself, it is one of the loveliest of water-plants. In the open air, plant it rather deep in a clean spot and in good soil, and see that the long and soft leaves are not injured either by waterfowl or any other cause. The Water Ranunculi, which sheet over our pools in spring and early summer with such silvery beauty, are not worth an attempt at cultivation, so rambling are they; and the same applies to not a few other things of interest. *Orontium aquaticum* is a scarce and handsome aquatic for a choice collection, but as beautiful as any is the Water Violet (*Hottonia palustris*). It occurs most frequently in the eastern and central districts of England and Ireland. The best example of it that I have seen was on an expanse of soft mud near Lea Bridge, in Essex. It covered the muddy surface with a sheet of dark fresh green, and must have looked better in that position than when in water, though doubtless the place was occasionally flooded. *Polygonum amphibium* and *P. Hydropiper* frequently flower prettily by the side of streams and ponds, while the Marsh Marigold (*Caltha palustris*), that "shines like fire in swamps and hollows grey," will burnish the margin with a glory of colour which no exotic flower could surpass. A suitable companion for this *Caltha* is the very large and showy *Ranunculus Lingua*, which grows in rich ground to a height of three feet or more. It is not scarce and yet not common—locally distributed, in fact. *Lythrum roseum superbum*, a beautifully coloured variety of the common purple Loose-strife, and *Epilobium hirsutum*, are two large and fine plants for the water-side.

WHAT TO AVOID.

In the selection of a few illustrations showing on what a mistaken principle, and with what deplorable taste, rockwork is generally made, my first intention was to have had them all engraved from drawings taken in various gardens, public and private; but as this course might have proved an invidious one, I have preferred to take most of them from our best books on Horticulture—the works of our highest authorities, Loudon, Macintosh, and others. From these the reader may glean some idea of popular notions on this subject, and it is scarcely needful to add that, if such ridiculous objects occur in our most trustworthy books, they must be yet more absurd in many gardens.

What to avoid.
Frontispiece of a book on Alpine Plants.

The first simple beauty is copied from the frontispiece of a small book on alpine plants, published not many years ago. Growing naturally on the high mountains, unveiled from the sun by wood or copse, alpine plants are grouped here beneath what appears to be a weeping willow—a position in which they could not possibly attain anything like their native vigour and beauty, or do otherwise than lead a sickly existence. The degree of contentment and delight felt by the artist for his subject is shown by his planting the ponderous vase in the centre of the group, and the introduction of the railing is quite beyond all praise. A few blacking pots or pieces of broken crockery are all that is needed to make the group complete.

One of the commonest forms which rockwork is made to assume is that of a rustic arch; and the following illustration, copied from Loudon, is less hideous than numbers that may be seen about London. Frequently they are formed of burrs, and occasionally of clinkers, but even if composed of the finest stone obtainable, they are utterly useless for the growth of alpine vegetation. How many Saxifrages, or Pinks, or Primroses, would find a home on such a structure planted in a part of the Alps highly favourable to vegetation? Probably not one, and should a few succulents establish themselves on its lower flanks, they would in all probability perish from heat and drought if their roots had not a free course to the earth beneath. Even persons with some experience of plant life may be seen sticking plants over such objects as these, as if their tender roots were capable of bearing as many vicissitudes of heat and cold as a piece of copper wire. The fact that plants push their roots far into masses of old brickwork is no justification for the rustic arch as a home for alpine flowers. If the cement, burrs, and clinkers permitted them even to enter it, they have nothing of any kind into which to descend. There is rarely an excuse for constructing such arches; where they occur, they should be completely clothed with Ivy or other vigorous climbers: the expense necessary to construct one would suffice for one of the simpler types of rock-garden already described.

What to avoid.
Rustic Arch (after Loudon).

The sketch on page 86, taken not long since at Hammersmith, shows something of the harsh, bare, and unnatural effect of structures of this sort. The grotesque figure who presides at the archway is in admirable keeping with the

What to avoid.
Rockwork in Villa at Hammersmith.

hybrid mixture of architecture and "rockwork" which is there exhibited.

The next scene is one in which a miniature representation of various mountains is attempted. Efforts of this kind usually end ridiculously, except when carried out at a vast expense. Let us succeed with a few square yards of stony mountain turf and flowers before we attempt to delineate *all* the mountains of a continent. A few hundred yards in length or even a single nook of many an alpine valley is often sufficient to impress the traveller with wonder and awe. We cannot therefore help admiring the boldness of those who try their hand even at a solitary alp.

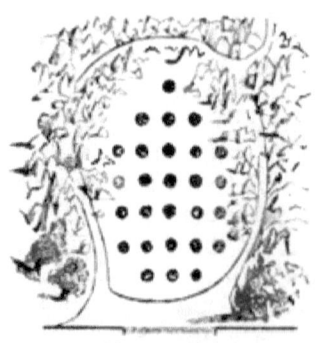

What to avoid.
All the Alps seen from the hall-door (after Macintosh).

The next illustration shows a rockwork and fountain in what we may call the true mixed style—huge shells, "cascades," and "rockwork." How any such object can be conceived to be in any sense ornamental is not easily explained, but it has been extracted as a model from a work of authority. In the fulness of time, no doubt, such abominations will be suppressed by act of parliament; but as many foolish persons will continue to erect them in the mean time, let us beg of them not in any way to associate them with alpine flowers. Even if it were possible to induce these to luxuriate on such a structure, they would merely serve to spoil the unity of the design.

What to avoid.
Fountain and Rockwork
(after Loudon).

Our next figure shows a truly laudable attack upon monotony. The tall stones are to the smaller ones as the Lombardy Poplar is to his roundheaded brothers of the grove. The front margin of this graceful scene consists of two rows of prostrate and one row of erect clinkers, and is much less irregular and more hideous than the engraver has had the heart to make it. The back wall is of a very common type, and precisely of that texture on which alpine plants will *not* exist. This cut is not extracted from the great books of Loudon or of Macintosh; it is a compara-

What to avoid.
" Infandi scopuli " (after ——).

tively recent improvement, and was sketched not long since in a botanic garden not one hundred miles from London.

Mrs. Loudon's design, while not so repulsive as some of the others, shows in its elevated nodding head the tendency to make such arrangements conspicuously offensive by raising them too high proportionately, and by so placing the stones that the rain cannot nourish the plants. Like the arches, such structures as this should in all cases be covered with Ivy, or some kindly veil of vegetation. It should be

What to avoid.
Rockwork (after Mrs. Loudon).

noted that when rocks or stones are properly placed in the rock-garden, they do not require any cementing, but are surrounded by and placed on moist stony earth or grit, inviting to every fibre of the root that descends. From this we may deduce the rule—Rockwork consisting of stones cemented together is utterly bad in all respects.

A distinction should, however, be here drawn between this variety and that in which a shell of artificial rock, so constructed as to resemble natural strata, is made to contain rich bodies of earth suitable for Clematises, Rhododendrons, etc.

A variety is occasionally seen bordering drives, often with large stones arranged in porcupine-quill fashion, and showing a dentate ridge of rocks springing up close from each side of the drive for a considerable length near the entrance gate. This may be described as the style dangerous for coachmen on dark nights, or, indeed, at any time when a swerve or tumble occurs. Such a position is the last that should be chosen for the rock-garden, especially as we live in an age when it is not desirable to combine it with any kind of fortification.

Without alluding to even half the genera, much less the species, of the ridiculous rockwork tribe, I have the pleasure of here presenting a plan of some, recently constructed on

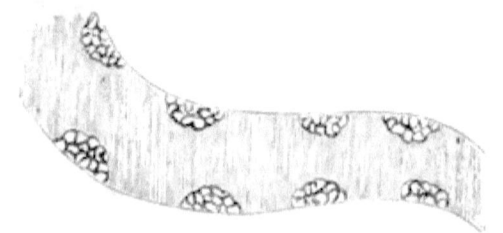

What to avoid.
Ground plan of "rockworks" recently made in a London park.

the margin of a stream in a great London park. It shows exactly what *not* to do with any rocks introduced near the margin of water. A poultry breeder, desirous of constructing a series of nests for aquatic birds, could scarcely have originated anything in baser taste. By turning to p. 28, something suggestive of Nature's work in this way will be seen, and that by no means a selected example. So far from these figures illustrating exaggerated or extreme instances, I should have no difficulty in finding many, even uglier and more unsuitable, in a few hours' walk near London. That such blemishes are not confined to obscure places, where the light of modern progress in these matters has not yet shone, is evident, as one of the most absurd

PART I.   *WHAT TO AVOID.*   89

sketches was taken in one of our greatest parks and another in one of the most popular of London public gardens.

No public garden should show anything in the way of rockwork that is not tasteful, and that is incapable of

What to avoid.
Sketched at Kew in 1872.

answering some useful end. This rule should particularly apply to botanic gardens. Better a thousand times content ourselves with the manifold good effects which we can produce with trees and shrubs and flowers on the level

What to avoid.
Sketched in the Botanic Gardens, Regent's Park, 1872.

ground, than add to the hideous piles of rubbish that go by the name of "rockwork" all over the country. And where these excrescences do occur in public gardens, if the finances or other circumstances will not permit of a proper rock-garden being made, the right thing to do is, to convey the offensive pile to the rubbish-yard some time when the ground is hard in winter, and labour plentiful. Few public gardens show worse examples of the traditional rockwork than Kew. Our sketch shows that on which the collection of alpine plants, etc., is shown in summer. It speaks for itself.

Lastly, among the features illustrating the way *not* to do it is the rockwork figured on page 89, which occurs in the Botanic Gardens, Regent's Park.

What a check to progress in this direction are such "rockworks" as these! And yet there is no way in which our public gardens would do more good than by growing well, in the open air, and arranging in a tasteful manner, the numerous brilliant flowers of the mountains of our own and other cold and temperate regions.

ON THE GEOLOGICAL ASPECTS OF ROCKWORK.

When rockwork has to be erected for horticultural purposes it will always be found that success will be attained just in the proportion in which some broad principles, based on a study of Nature's own work of that kind, have been followed.

Every lover of Nature must have envied her power of adorning rough stony nooks by means of a few of the commonest plants; a fern or two and a little moss convert a few weather-beaten rocks into objects of beauty, which it has often been attempted to imitate — usually without success. And success *is* attainable in almost every case if sufficient attention be only paid to the rules which, it will be seen, are as sacred to the physical agents which model our scenery as they ought to be to every landscape gardener. It is a trite observation to say that what pleases us in Nature is the perfect fitness of things which pervades all her belongings. The most rugged, abrupt, and even grotesque rockmasses, when untouched by man, never repel us by a sense of incongruity; they may be pleasing or awful, as the case may be, but they do not strike us as being out of place. Who, on the other hand, has not seen a lovely view marred by some unintelligent human hand, whether its work took the form of a quarry, a statue, or a vase? The secret of the difference lies in the words *weather-beaten*: rain, the chief rock-sculptor, working uniformly, slowly, and gently, leaves to each stone which it is fashioning its proper character, models it according to its peculiarities of composition and structure—in short, uses it fitly; while men with the most artistic pretensions, and armed with ruthless tools, too often misuse their materials—engaged, as they are in their igno-

Granite tor.

rance—in the hopeless task of making "silk purses out of sows' ears."

The first great rule which it behoves constructors of rockwork to look to is one very easily followed and constantly broken,—it is, that "your rockery should be characteristic of the part of the country in which it stands." That is to say, use chalk at Brighton and sandstone at Tunbridge, granite on Dartmoor, and trap near Edinburgh. This seems obvious enough; but the experience of every one must include cases where the reverse rule appears to have been acted on. Some "artists" have even carried their Philistinism in this respect so far, that the more they have succeeded in giving to their rockwork the appearance of a miscellaneous collection of mineralogical specimens from all parts of the world, the better they have been pleased. The familiar burnt brick of the South of England, and the slag and painted coke of the northern coal districts are better than these.

It is needless to point out in detail what rocks are suitable for rockwork in the different parts of Britain; a walk in the country will show you the rocks, and a glance at any geological map will tell you their names.

The second rule not to be departed from is one not so easy to adhere to as the last, but quite as important, viz.:— "The form of your rockery should be that which in Nature is assumed by the particular kind of rock of which it is composed." In order to appreciate the amount of careful observation which this rule renders necessary, we must consider what are the various agencies which together bring about on rocks the result which geologists know by the name of "weathering." Nature's mode of making her rocks weather-beaten requires such an amount of time, that we cannot attempt to imitate her in that respect; but if we cannot use her means, we can copy her results. Now, the weathering of a rock depends, before all things, on the structure of that rock, on its composition, and on the manner in which it is exposed to sun, rain, frost, wind, and the atmosphere itself, which are the great weathering and rock-carving agents. On many rocks water acts mechanically only; or, to be more accurate, its power of dissolving some rocks, such as quartz, is so limited, even when, as is almost always the case, it is charged with carbonic acid, that it is inappreciable, and may for practical purposes be left out of the reckoning. On a great mass of quartzite rock, for instance, the effect of rain would be of this kind. It could scarcely dissolve any of it away; but it would insinuate itself into every crevice and fissure and crack with which such hard rocks abound near the surface, and thence, by the help of frost, it blasts to shivers, winter after winter, layer after layer of this tough rock, just in the same manner as it blasts the water-pipes of our houses. By observation it is found that every rock affects a more or less peculiar kind of fracture; so that in bursting splinters from them, as has just been shown, the lines of fissure are not arbitrary and accidental. They, like everything else in Nature, form part of a plan. Hence a particular class of form is the result for each rock of this purely mechanical action of the weathering agent. In the case of quartzite, for instance, the fracture is "conchoidal," or shell-shaped, concave and wavy; this, on a large scale, gives rise to peaks with somewhat hollow sides, ridged with sharp serrated edges. This may serve as an example of simple weathering on a homogeneous, hard, and practically insoluble rock. Let us see what takes place with more complex rocks, of which granite may serve as a representative. This rock is made up essentially of three minerals—quartz, felspar, and mica in various

Chalk.

proportions. Now here the water with its carbonic acid will act not only mechanically, as in the case of quartzite, but as a powerful solvent and disintegrator. The fissures in granite are large and continuous, taking the form of immense joints, which cross and recross each other, often, but not always, in a regular manner; but besides these larger lines of weakness, which affect the whole rock, there are those minute lines which separate the constituent minerals from one another. Into all these the water trickles, decomposing the granite along the joints and cracks, "widening them, and rounding off the angles of their intersections, and ultimately only the harder masses, or the hearts of the blocks defined by the joints, remain as solid crystalline granite, some—though little—of the quartz is dissolved away by the water; the iron," which is usually present in small quantities in granite, "becomes oxidized and weakens the rock; but it is chiefly the felspar that is decomposed by the action of carbonic acid, its alkalies are removed, and its residue is washed away in the form of fine china clay... The quartz crystals remain as sand; the mica remains, but is less observable, and is partially decomposed." (Professor Rupert Jones.) It is by processes such as that described that the many fantastic shapes assumed by granite rocks have been arrived at, whether they be those of the curious balanced "Logging" stones of Cornwall or Brittany, the bare rounded tors of Devon, or the grey sterile mountain tops of Aberdeenshire. All felspathic rocks of eruptive origin, such as porphyries, are moulded into the shapes which they now exhibit in the same way as granite, and such also is the case with those sedimentary rocks which consist to a considerable extent of felspar, such as many of our gritstones. In these, however, a greater uniformity of weathering is caused by the regular lines of bedding which take the place of the horizontal joints of the former class of rocks. The vertical joints are similar in both. In igneous rocks, such as basalt and greenstone, the jointing and fissuring is often of such a kind as to give rise to very striking effects, very various in their appearance, though probably closely allied in their origin. Thus, from the simple dark brown, or black trap, without apparent structure, forming shapeless masses of a rounded, somewhat unpicturesque, outline, there is but one step to the bold semi-columnar escarpments of trap, which are so conspicuous in Northumberland and in many parts of Scotland; from these to the wonderful assemblages of rigid geometrical pillars of Staffa and the Giant's Causeway, with all their suggestiveness to rock-builders, the transition is shorter still; whilst in many parts of the three countries, we have examples of trap weathering into a mass of many-coated spheres of every size, decomposing layer by layer, with only a small core of the untouched rock in the centre of each ball. It is a noteworthy fact that basalt in this spheroidal condition weathers and decomposes much more rapidly than it does in the prismatic or columnar state. Rocks such as those we have been considering (with the exception of the grits and quartzite) have all been thrown up in a molten or pasty condition, which precluded their being subject to many of the rules which water-deposited rocks are bound by. Their structure is in a great measure the result of cooling; and although they frequently have a bedded appearance, they are not under the rigid sway of dip and strike, which in other rocks

Limestone.

is all-powerful in producing, or rather in preparing, the structure of a country. Indeed, in the great majority of cases it is the advent of the eruptive rocks which has given the sedimentary deposits their present positions, or what is technically called their "lie." Few of the latter, whether sandstones, limestones, shales, clays, or sands, are now lying in the horizontal positions in which they were formed, especially in much-disturbed and dislocated Britain. Great geological operations have taken place since then, and have squeezed, tilted up, and broken these beds of rock into every shape. And it will be obvious to all that had it not been for these great changes, the edges of these rocks could never have been brought under the influence of rivers and glaciers to carve them on the large scale into hill and dale, and of rain more delicately to "weather" and ornament them. It is therefore very necessary to observe the dip, or general mode of lying of the beds of any district which it is desired to make use of for rockwork purposes. The writer has seen a large rock-garden in the north of England which was laid out with great care and at vast expense, and which is spoilt by one apparently small but fatal oversight—the dip of the beautifully-arranged rockery-blocks is westerly and strongly marked, while the dip of the real "live" rock immediately beneath is due east. Now this seems a small thing to find fault with; and it is true that an uneducated eye might be well pleased, in ignorance of the defect. But consider that this easterly dip in that part of the country is the *raison d'être* of the shape of the hills and valleys which make its beauty: without it the fine slope on which this garden stands would not be in existence—the entire district would be altered, to say nothing of the fact that, were it not for this dip and the vast industries which it fosters, the wealth which built the rock-garden would have been elsewhere. "Follow Nature in all things" is the only safe motto for the landscape gardener. It would be tedious and perhaps not very useful to enumerate the different kinds of water-bedded rock which can in Britain be used for rockwork. A glance at the chief members will suffice.

Of the grits we have already spoken, and their mode of weathering is that of the entire class of sandstones, coarse and fine grained, massive and flaggy. With regard to the latter, it may be allowable to point out, for special reprobation, a mode of rock-building which seems to be gaining favour in many districts. It consists in placing a number of broken flagstones on end, and in every position relatively to one another; the result is peculiarly hideous, and resembles no possible combination of Nature's art, since the flags, at whatever angle they may be dipping, must be always parallel among themselves, except in the case of the arrangement known as "false bedding," which is one not likely to be successfully imitated. Sandstones are, as a rule, peculiarly adapted for rockwork by the forms they assume on weathering, by their great frequency, and by the great variety of their colours. From dark brown to bright red, from red to yellow, from yellow to white, thence through every tint of grey to blue and purple, the choice of colouring is great indeed in these rocks. They are found every-

Old Red Sandstone.

where—as hard grits in the old Silurian and Cambrian districts, as great rugged crags throughout the Carboniferous regions, forming the well-known Old Red and New Red sandstones, more sparsely distributed among the Oolites, but forming occasional bands of striking character among the sands and clays of the Wealden (witness the " Greys "

7

of the Lover's Seat and other marked natural rockeries in the neighbourhood of Hastings and Tunbridge Wells), and in the much more recent tertiaries appearing occasionally, as in the sand of Brussels, as lines of grotesque fistulous masses running through incoherent sand, very much as flints lie in our Upper chalk.

Many sandstones and grits pass gradually into more or less coarse conglomerates, that is to say, rocks formed of rolled pebbles and blocks of stones derived from other preexisting formations. Of such conglomerates there are many examples in Britain, and they are often very suitable for rockwork, owing to the uneven weathered surface which is the result of the different sizes of the pebbles, and occasionally of their different hardness, and which causes them to be dislodged unequally. The Permian conglomerates, in many places of Central England, are great additions to the natural beauty of the scenery, and have been frequently taken advantage of for the ornamentation of grounds.

Under the name of limestone must be included a very large number of rocks different in texture, hardness, and general aspect, but having this in common—that they are chiefly composed of carbonate of lime. The result of this composition is that more than any other rocks are they liable to the solvent as distinguished from the disintegrating action of water charged, as rain-water always is to some extent, with carbonic acid. This action we see displayed on a large scale in the great stalactite-lined caverns in the Carboniferous limestone of the North of England, or in the sand-pipes running deep into the chalk of the South country. On a smaller scale, the effects of this dissolving power are marked on every exposed face of limestone of every age, and help to make them everywhere worthy of the attention of the rock-gardener. In some instances thin beds of hard limestone are weathered into a curious honey-combed state, the exposed parts being of a lighter colour than the inner stone; in others the face of the beds present the appearance of a clumsy balustrade of the Louis XIV. style, the interstices having been gradually eaten away by the water running down the original lines of upright joints. Sometimes the most peculiar forms are assumed in this manner by limestones, and each kind has its own special characteristic shape, to be known only by constant observation; but perhaps no rock equals the great Magnesian limestone of Durham in the eccentricity or in the multiplicity of its disguises. This limestone is of a yellowish colour, and its structure is wonderfully diversified, sometimes hard and compact, sometimes friable, often concretionary and botryoidal, occurring as a mass of radiated concentric spheres of all sizes, generally crystalline, often as a distinct breccia or agglomeration of angular fragments held together by a cement of similar material. A walk along the coast of Durham, from South Shields to Roker, will show to what vagaries of weathering and denudation this extraordinary variety of conformation has given rise. The high cliffs are in places worn into deep caves, in others slender pillars of rough rock have been separated from the main mass, and stand solitary on the beach, while larger islands of rock stand out at sea, through which arches of every size and shape have been excavated. No rock can be better suited for rockwork if used judiciously, and it is moreover known that its chemical composition is such as to be very beneficial to rock-plants. These magnesian limestones are called Dolomites, and it is notable that their fantastic shapes are by no means confined to England, since no mountain-range is so remarkable for abruptness and startling variety of configuration as that in the Italian Tyrol known as the Great Dolomites. Besides the hard old stony limestones, of which we have spoken, there are in England a number of other kinds, from the oolitic limestones to the chalk, which can occasionally serve the landscape gardener's purpose. Their appearance is too well known to need description here. In the newer geological series there are frequently beds of a light porous limestone, very similar in appearance to the sinter which is deposited by petrifying springs. In many places this is called "ragstone," and it is extremely well adapted for our purpose; their distribution is, however, very local in Britain, so that, according to our theory as to æsthetics of rockery, they cannot be very widely used. Abroad, in Tertiary districts, they are far more common, especially on the shores of the Mediterranean, both on the European and on the African side.

Schists and shales may, for the purposes of rock-building, be considered together; the former being simply the hardened and altered form of the latter. Their weathered appearance, where exposed, varies very much with the angle of their dip and with the degree of crystallisation to which they have attained. Some schists are quite as crystalline as granite, and they then weather in the same manner, with this proviso, that the lines of foliation, or lamination, direct the operation. Where such beds are highly inclined, as on the south-west coast or in Brittany, a curious appearance is often seen, which may be called the "Artichoke form," as

Mica Schist.

it exactly resembles the mode of arrangement of the Artichoke leaves. At lower inclinations, schists and the harder shales do not form striking features; but, by offering slight rocky elevations, above a more or less level ground, with distinct "craig and tail shapes," they can be made highly effective in rock-gardening where they occur naturally. This has been done with the greatest success in the Central Park, New York. The softer shales may be dismissed as rockery materials, except for the purpose of forming the lower of the two beds of rock essential to the construction of a good waterfall or of an overhanging crag. While on the subject of waterfalls, it may be as well to remind the landscape gardener that, with very few exceptions, the rocks forming waterfalls in Nature dip *up-stream*, and this holds good for great and small falls alike. The clays and sands need not detain us; where these unrocky materials prevail, the rockmaker is clearly entitled to do the best he can to try and imitate the rock-masses of more favoured districts. But even then he should be bound by what we will call our third rule, which flows naturally from our other two, enounced above: "In no case should your rockery be constructed in a manner contrary to the broad geological laws to which all rocks are subject in their natural state."

In this brief survey of a large and interesting subject, it has only been intended to suggest some points for the consideration of rock-builders, and to show that success in their art, as in every other, is to be attained only by strict and careful observation and study of Nature's own models.

G. A. LEBOUR,
F.G.S. London and Belgium, F.R.G.S., etc.

"Excelsior!"

PLANT-HUNTING ON THE MOUNTAINS.

"The best image which the world can give of Paradise is in the slope of the meadows, orchards, and corn-fields on the sides of a great Alp, with its purple rocks and eternal snows above; this excellence not being in any wise a matter referable to feeling or individual preferences, but demonstrable by calm enumeration of the number of lovely colours on the rocks, the varied grouping of the trees, and quantity of noble incidents in stream, crag, or cloud, presented to the eye at any given moment."—*Ruskin*.

As many lovers of alpine plants have no opportunity of seeing them in a wild state, I have thought it worth while to include a few notes of my first short excursion in a really alpine country, which may serve to give some notion of such a region to those who have no better means of becoming acquainted with it. They relate no exciting accounts of attempts to mount any peaks that happen to be a few hundred feet higher than those of easy access, but only deal, in passing, with one of the many texts that may be read in the great book of the Alps.

My first day's work was devoted to the ascent of the Grande Salève, which, though not a great mountain, and with green meadows instead of snow at its top, is nearly

5000 feet high, and affords a good opportunity of commencing training for more serious work. The limestone chain, to the highest point of which we have to walk, is situated a little to the south of Geneva, and has vast escarpments looking toward that town. It will afford us our first introduction to alpine flowers, and a magnificent view of the mountains around. We are on the banks of Lake Leman, and valleys, hills, and far-off mountains all glow with the genial sun of a clear June morning. A few miles' drive through the clear sparkling air brings us from the margin of

An Alpine Lake.

the lake to the foot of the mountain before six o'clock, and then we gradually and pleasantly begin the ascent, through the last patches of meadow land, for the most part very like English meadow land, but much more gay with flowers. Bright Pinks, blue Harebells, Sages, and various Pea-flowers, make the scene gay with colour, and the air is full of the voices of innumerable insects, for which the long grass is a forest. Soon we pass the cultivated land, and enter on the hem of an immense belt of hazel and low wood, with numerous little green and bushless carpets of grass here and there, which cuts off vine, and corn, and meadow, from the slopes of the mountains. Here, at half-past six in the morning, the nightingale is singing, while white-headed eagles float aloft, now over the lake, and now over plain and hill, sometimes on motionless wing, and yet rapidly and silently gliding along on the look-out for prey. From floating bird in glowing air perfumed by wild Lily-of-the-valley, the white bells of which may be seen leaning out of its tufts of tender green at the base of the bushes, to the flower-clad

In the woody region.

heaps of stone, and in every peep which the eye obtains through the bush and wood to the villa-dotted margins of the lake, the scene is one of unalloyed beauty and abounding life—

> "A populous solitude of bees, and birds,
> And fairy-formed and many-coloured things."

Some magnificent gorges and precipices are gradually reached and exposed to view, every crevice having some plant in it, and all the ledges being clothed with the greenest grass or bushes, but as yet few of such as are generally termed alpine plants are seen. Many of the most delicate and minute of these would grow well in such a position, but the long and luxuriant grass and low wood would soon overrun and destroy them. The low tree gets a vantage ground on the shattered flanks of the mountain, and retains it. But in such positions we find numbers of beautiful flowers that may be termed sub-alpine, and occasionally plants, that are found of very diminutive size near the crest of an alp, are here met with much larger and taller. The plants that occur in such places should have a peculiar interest for all who love gardens, because they flourish in a temperature nearly like that of the greater part of these islands. Every copse, shrubbery, thin wood, or semi-wild spot in pleasure-grounds, throughout the length and breadth of the land, may grow scores of these plants, that now rarely or never find a suitable home in our gardens.

That fine rock-plant, *Genista sagittalis*, with curious winged stems and profuse masses of yellow flowers, forms the very turf in some spots. Dwarf neat bushes of *Cytisus sessilifolius* become very common; it is well worthy of cultivation: and soon I gather my first truly wild Cyclamen. The Lily-of-the-valley forms a carpet all under the brushwood. The Martagon Lily shoots up here and there among the common Orchids and Grass, and I begin to enjoy, for the second time during the year, the fragrance of the Hawthorn Bush. The Laburnum is mostly past; but on high precipices you may yet see bushes of it in flower. The great yellow Gentian begins to be very plentiful everywhere, and *Globularia cordifolia* is in dense dwarf sheets here and there, showing its latest flowers. *Anthericum Liliago* is very plentiful and pretty; and we see all this by the side of a well-beaten path, from which, no doubt, every rare plant has been gathered. Trifolium, Dianthus, Melampyrum, Anthyllis, and Euphorbia struggle for the mastery wherever a little grass has a chance to spread out, and every chink and small hole in the rocks where a little decomposed mould has accumulated supports some vegetation.

After a walk of three hours, we reach the top, having often stopped to admire the magnificent and varied views. From the bottom the visitor might have expected a stony barren mountain-top, a contracted space, with stunted (if rare)

Pine Woods, Glacier, and Alpine Village.

forms of vegetation; but it is an immense plateau, stretching miles in length, and covered with the greenest and freshest verdure. The best meadows of England, or even the Green Isle, could not vie with it in these points, while the grass is gay with flowers to which they are strangers, and here and there young plants of the great yellow Gentian, with their large and handsome leaves, form the "foliage plants" of the region. Trees there are none; but occasionally the Hazel, Cotoneaster, and other shrubs form a little group, and perhaps enclose some spot, so that the cattle that are driven up here in the summer months cannot eat down the flowers there. The mountain is of a limestone formation, but now and then we meet with a great block of solid granite, a remembrancer of the days when enormous glaciers from the far-off Mont Blanc range stretched to this, and when the rich and pleasant valley of the Alps was not. In several places there is a large expanse of well-worn rock, a level well-denuded mass, with cracks in it, in which *Polypodium Robertianum* and other Ferns grow luxuriantly. The surface is indented with roundish hollows, as if great lizards and salamanders had left their impress on it; these have in the course of ages become filled with a few inches of mould from decomposed moss, etc., and in them grow Vacciniums, Saxifrages, and Ferns, quite as well as if the "most perfect drainage" were secured. I tore up some flakes of plants here as easily as if they had been carefully detached from the rocks before, so lightly did they grow in the smooth hollows.

I was of course very glad to meet with my first silvery Saxifrage in a wild state, having long held that these Saxifrages, so often kept in pots even in botanic gardens, require no such attention, and may be grown everywhere in the open air with the greatest ease. The specimens which I found here fully confirmed this opinion. They grew in every conceivable position, at the bottom of small narrow chasms, under the shade of the bushes, in little thimble-holes on the surface of the rocks, in a tiny and sometimes flaccid condition from the drought; and here and there among *Festuca glauca* and *Asplenium Trichomanes*, where the accumulated soil was a little deeper.

The vernal Gentian is known to many as the type of all that is charming in alpine vegetation: its vivid colour and peerless beauty stamp themselves on the mind of the dullest traveller that crosses the Alps as deeply as the vast and death-like wastes of snow, the ever-darting silvery waterfalls, or the high, dark, plumy ridges of pines, though it be but a diminutive speck compared to any of these. It is there a hardy little gem-like triumph of life in the midst of death, buried under the deep all-shrouding snow for four, six, or even eight months out of the twelve, and blooming during the brightest summer days near the margin of the wide glaciers, and within the sound of the little snow cataracts that tumble off the high Alps in summer. But it is not confined to such awful, if attractive, spots; it descends to the crests of comparatively low mountains like this, where the sun's heat has power to drive away all the snow in spring, and where the snow is quickly replaced with boundless meadows of the richest grass, that form a setting for innumerable flowers. Among these the "blue Gentian" occurs, and blooms abundantly late in spring, while acres of the same kind lie deep and dormant, under the cold snow, on the slope of the high neighbouring alp for months afterwards. It also ventures into non-alpine countries, being found in Teesdale and Galway. This brilliant Gentian is very plentiful in the pastures here, but it is now passed out of flower, and the seed-vessels, very full and strong, are to be seen among the taller herbage. A few weeks earlier this plant was in perfection—now the grass makes it difficult to see its leaves, and somewhat obscures the dwarf silky Cudweed, which seems placed to form a silvery bed for the Gentian. Alpine travellers, botanists, and horticulturists, say that this lovely plant and its fellows cannot be cultivated, and Dean Close regretfully echoed this in describing in 'Good Words' his passage over the Simplon. This idea is quite erroneous.

Having arrived at the summit, let us sit down and survey the varied and magnificent prospect around. On one side we have the Jura range, and the wide sunny valley cultivated in every spot below the town of Geneva, and, between the Jura and our position, the lower part of the lake of Geneva, scarcely fluttered by the light breeze, the countless pleasant spots along its famous shores, and issuing from it the blue waters of the Rhone. Below the town it flows for some distance before being joined by the Arve, and from the summit of this mountain both may be seen wending their way to the meeting place—the one a dirty ash colour, the other almost a porcelain blue. By turning to the other side, another beautiful and well cultivated valley is seen, and beyond it a round isolated mountain, which from Geneva looked as tall as some of the giant ones, but which now seems a mere Primrose Hill compared with others to be seen from this spot. Many green and well-pastured mountains lie beyond, with dark clouds of Pine woods on their sides and summits. Others still higher, and with the

View of a part of distant range.

verdure less visible, are behind, and above all a great, bony, steep-scarped, dark range stretching all across the view. The hollows between the angular-pointed ridges are white with vast fields of snow, while others seem variegated with it in narrower bands, against the dark framework of the mountains. Clouds cap this far-off region—round, high, silvery clouds, contrasting with the deep blue sky above, and the wide range of mountains with the deep snow-seams down their dark sides below. A few minutes afterwards a break has occurred in these great "cloud-lands," and something reveals itself among them lit up with the hues of the silvery woolpacks around, and yet not of them, for there is here and there a dark spot suggestive of solid earth, and you ask yourself, can it be that that is a mountain? Yes, that is the tall old father of all the magnificent mountains around, his head silvery with age, while his eldest sons are merely beginning to show the silver here and there, and the younger ones have not a trace of anything but the fresh hue of youth.

The indescribable variety and beauty of the country traversed on descending the other side of the Salève, and the margins of calm, blue, celestial-looking Lake Leman, with vast ranges of snowy mountains beyond its broad expanse, give the young traveller a very rose-coloured impression of the Alps, which forty-eight hours' journey from Geneva was quite sufficient to modify in my case. The country has every conceivable variety of attractive pastoral scenery, and, better still, the human beings in it seem to partake of the

Chillon.

felicity which appears to be here the lot of all animated nature. Their cottages and houses, nestling in nooks in the sweetest of flowery fields, and carved out of the abundant wood of the region, snug gardens, fields of emerald green, vine-clad slopes, happy-looking villas, numerous flocks, and high ridges of mountain-lawn, with noble groups of dark Pines, in vast natural parks, form pictures of which the eye never wearies. The Castle of Chillon comes in to make the scene more interesting, associated as it is with thoughts of Rousseau and the author of 'The Prisoner of Chillon.' The too rapid rail allows but a moment to see the castle and the neighbourhood of varied loveliness by which it is surrounded, and a glimpse at the solitary little islet—

> "A small green isle, it seemed no more,
> Scarce broader than my dungeon-floor;
> But in it there were three tall trees,
> And o'er it blew the mountain breeze,
> And by it there were waters flowing,
> And in it there were young flowers blowing,
> Of gentle breath and hue."

But we must leave these Arcadian spots, and enter the Saas valley.

Compared to the enchanting shores of the lake I had passed the day before, this dark valley, with its deeply worn river-bed, and vast sides of gloomy rock, looked anything but a cheerful introduction to the Monte Rosa district; but fortunately I had other resources than those of the landscape or the sky, and as yet the weather permitted of enjoying them, for here were countless tufts of the interesting Cobweb Houseleek (*Sempervivum arachnoideum*), a not common, though always admired, inhabitant of our gardens. It was the first time I had ever met with it in a wild state, and cushioned in tufts, over the bare rocks, in the spaces between the stones that here and there had been built up to support the side of the pathway, and in almost every chink, I could have gathered thousands of plants of it. Although some of the Houseleeks are among the most interesting and singular of all dwarf plants, many persons do not grow a single kind, except it be the common one. They are the succulent plants of the Alps; their geometrically carved little rosettes may be compared to miniatures of the great stately Agaves of America. Some have rosettes as large as a saucer; some are small enough to be covered with a thimble; they vary in the hue of their leaves from a decided glaucous tone to light green; some are ciliated at the margins of the leaves, while the Cobweb one is white from a densely interwoven cottony down. They are amongst the hardiest of all plants, enduring any weather, and living even in smoky London, where many plants which people generally think much more hardy and vigorous quickly perish. There is not a window-sill in London to which the light of the sun can occasionally

penetrate on which they may not be grown either in pots or boxes, while in all open gardens they merely require to be kept free from weeds and "left to nature;" though even in our largest scientific gardens it is common to see them grown in pots, which they no more require than does a young oak. To the cottage or garden of the poorest they will lend an interest, and any of our largest gardens would be improved by their presence, if suitably arranged. Generally it is rare to see them in cultivation, but lately a fine species, S. calcareum, has come into use as an edging plant.

Next our pretty old friend, the Hepatica, came in sight, peeping here and there under the brushwood, but rarely in such strong tufts as one sees it make in our gardens. In a wild state it has, like everything else, to fight for existence, and is none the worse for it. To meet the little king of all our early spring flowers in his old wild home would have rewarded one for a day's hard walking in these solitudes. This plant had many interesting companions; not the least attractive and welcome being the Helvetian Selaginella, occasionally seen in fern collections in this country, which mantled over the rocks in many places, pushing up little erect fruiting stems from its green branchlets. It is hardy and well suited to gracefully accompany the smallest flowering rock-herbs.

The scenery now began to get very bold and striking, and, after a walk of nearly two hours, we reached a village with a very poor inn, where we had some black bread and wine. By this time a slight misty rain had begun to fall, and bearing in mind the long and toilsome valley we had to traverse before reaching a place where we could rest for the night, we resolved to use our legs as rapidly as possible, and practically shut our eyes to all the interesting objects around us. A soaking rain helped us to carry out this part of the plan. With rapid pace and eyes fixed on the stony footway, on we went, the valley becoming narrower as we progressed, and in some parts dangerous-looking from almost perpendicularly rising hills of loose stone. Presently a little rough weather-beaten wooden cross was passed beside the footway. "Why a cross here?" said I to the guide. "That great stone or rock you see, killed, on its way down, a man returning with his marketings from the valley," he replies.

Poor fellow! he must have formed but a small obstacle to that ponderous mass—hard as iron and big as a small cottage—which fell from its bed with such impetuosity that it leaped from point to point, and at last right over the torrent-bed, resting on a little lawn of rich grass and bright flowers on the other side.

Ten minutes afterwards we came to a group of three more rough wooden crosses, almost projecting into the pathway, and loosely fixed in the stones at its sides. They marked the spot where two women and a man had been buried by an avalanche. "And how," said I, "do you recover people's bodies who are thus overwhelmed?" "We wait till the snow melts in spring, and then find and bury them." It is no exaggeration to state that in many places along this valley these wooden crosses, marking the scene of deaths from like causes, occurred so thickly as to remind one of a cemetery. A railway collision would seem to offer capital chances of escape compared to what one would have in case of being in the way of any falling matter in these parts.

An Alpine Pathway.

We have all heard of the merry Swiss boy, but few of us have an idea of the hard and fearful nature of the lives of the peasantry of the elevated parts of this country. In the wide valleys and level land about the lakes life is as easy as need be; but where man creeps up to occupy the last tufts of verdure that are spread out where the Alps defy him with fortifications of rock and fields of ice and snow, there his life is not an enviable one. Perhaps its hardships make it none the less dear. Even the procuring of the necessaries of life renders the inhabitants liable to dangers of which in our own country we have no experience; almost every commodity used has to be dragged up these valleys on the backs of men or mules from the villages and towns in the Rhone valley; while in their dwellings, made of stems of the ever-abundant pine, and usually placed on spots likely to be free from danger from avalanches, they are sometimes buried alive.

The following description by Mr. Ruskin of one of their sad little groups of houses is fearfully true, and as it is perhaps desirable that we should know a little about the people as well as the plants, it may not be out of place here :—

" Here, it may well seem to him, if there be sometimes hardship, there must be at least innocence and peace, and fellowship of the human soul with nature. It is not so. The wild goats that leap along those rocks have as much passion of joy in all that fair work of God as the men that toil among them. Perhaps more. Enter the street of one of those villages, and you will find it foul with that gloomy foulness that is suffered only by torpor, or by anguish of soul. Here it is torpor—not absolute suffering—nor starvation or disease, but darkness of calm enduring; the spring known only as the time of the scythe, and the autumn as the time of the sickle, and the sun only as a warmth, the wind as a chill, and the mountains as a danger. They do not understand so much as the name of beauty, or of knowledge. They understand dimly that of virtue. Love, patience, hospitality, faith—these things they know. To glean their meadows side by side, so happier; to bear the burden up the breathless mountain flank unmurmuringly; to bid the stranger drink from their vessel of milk; to see at the foot of their low deathbeds a pale figure upon a cross, dying also, patiently—in this they are different from the cattle and from the stones, but in all this unrewarded as far as concerns the present life. For them there is neither hope nor passion of spirit; for them neither advance nor exultation. Black bread, rude roof, dark night, laborious day, weary arm at sunset; and life ebbs away. No books, no thoughts, no attainments, no rest; except only a little sitting in the sun under the church wall, as the bells toll thin and far in the mountain air; a pattering of a few prayers not understood, by the altar rails of the dimly gilded chapel, and so back to the sombre home, with the cloud upon them still unbroken —that cloud of rocky gloom, born out of the wild torrents and ruinous stones, and unlightened, even in their religion, except by the vague promise of some better thing unknown, mingled with threatening, and obscured by an unspeakable horror—a smoke, as it were, of martyrdom, coiling up with the incense, and, amidst the images of tortured bodies and lamenting spirits in hurtling flames, the very cross, for them, dashed more deeply than for others, with gouts of blood.

"Do not let this be thought a darkened picture of the life of these mountaineers. It is literal fact. No contrast can be more painful than that between the dwelling of any

An Alpine Village.

well-conducted English cottager and that of the equally honest Savoyard. The one, set in the midst of its dull flat fields and uninteresting hedgerows, shows in itself the love of brightness and beauty; its daisy-studded garden-beds, its smoothly-swept brick path to the threshold, its freshly sanded floor and orderly shelves of household furniture, all testify to energy of heart, and happiness in the simple course and simple possessions of daily life. The other cottage, in the midst of an inconceivable, inexpressible beauty, set on some sloping bank of golden sward, with clear fountains flowing beside it, and wild flowers, and noble trees, and goodly rocks gathered round into a perfection of Paradise, is itself a dark and plague-like stain in the midst of the gentle landscape. Within a certain distance of its threshold the ground is foul and cattle-trampled; its timbers are black with smoke, its garden choked with weeds and nameless refuse, its chambers empty and joyless, the light and wind gleaming and filtering through the crannies of their stones. All testifies that, to its inhabitant, the world is

An Alpine Waterfall.

labour and vanity; that for him neither flowers bloom, nor birds sing, nor fountains glisten; and that his soul hardly differs from the grey cloud that coils and dies upon his hills, except in having no fold of it touched by the sunbeams."

Soon the rain began to be mingled with an occasional wet flake of snow, which in another half-hour was descending in a regular heavy fall; and as we gradually ascended, soon every surface was covered with it, except that of the torrent beneath, which roared away with as much noise as if the waters of a world, and not those of one hollow in a great range, were being dashed down its wonderfully picturesque bed—sometimes cutting its way through walls of solid rock of great depth, at others dashing over wastes of worn and huge stones, carried down and ground by its action. Often we crossed it on small rough bridges of pine-wood, fragile looking, and heavily laden with fresh-fallen snow, not offering a very agreeable

An Alpine Stream.

passage to the nervous. The hissing splash of many cascades accompanied the tumult of the river-bed—many of these born of the melting snow and previous heavy rain, the main ones much swollen by it. The air being full of large downy flakes of snow, the pines on the white mountain side began to look quite sharp-coned from the pressure of its weight on their branches.

We had by this time evidently got into a region abounding with flowers, as every one of the caves was literally lined with the pretty little yellow *Viola biflora*. Every cranny was golden with its flowers; every seam between the rocks and stones enlivened by it. On entering one of these caves, I saw some crimson blooms peeping from under the snow about the roof or brow. They were those of the first Alpine Rose (*Rhododendron ferrugineum*) I had ever seen wild. Occasionally, pressed by the snow, the handsome flowers of a crimson Pedicularis might be seen; and in almost every place where a little soil was seated on the top of a rock or stone, so straight-sided that the snow only rested on the top, the beautiful, soft, crimson, white-eyed flowers of *Primula viscosa* were to be seen. It grows in all sorts of positions—wherever, in fact, decomposed moss, etc., forms a little soil. In dry places it is smaller than in wet ones, and is usually particularly luxuriant on ledges where a gradual or annual addition of moss or soil takes place, so that the tendency of the stems to throw out rootlets is encouraged.

Several hours in falling snow, feet saturated with deep snow-water, and extremities beginning to chill, notwithstanding the hard walking, make Saas, and Saas only, the one object to attain. To gain it, we passed through one or two small hamlets, the inhabitants of which were as much surprised as ourselves at the sudden and heavy fall of snow early in June, and eventually reached Saas just as evening was falling. By this time nearly a foot of snow had fallen on the corn, already far advanced in the ear. Unhappily we found the hotel closed, as the tourist season had not yet commenced. Standing on its threshold, thoroughly soaked with snow, waiting till somebody came to open it, and finding a hotel in such a region and on such a day without an inmate or a fire, was cold comfort indeed. Among the first of those who came to see us was the curé, who wondered how we got there in such weather; and he immediately set to work to dispel the hunger and the cold by instructing a maid to make a fire with all haste, and by ordering dinner.

As the country for miles around was covered with a dense bed of snow, my hopes of seeing the plants of the high Alps in this region were over, and rather than return by the same long and dreary valley, I determined to cross the Alps and descend into the sunny valleys of Piedmont, where we should, at all events, probably see some traces of vegetable life.

Next day we set out for Mattmark, nearly nine miles from Saas, more than 7000 feet higher than the sea-level, and above the level of the pine or any exalted vegetation. Only a few spots under ledges, etc., were bare, but we found many ordinary and well-known plants, as well as the rare *Ranunculus glacialis*, in full beauty, some of the flowers measuring nearly an inch and a half across. Near where we found this, a great sea-green arch shows the end of a large glacier, apparently a wide and deep river of ice beneath a field of snow, except where in places it is riven into glass-green crevasses. We have to skirt this field of ice to reach Mattmark, where there is a lake, the overflow from which passes right under the glacier. Although all surfaces were rendered pretty much alike by the snow, the scene was a striking one. Within a few steps of the lonely hotel there stand several enormous boulders, so large that, but for the frequent evidence of the great masses borne onwards by glaciers, it would be difficult to believe that any such agency had brought them there.

Lloydia serotina we met with in great abundance in the region of the glacial Ranunculus, and also *Androsace Chamæjasme*, the still rarer *A. imbricata*, and the mountain form of *Myosotis sylvatica*. By scraping off the snow here and there, we could see the very pretty little *Pyrethrum alpinum*, reminding one of a Daisy with its petals down in bad weather. Several not common Saxifrages, and a few Sempervivums, *Geum montanum*, *Linaria alpina*, very dwarf, but with the flowers much larger than usual; *Gentiana verna*, abundant; a pink Linum, *Polygala Chamæbuxus*, *Loiseleuria procumbens*, *Androsace carnea*, *Senecio uniflorus*, with deep orange flowers, and the most silvery of leaves an inch or so high; and the beautiful *Eritrichium nanum*, from half an inch to an inch high, and with cushions of sky-blue flowers —were among those not hidden from us by the snow.

Next morning we were up early to cross the pass of Monte Moro into Italy; the snow was very deep, and we were the first tourists who had crossed during the year. The snow was eighteen inches thick even in the lower parts of our three hours' walk, so that it was impossible to gather any specimens; and this was unfortunate, as the neighbourhood of the little lake of Mattmark, between two glaciers, is said to be very rich in plants. However, there was quite enough to do to ascend Monte Moro, with its deep coating of snow; in fact, it was hard work, and consequently took more time than usual. Arrived at the cross which marks the top, a new and magnificent prospect bursts upon us—the white clouds lie in three thin layers along the sides of Monte Rosa, but permit us to see its crest, while the great mountains whose snowy heads tower around it are here seen in all their majesty. On the Swiss side nothing but snow is seen on peak or in hollow; on the Italian, a deep valley has wormed its way among the magnificent mountain peaks, crested with sun-lit snow and dark crags, and guarded by vast ice rivers and unscaleable heights. We can gaze into this valley as easily as one does from a high building into the street below; and, crouched on the sunny side of a vertical cliff, to gain a little shelter from the icy breeze that flowed over the pass, view its quiet signs of life and green meadows, and above their highest fringes the vast funereal grove of pines on every side, guarding, as it were, the green valley from the vast and death-like wastes of snow above it. A grander scene it would be difficult to find, even in the most remarkable alpine regions of the world. Its effect was much enhanced by the quantity of snow that had just fallen and covered up thousands of acres of the higher ground. The contrast between the valley flushed with verdant life and the great uplands of snow was most imposing.

We had several miles to descend through the snow before a trace of vegetation could be seen, when fairy specimens of the nearly universal *Primula viscosa* began to show their rosy flowers here and there on ledges, where they were pressed down by the snow; and by clearing little spaces with the alpenstock, we found the ground nearly covered with them. Then the glacial Ranunculus began to make its appearance in abundance. Another rare and minute gem was here in quantity—the silvery *Androsace imbricata*, growing on the hollowed flanks of rocks—the tufts, not more than half an inch high, sending roots far into the narrow chinks. These having a downward direction, the water could reach the roots from above. One plant was gathered in the hollow recess of a cliff, with at least one hundred little rosettes and flowers, forming a tuft three inches in diameter, all nourished by one little stem as thick as a small rush, and which was bare for a distance of two or three inches from the margin of the chink from which it issued. The tuft, bloom, and minute silvery leaves suspended by this were in all probability as old as any of the great larches in the valley below.

The *Androsaces*, with very few exceptions, which have not until quite recently been successfully cultivated, are, as it were, the very humming-birds of the vegetable kingdom. Their silvery rosettes are more delicately chiselled than the prettiest encrusted Saxifrage; their flowers have the purity of the Snowdrop, and occasionally the glowing stains and blushes of the alpine Primulas. They are the smallest of beautiful flowering plants, and they grow on the very highest spots on the Alps where vegetation exists, carpeting the earth with wondrous loveliness wherever the sun has sufficient power to lay bare for a few weeks in summer a square yard of wet rock-dust.

The icicle-fringed cliffs, on the concave sunny faces of which the only traces of vegetation seen about here were found, and the rocky precipices seen from the spot, make all this diminutive enduring flower-life the more interesting and remarkable.

> "Meek dwellers mid yon terror-stricken cliffs!
> With brows so pure, and incense-breathing lips.
> Whence are ye? Did some white-winged messenger
> On mercy's missions trust your timid germ
> To the cold cradle of eternal snows?
> Or, breathing on the callous icicles,
> Bid them with tear-drops nurse ye?
>   Tree nor shrub
> Dare that drear atmosphere; no polar pine
> Uprears a veteran front; yet there ye stand,
> . . . . unblanched amid the waste
> Of desolation.   *Mrs. Sigourney.*

A very pretty dwarf Phyteuma, with blue heads, was found on the rocks here, and as we got down the mountain, *Geum montanum*, with its large yellow flowers, gilded the grass somewhat after the fashion of our Buttercups. *Sempervivum Wulfenii*, a large kind, was in flower, and the fine *Saxifraga Cotyledon* was also coming on. One specimen found had a rosette of leaves eight inches across. *Pyrethrum alpinum* here takes the place of the Daisy, and is full of flower. *Arnica montana*, so well known as a medicinal plant, is in

great abundance, and very luxuriant, looking like a small single Sunflower. *Silene acaulis* is everywhere, and no description can convey an idea of the dense way in which its flowers are produced. Starved between chinks, its cushions are as smooth as velvet, one inch high—though perhaps a hundred years of age—so firm that they resist the pressure of the finger, and so densely covered with bright rosy flowers that the green is totally eclipsed in many specimens. These flowers barely rise above the level of the diminutive leaves.

Soon we reached the meadow-land towards the bottom of the warm valley, and found this Piedmontese meadow almost blue with Forget-me-nots and strange Harebells, enlivened by orchids, and jewelled here and there with St. Bruno's Lily (*Paradisia Liliastrum*). This is one of the very best of all herbaceous or border plants, but I never saw it in such perfection as here in the fresh green grass. The flower is nearly two inches long, of as pure a white as the snows on the top of Monte Rosa. Each petal has a small green tip, like the spring Snowflake, but smaller and purer, and golden stamens adorn the interior of the flower. The pleasure of finding so many beautiful plants, rare in cultivation, growing in the long grass under conditions very similar to those enjoyed in our meadows, was greater than that of meeting with the more diminutive forms on the high Alp, verifying, as they did, the conviction which I had long entertained, that no flowers grow in those mountain meadows that cannot be grown equally well in the rough grassy parts of many British pleasure-grounds, woods, and copses!

From the top of the pass, in addition to the great glacier, two remarkable objects were seen—one an island, called the Belvedere, which breaks the descending ice river, dividing it into two branches, so fresh and green and garden-like as to seem quite out of place in such a position; the other a great moraine, so formal in outline that to the inexperienced it actually looked like a large embankment, the recent work of some railway company about to open the valley. But it, like all its fellows, is simply one of those colossal accumulations of rocks and grit borne down for ages by the great ice river and deposited along its flanks.

Next day we explored the Belvedere between the two branches of the glacier, and then turned to the left and traversed a great deal of the mountain above Macugnaga up to the line of snow, but, strange to say, found both the Belvedere moraines and mountains a desert, so far as rare alpine plants are concerned. *Soldanella alpina* was extremely abundant. The great bearded seed-heads of the fine alpine Anemone was a marked feature of the meadows in some places. The yellow alpine Anemone was not uncommon

The limit of life.

Where the birds dare not build, nor insects wing
Flit o'er the herbless granite."

higher up. The little two-leaved Lily-of-the-valley grew along with the common one in the lower fringes of the woods. The dwarf *Loiseleuria procumbens* half covered the mountains. The white-flowered *Ranunculus aconitifolius* was very common in the tall grass; this is the wild form of the double flower known in English gardens as the " Fair Maids of France." The sky-blue *Campanula barbata*, with delicate downy hairs about the margins of its bells, was very common, and the sweet *Primula viscosa* was everywhere. Coming over the pass of Monte Moro, it was in perfect condition and full bloom, and yet so small that a shilling would cover the entire plant. In lower spots on the opposite side of the valley single leaves of it were nearly three inches across and five inches long ! This will help to

Alpine Larch-wood.

show the fallacy of supposing that, because a plant is found in almost inaccessible places and hard chinks of cold alpine rock, we must attempt the nearly impossible task of imitating such conditions, or give up the culture of such an interesting class of plants.

The cliffs here rise in some parts like a vast wall to a

height of 8000 feet—stupendous and beautiful towers of rock and sunlit snow, perfectly lifeless, but reverberating now and then with small tumbling avalanches of the recently fallen snow. Above the village of Macugnaga, as in many other parts of the Alps, some of the Larch-woods are beautiful from the evidences of the struggle for life. Once the breath of summer has passed over the earth, the dwarf herbage is all freshness and life—the smallness and feebleness of the minute vegetation preventing us from seeing the stamp of the destroyer. The winter snow weighs down the little stems, and then when in spring their successors come up in crowds, the earth is covered with a carpet as if winter would never come again. But not so with the trees. Many lay prostrate, dead, barked, and bleached nearly white among the flowers that crowded up around them. Others were in the same condition, but leaning half erect amidst their fresh green companions: others were dashed bodily over the faces of cliffs: others had their heads and trunks swept over the cliffs by the fierce mountain storms, but holding on by their roots, and, assuming the quaintest and most contorted shapes, endeavoured to lift their living tops above the rocky scarp from which in their pride of youth they had been cast. I never in any wood saw anything so wildly and grimly beautiful as this. It suggested that it would be an improvement to sometimes allow old dead trees to remain in woods planted for ornament only.

We next resolved to descend into the plains of Lombardy, cross the lakes of North Italy, go as far as Lecco on the lake of Como, ascend Monte Campione, and find *Silene Elisabethæ*, a plant as rare as beautiful, and any like subjects which that region might afford. The long and ever-varying Val Anzasca, which runs from the foot of Monte Rosa to the great road from the Simplon, is unsurpassed for the grandeur, beauty, and variety of its scenery. We started from the Hotel Monte Moro at half-past three in the morning, when several of the highest peaks were illumined by a ruddy light, and all the lower ones were in the dull grey of daybreak. Almost every step revealed a fresh prospect of the mountains. The beauty of the orange Lily in the grass was something quite remarkable. Not growing higher than the grass, and in single specimens, not tufts, the effect was not what we are accustomed to see in Lilies. By looking over a ledge now and then, those small alpine meadows, apparently stolen from the vast wilderness, were seen thinly studded with large fully-expanded Lily blooms, every flower relieved by the fresh grass. *Asplenium septentrionale* was extremely abundant. Of flowers we saw but few, for the taller tree vegetation cuts off the view and runs up and

Cascade in a high wood.

clothes the secondary mountains to the very summits, except where grass that is like velvet spreads out as if to show the small silvery streams, which soon hide in the woods, and by-and-by are seen in the form of cascades falling over wide precipices, to be again lost in deep, wet, tortuous, stony beds, and presently forming larger cascades near the path of the traveller, who is obliged to cross them by bridges. Then lower down they break and shoot perhaps for three hundred feet, till they join the main stream of the valley below, which has cut itself an ever-winding, diving, and foaming bed between terraces, and cliffs, and gullies of rock, affording scenes of infinite beauty and variety.

We walked twelve miles down the valley before breakfast, and every step revealed a new charm. Before us, a great succession of blue mountains; on each side, mountain slopes green to the line of blue sky; behind, all the glory of the Monte Rosa group, in some places flat-topped and of the

Alpine Road through Cliff.

purest white, like vast unsculptured wedding cakes—in others, dark, scarred, and pointed to the sky, like some of the aged pines on their lower slopes, standing firmly but with branch and bark seared off by the fierce alpine blast. Lower down, the valley begins to show pleasant signs of human life, with really well-built and clean-looking houses; the slopes of the hills are frequently terraced, to give the necessary level for pursuing a little cultivation; and the churches are large and well decorated in the interior. Vines begin to appear, and for the most part are trained on a high loose trellis from five to seven feet above the surface of the ground, so as to permit of the cultivation of a crop underneath. The trellises are frequently held up by flat thin pillars of rough stone, which support branches tied here and there with willows. It seems a good plan for countries with a superabundance of light and sun. From nearly every rock and cliff along the valley spring the pretty rosettes and foxbrush-like panicles of flowers of the great silvery Saxifrage. But, beyond doubt, the charm of the valley is its ever-varying and magnificent scenery. Nothing can surpass many of the prospects from the lower parts, where you get a foreground of Italian valley vegetation—

Island in Lake Maggiore.

the deep-cut river bed below, the ascending well-clothed mountains to the right and left, and then up the valley the higher pine-clad slopes, all again crowned by the majestic mountain of the rosy crest. The most passionate and unreasoning love of country would be excusable in the inhabitants of these happy spots, enriched with the vine and other products of the south, sheltered by evergreen, and walled in by arctic hills.

We will hasten by the streams that feed Lake Maggiore,

and stop for a while near the islands on its fair expanse. Mountains with dense green woods creeping to their very tops are reflected in the transparent water in which they seem to be rooted, so near do they rise from its margin, and only showing their stony ribs here and there, where a deep scar or scarp occurs too precipitous for vegetation.

The isles look pretty, but not beautiful, because of the rather extensive and decidedly ugly buildings and terraces upon them; but they are only specks in a great natural garden, which even if dotted with smoke-polluted towns, like those in the North of England, would still be lovely. Brockenden is quite right when he says of one of the islands, "It is worthy only of a rich man's misplaced extravagance, and of the taste of a confectioner." The Maiden-hair fern is abundant on the islands. The vegetation here and on the margins of the lake is often of a remarkable and interesting character, quite sub-tropical in some places; but as our business is with alpine and rock plants only, we must pass all this by, and hasten on to the shores of Como. When approaching Isola Madre, the first thing that struck my attention was a plant like a greyish heath, covered with light rosy flowers, growing out of the top of a wall. It proved to be an old friend, the Cat Thyme, and in beautiful condition; as grown in England, nobody would ever suspect it to be capable of yielding such a bright show of flowers. *Trachelium cæruleum* grows very commonly on the walls, and so does the Caper, a noble plant when seen issuing from a wall and bearing numbers of its large blooms.

Arrived at Lecco, the next object is to hunt for the handsome Catchfly on the crest of Monte Campione, and we start at three o'clock in the morning, as it is desirable to get up a little out of the warm valleys before the dew has been dissipated. Soon we find ourselves on the spur of a mountain, on which Cyclamens peep forth here and there from among the shattered stones—sometimes a solitary bloom or two, at others handsome tufts, where the position has favoured free development, and now and then springing in a miniature condition from some chink, where there was very little nutriment or root-room to be obtained. Lower down we met with the neat *Tunica Saxifraga* on the tops of walls, and it continued to appear for some distance higher up, rarely looking so pretty as when well cultivated. The Maiden-hair fern does not ascend up the mountain sides, nor even find a home in the villages up the valley, though in the town of Lecco it adorns the mill wheels and moist walls near watercourses with abundance of small pretty plants adhering closely to the wall, and dwarf from existing on moisture or very little more. The pretty *Coronilla varia* is often seen low down; and what can form prettier tufts, or fall more gracefully over the brow of rocks? As we ascend, the fine flowers of *Geranium sanguineum* are everywhere seen, and *Horminum pyrenaicum* begins to show itself here and there, becoming more abundant as the mountains get higher, and growing to the top. It is barely worthy of cultivation; a pinkish variety was noticed in several places. The Privet, in a very dwarf and floriferous condition, adorns the rocks in abundance; while Aconites, Lilies, etc., are occasionally seen. The orange Lily is a great ornament hereabouts. I saw on one of the topmost and most inaccessible cliffs of the mountain one of its bold flowers like a ball of fire in the starved wiry grass, and small plants of it growing on a narrow ledge. The Martagon Lily is also abundant, though not so effective. Dwarf Cytisuses are great ornaments to the rocks, and here and there the leaves of Hepatica are mingled with those of Cyclamen, suggesting bright pictures of spring in these localities. The Cyclamens are deliciously sweet, and the great spread of *Erica carnea*, seen in all parts, must afford a lovely show of colour in spring.

The ground is rocky, and we think we have taken leave of all the meadow-land, when the hills again begin to break into small pastures, where Orchises, Phyteumas, Arnica, Inula, Harebells, and a host of meadow plants, struggle for the mastery. Soon we come to great isolated masses of erect rock, whose surface is quite shattered and decayed in every part; and, after half an hour among these, see far up rosettes of the blue flowers of *Phyteuma comosum*, projecting about two inches from the rock. The rosettes are as wide as the plant is high, and much larger than the leaves, which are of a light glaucous colour. We ascend far above these rocks, and find the mountain-side has broken into wide gentle slopes, park-like with birch and other indigenous trees here and there, but for the most part a great spread of meadow-land, adorned in every part with a glorious company of flowers. Conspicuously beautiful was the St. Bruno's Lily, growing just high enough to show its long and snow-white bells above the grass. It should be called the Lady of the Meadows, for assuredly no sweeter or more graceful flower embellishes them. In every part where a slight depression occurred, so as to expose a little slope or fall of earth on which the long grass could not well grow, or along by a pathway, *Primula integrifolia* was found in thousands, long passed out of flower.

In wandering leisurely over the grass, an exquisite Gen-

Scene in the higher Alps.

tian, of a brilliant, deep, and iridescent blue, came in sight. At first we thought it was the fine *Gentiana verna*, but on taking up some plants, it proved to be an annual kind, quite as beautiful and brilliant as either *G. bavarica* or *G. verna*, gems as they are. Wherever a boulder or mass of rock showed itself, *Primula Auricula* was seen, often in the grass and always on the high rocks and cliffs. A species of Pedicularis, with deep rosy shining flowers, is a fine ornament, and ascends to the very highest points. A showy Epilobium and Dentaria are also seen among the taller vegetation, while the compact little blue Globularia creeps from the surrounding earth over every rock. As we mount, the mist of the higher points begins to envelop us, and hide the lovely and ever-varying scenery below and on all sides, except now and then when the breeze clears the vapours away.

As the upper lawns are reached, the extraordinary nature of the mountain begins to be seen through the increasing mist. Lower down, and indeed in all parts, erect, isolated masses of rock are met with; but towards the great straight-sided mass that forms the central and higher peak, huge *aiguilles* are gathered together so thickly that, dimly seen through the mist, they seem like the ghosts of tall old castles and towers creeping one after the other up the mountain-side. The highest point, formed by a most imposing rock of this description, has never yet been ascended. Lower down, cliffs of the same nature and great height form one side of the mountain, their giant and weird appearance being much heightened by the mist which completely hid the valley and made them seem baseless.

Hereabouts we came upon some little tufts of the most diminutive and pretty *Saxifraga cæsia*. In little indentations in rocks it sometimes looked a mere stain of silvery grey like a Lichen; on the ground, it spread into dwarf silvery cushions, from one to three or four inches wide. It seemed quite indifferent as to position, sometimes growing freely along, and even in, a channel, the sides and bed of which are a mass of shattered rocks, and which is in winter a stream and a torrent after heavy rains and thaws. I found one plant as circular and as wide as a dessert plate, a mass of Lilliputian silvery rosettes, each about the eighth of an inch across, and formed of from fifteen to twenty-five diminutive leaves, and hundreds of rosettes going to form a tuft about an inch high.

This is one of the brightest little gems in the large Saxifrage family, which affords a greater number of distinct plants worthy of cultivation in the rock-garden than any other at present known to us. These plants grow upon the mountain tops, far above the abodes of our ordinary vegetation, not only because the cool, pure air and moisture are congenial to their tastes, but because taller and less hardy vegetation dares not venture there to overrun and finally extinguish them. But though they dwell so high in alpine regions, they are the most tractable of all plants in British gardens, and with but little attention grow away as freely as

PART I. *PLANT-HUNTING ON THE MOUNTAINS.*   135

our native lowland weeds in gardens where Gentian and alpine Primula and precious mountain Forget-me-not sicken and die. They are evergreen, and more beautiful to look upon in winter than in summer, so far as the foliage is concerned, and their foliage is beautiful exceedingly. But unlike many other things which have attractive leafage or a peculiar form and habit, they flower as freely in the

The limit of the Pines.

early summer as if they were herbaceous and uninteresting, instead of being evergreen and of exquisite chiselling.

· One would think that coming from habitats so far removed from all that is common to our phlegmatic and monotonous skies, it would be impossible to keep these little stars of the earth in a living state; and reasonably enough, as it would be easier to imitate the temperature of the hottest ravines of Borneo, or the clime where the unearthly-looking Welwitschia grows, than to produce in any way known to us even the faintest imitation of such a climate as theirs. But that is needless, as they can grow no better on their native hills than they do even within large towns and cities in the United Kingdom. Our climate suits them to perfection, and they are the chief glory of the cultivator of alpine plants. Hitherto they have been but very little appreciated. They are usually grown in pots, where people cannot see

View on the Simplon Road.

half their loveliness, and in which they sicken and dwindle. Not so when planted in the open air. In autumn, when most plants and trees are making themselves quite melancholy-looking before the approach of darkness, winter, and frost, and casting off their soiled robes, the Saxifrages are expanding their compact little rosettes, and glisten with silver and emerald when the rotting leaves are hurrying by before the stiff, wet breeze.

The Lion's-paw Cudweed is very abundant on Monte Campione. Daphne and Rhododendron in small quantities, and the pretty little *Polygala Chamæbuxus*, often crop out in a very diminutive state, much less beautiful than when in cultivation. A blue Linum, probably *L. alpinum*, is very common; the rare *Allium Victoriale* we found sparsely on high rocks; and *Dryas Octopetala* abundantly in flower, with *Anemone alpina* in a very dwarf state; while pale flowers of the common *Gentiana acaulis* looked up singly here and there. In the higher and barer parts of the meadows, *Aster alpinus* was very charming, not in tufts or masses, but dotted singly over the turf. Having climbed so high for the chief object of our ascent, we failed to find it there after a long search, and, disappointed, were descending the mountain down a long and rocky chasm formed of a vast bed with banks of shattered rock, when, much to our pleasure, a little plant with a few leaves was discerned growing from a chink on a low mass of rock. By carefully breaking away portions of this we succeeded in getting the plant, roots and all, out intact, and by very diligent searching, found a few more specimens of it. It was not yet in flower, but pushing up the stem preparatory to it. Then a long trudge down mountain, valley, and hilly road brought us home to our quarters at half-past nine, after a long and interesting day of nearly twenty hours' walking. A description of the scenery from the top of this mountain is better not attempted, and, indeed, for several hours near the top we could not see many yards before us because of a white mist. But at one time, when as high as we could go, we saw through a rent in the mist the far-off country below —lake, hills, and villa-dotted lowlands, warmed by a bright sun, and happy-looking as Eden, "when o'er the four rivers the first roses blew."

With a few words on the vegetation of some parts of the Simplon great range these notes will end. The chief feature of the smaller vegetation alongside the great Simplon Road is the foxbrush-like flowering pyramids of the great *Saxifraga Cotyledon*. The little *Campanula cæspitosa* is very abundant and pretty in some spots, and on the highest parts of the road, wherever the ground near it breaks into anything like turf, the vivid blue of the vernal Gentian sparkles amongst bright yellow Potentillas and Ranunculi.

It is pleasant to meet with it in flower weeks after one has left it in full flower in England in April, and seen it bear seed on mountains about 5000 feet high. About the end of June it was in fresh and perfect condition here, and likely to remain so for some time to come. Observe the capabilities of the plant, and the changes that it endures without losing health in any case. In perfect health in England, without a covering of snow through the winter,

A Glacier.

and flowering strongly in early spring, it flowers here in the month of June, and higher up in July.

Let us ascend one of the highest mountains of the range a little way, climb upwards for two hours, passing the limits of the pines, till we get at the base of the bed of an enormous glacier, a vast high field of snow apparently, which fills the upper portion of a wide gap between two mountains. Here and there you see flakes of it like green glass, and its face, where the water wears itself arches in issuing from beneath the slowly melting mountain of ice, is also of that tint. The wide expanse of ground which we are traversing is simply a mighty bed of shattered rock, which at a remote day was carried down by this colossal, ever-gathering and ever-levelling machine, and it is now covered with a scanty vegetation of alpine Rhododendron and high mountain plants.

Everywhere, and very pretty, is the mountain form of the Wood Forget-me-not, but no trace of the true *Myosotis alpestris*. Apparently the white form of the Wood Forget-me-not is very abundant among the blue, but upon looking closer, the simple-looking white flower growing amongst the Forget-me-nots is seen to be a white Androsace. Everywhere the large white flowers of the mountain Avens are covering the surface; but as we are in such rich ground, we had better confine ourselves to plants not British, and—climb. That exertion is above all things necessary; the vast slopes of shattered rock seem interminable—an hour's hard work brings you to a point that you thought you could reach in five minutes, and this point, instead of proving the resting-place and exploring-ground you had expected it to be, merely shows you that still the wide and mighty mass of shattered rock creeps higher and higher, far beyond your powers of approach, until at last the wall of ice, "durable as iron, sets death like its white teeth against us." On a great ridge beneath it are some scattered fragments of vegetation rooting deeply among the stones and gaining a scanty subsistence from the sandy grit which results from the decomposition and friction of the fields of brittle rock. The opposite-leaved Saxifrage is a perfect mass of flower; you cannot see anything but flowers on its dense cushions, here as beautiful in this awful solitude as the choicest flowers of climes genial enough for the humming-bird. Here and there a large yellow flower is seen, which proves to be *Geum reptans*, a fine plant, from three to six inches high. Presently, while admiring the great beauty of the crimson Saxifrage here, within a few feet of wide beds of snow, that lie on each side of the ridge on which we stand, what appears a giant specimen comes in sight; the flowers are much larger, so that instead of little cushions made up of a multitude of blooms, we see the individual cup-like blooms standing boldly up, of much deeper hue, and the leaves also grown large and distinct. It is the noble *Saxifraga biflora*, and I hope nobody will object to calling noble that which only grows about half an inch high! It is raining heavily, and the place is anything rather than cheerful, but it is a very great pleasure to gather this plant here, and also *Linaria alpina*, more familiar to me, and so beautiful here that I can hardly hope to give the reader an idea of it. Many alpine plants are prettier in cultivation than in a wild state, for instance, *Polygala Chamæbuxus*, which grows here—just venturing out one or two little shoots and flowers at a time. Not so *Linaria alpina*, which grows and flowers well in sandy soils and moist places at home, and gets so strong that its glaucous leaves form quite a strong tuft, almost high enough for an edging plant, but which here shows its rich orange and purple flowers, gathered in dense tiny tufts here and there among the stones, without any leaves being perceptible. It

A Glimpse at the Home of the two-flowered Saxifrage.

is infinitely more lovely here than in cultivation, though its beauty in either case is of a high order. The very dwarf and pretty little *Campanula cenisia* was abundant among the higher plants, its tufts of very light green growing among the débris. By turning over the stones, plants with good roots could be got out. One solitary tuft of *Ranunculus alpestris* was met with by the side of a little rivulet; it was a roundish specimen, about six inches in diameter, and quite pretty where "specimens" are rare, and where one thing struggles with another in the grass.

Descending, the ground, becoming more level, begins to form an undulating basin between two ranges, and here the short grass is perfectly jewelled with dwarf alpine plants and flowers. The silky-leaved and very dwarf *Senecio incanus* occurs in thousands, the Cudweeds too are abundant, while a few inches above the dense silvery turf formed by such plants, the large and beautiful purple flowers of *Viola cal-*

carata form, not quite a sheet of colour—for the flowers occur singly, and are separated one from the other by bits of green and silvery turf—but sometimes the eye is brought nearly level with the surface of a bank dotted with blossoms, and the effect is lovely. It is not the effect of "massing" flowers, but that of "shot" silk. The flowers of this Violet were generally very large—I measured several an inch and a half across, while the plants from which they sprang were almost inconspicuous, and generally I had to use the flower stem as a guide to the minute rosette of leaves in the grass. A still more beautiful effect, and perhaps more so than I have seen either in flower-garden or wild, was observed when tufts of *Gentiana verna* occurred pretty freely amongst this Violet, the vivid blue of the Gentian in patches amongst the groundwork of the Violet. In quite a valley of Gentians—a little lawn at an elevation of about 7000 feet—I noticed some growing in a watery hollow. I had almost passed them by when I chanced to look closely down to admire their deep, vivid, and exquisite blue, and saw that they were large tufts of *Gentiana bavarica*. The little Box-like leaves were in compact tufts, and the flowers were larger, of a deeper and more beautiful blue, than *G. verna*, which is saying a great deal. I have one specimen now with thirteen perfect blooms—a by no means selected specimen—in a single close tuft, not more than an inch and a half across.

There were spots near at hand, where *G. verna* formed a turf of its own, and yet it was not so beautiful as *G. bavarica*, which was growing exactly in positions that would suit the Bog Bean and the Marsh Marigold. Attempts to cultivate *G. bavarica* in England have hitherto been a failure. It is very rarely seen with us even in botanic gardens, and, when it is seen, is usually yellow and in poor health. A few words, then, about the position in which I found it in such perfection may prove useful. A little mountain streamlet diverges from its channel and spreads over the surface of the ground for twenty or thirty yards across, not destroying the grass, but simply showing itself in trickling patches here and there. On the little hillocks of grassy earth that stood a few inches above the water, I found the plant in very good condition, the roots certainly in the water, and the "collar" of each plant very little above it. Somewhat lower down the waters gathered together again, leaving the sides of that marshy spot and the intermediate ground perfectly green, but very wet, and here and there dotted with clusters of blue stars, to which in brilliancy of tone the choicest gems ever seen were but dull and earthy. In walking on this green spot the water hissed and bubbled up around. Here the specimens were very fine, the pretty little close-growing tufts of light green leaves clearing spots for themselves in the longish grass. The slightest impression made here immediately became a small pool, and in no place did I find the plant but where the hand, if pressed into the grass, became immediately surrounded by water. A few steps away and *Gentiana verna* was everywhere in full beauty on dry banks; but in no case did either species manifest a tendency to invade the ground of the other. In fact, proof was there that *G. bavarica* is a true bog-plant. And what a beautiful companion for the Wind Gentian, the Water Violet, the fine white Bog Arum, the moist-peat-loving *Spigelia marilandica*, and the early Myosotis *(M. dissitiflora)*, which loves a bog, *Rhexia virginica*, the little creeping Bell-flower and like plants! Why, it is worth our while to make a little bog, with a surface of Sphagnum and dwarf bog plants, for the sake of growing this exquisite Gentian.

Scene in the Rocky Mountains.

MOUNTAIN VEGETATION IN AMERICA.

The passage of the great American desert which is crossed on the way from New York to San Francisco is, perhaps, the best preparation one could have for the startling verdure and giant tree-life of the Sierras. Dust, dreariness, alkali—the earth looking as if sprinkled with salt; here and there a few tufts of brown grass in favoured places; but generally nothing better than starved wormwood, that seems afraid to put forth more than a few small, grey leaves, represents the vegetable kingdom in the plains of the desert region. Where the arid hills—showing horizontal lines worn by the waves of long-dried seas—are visible, a few thin tufts of alders and poplars mark their hollows; while willows fringe the streams of undrinkable water which course through the valleys. A better idea of the country can scarcely be had than by imagining an ash-pit several hundred miles across, in which a few light-grey weeds, scarcely distinguishable from the parched earth, have sprung up, regardless of drought.

As the train ascends the Sierra, it passes through darkribbed tunnels of long covered sheds, which guard it from the snow in winter. Dawn broke upon us as we were passing through these; and, looking out, we saw such a change from the Salt Lake scenery as one experiences in passing from a hot dusty road to a cool, green, ferny dell. Dust, alkali, dreariness, harshness of arid rock and hopelessness of barren soil, are seen no more. Near at hand a giant pine rushes up like a huge mast, while in the distance

Isolated Rocks in Rocky Mountains.

great pines grouped in stately armies of tree grenadiers, filling the deep valleys and cresting all the wave-like hills till these are lost in the distant blue.

To the western slopes of the great chain of the Sierras one must go to see the noblest trees and the richest verdure. There every one of thousands of mountain gorges, and the pleasant and varied flanks of every vale, and every one of the innumerable hills, are densely populated with noble pines and glossy evergreens, like an ocean of huge land waves, over which the spirit of tree-life has passed, creating giants. The autumn days I spent among these trees were among the happiest one could desire. Every day glorious sunshine, and the breeze as gentle as if it feared to hurt the long-dead trees standing here and there leafless, and often perhaps, barkless, but still pointing as proudly to the zenith as their living brothers. Wander away from the little rough dusty roads, crossing, perhaps, a few long and straight banks of grass and loose earth—the remains of dead monarchs of the wood, now rendered back to the dust from which they once gathered so much beauty and strength—and fancy willingly reminds us of the mast-groves of the Brobdingnags. There is little animal life visible, with the exception of a variety of squirrels, ranging from the size of a mouse to that of a cat, the graceful Californian quail, and occasionally a hare or a skunk. Everywhere vegetation is supreme, and in some parts higher effects are seen than in the most carefully-planted park or pleasure-grounds in the most favoured climate. This results not more from the stately pines (not often crowded together as in the Eastern States, but with perfect room for development, and often near the crest of a knoll, standing so that each tall tree comes out clear against the sky) than from the rich undergrowth of evergreens with larger leaves that form a smaller forest beneath the tall trees. Grand as are the pines and cedars (Libocedrus), one is glad they do not monopolize the woods; the evergreen oaks are so glossy, and form such handsome low trees. One with large shining leaves, yellowish beneath, and long acorns in thick cups, covered with a dense and brilliant fringe of fur, was the most beautiful oak I ever saw; but most of the evergreen oaks of California, whether of the plains or hills, are very ornamental trees. One day, in a deep valley darkened by the shade of giant specimens of the Libocedrus, I was astonished to see an Arbutus, about sixty feet high, quite a forest tree. This is Menzies' Arbutus, commonly known by the old Mexican name of the "Madrona"; and a very handsome tree it is, with a cinnamon-red stem and branches. Here and there, too, the Californian laurel (Oreodaphne) forms laurel-like bushes, and tends to give a glossy, evergreen character to the vegetation. Shrubs abound, the Manzanita (Arctostaphylos glauca) and the Ceanothuses being usually predominant; while beneath these and all over the bare ground are the dried stems of the numerous handsome bulbs and brilliant annual flowers, that make the now dry earth a living carpet of stars and bells of brilliant hues in spring and early summer.

On the very summit of the Sierra Nevada the vegetation is not luxuriant; there, as elsewhere on high mountain chains, is the frost that burns and the wind that shears. When you see a solitary pine that has been bold enough to plant itself among the boulders and rocks of the high summits, it is usually so contorted that it looks as if inhabited by demons; while here one has succumbed to the enemy, and a few blanched branches stick from a great,

Mountain Woods of California.

dead, barkless base, lapped over the earthless granite. But go a little lower down the mountain, and most probably you will find a noble group of Piceas, startling from the size and height of their trunks, though looking much tortured about the head by the winds that surge across these summits—the mast-heads of the continent. Snow falls early and deep on the Sierras, and the stems of the higher trees are often covered with it to a depth of from six to twenty-five feet. Near the railway, and near frequented places, thick stumps of pines, six to fifteen feet high, may be noticed; these are the trees which have been cut down when the snow was high and thick and firm about the lower part of their stems. But if the nights are bitterly cold, the sun is strong in the blue sky far into the winter months, so that the snow is melted off the tree tops, and the leaves of the pines live, in golden light, throughout the winter. All the pines that grow near the summit must resist the most piercing cold.

The golden light of the sky and the blue of its depths, and the purity of the fresh mantle of snow, are not more lovely in their way than the robe of rich yellow lichen with which the stems and branches of the pines are clothed. Imagine a dense coat of golden fur, three inches deep, clothing the bole of a noble tree for a length of one hundred feet, and then running out over all the branches, even to the small dead twigs, and smothering them in deep fringes of gold, and some idea may be formed of the glorious effect of this lichen (Evernia). It is the ornament of the mountain trees only; in the valleys and on the foot-hills it is not seen.

Those who have not visited the high lands of California can have no idea of the size and majesty of the trees there. It is a mistake to suppose the Sequoia (Wellingtonia) is such a giant among them; several other trees grow nearly or quite as high, and it is very likely that in such a climate all the pines known in gardens would attain extraordinary dimensions. There was a small saw-mill near where I stopped for some days, and several yokes of oxen were constantly occupied in dragging pine logs to it. The owner never thought of bringing anything smaller to this than a log three or four feet in diameter in its smallest part, and usually left one hundred feet or so of the portion of the tree above this on the ground where it fell, as useless. What is it that causes the tree-growth to be so noble there? There can be no doubt that the climate is almost the sole cause. Soil has very little to do with it. I have frequently noticed the trees luxuriating where there was not a particle of what we call soil, and, indeed, in places where twenty-five feet or so of the whole surface of the earth had been washed away by the gold-miners. A bright sun for nearly the whole year and an abundance of moisture from the Pacific Ocean explains the matter. This should draw our attention to the fact that, in ornamental planting, and especially in the planting of coniferous trees, we pay far too much attention to supplying them with rich and deep soil and far too little consideration to the capabilities of the climate in which we have to plant.

There is a foot or two of snow in some places on November 15, 1870; but the time for very deep snow has not yet come, and we are fortunately in time to see a patch of alpine plants here and there before they are tucked in under their wintry shroud. What are these brown tufts like withered moss among the rocks and boulders on exposed spots, some of them cushioned low and flat; others looking as if moss had assumed a shrubby habit and died full of years, at three inches high perhaps, on a gouty stem nearly as thick as the finger? These are little Phloxes, withered almost beyond hope by the heats of summer; but pull up one, and the old roots are seen sending out a mass of fragile feeders in the snow-moistened earth, and in the very centre of each juniper-like truss of prickly leaves may be discerned a small speck of green. When the twenty feet of snow melts in spring, and the glowing sun warms the saturated earth, these mites of Phloxes will be to the now arid solitudes as blossoms to the crabbed apple tree. The dead moss will change to bright, shining green, and presently this will be perfectly obscured by as fair a host of flowers as ever fretted over the small herbs on Swiss or Tyrolese Alp. There is as much difference in size between our common border Phloxes—the parents of which are wild in the middle and southern states of America—and the diminutive mountain Phloxes I speak of, as between swans and humming-birds. The alpine Phloxes of the Rocky Mountains or the Sierras are as indispensable to the choice collection of alpine plants as the Gentians or the Primroses. Very few of them have been introduced.

Everywhere on bare places there are tufts of dwarf, bush-like Pentstemons, from 2 to 5 inches high, and bearing nearly the same relation to the tall Pentstemon of our gardens as the tiny Phloxes do to the border Phloxes. The Pentstemons are among the most beautiful of rock-plants, their colours being of a more refined and delicate character than those of the tall varieties, richly toned as these are. Indeed, no flowers possess such indefinable shot-silk-like blues and purples as these. I secured a stock of two diminutive new species. Like the little Phloxes, many of these have woody stems, probably as old as some of the pines near at hand,

and have embellished these lonely heights for ages unadmired, unless the "grizzlies" or the woodpeckers delight in such objects.

It might perhaps be thought that, however well the minute plants thrive among rocks and boulders, the giant pines would require good soil, or at all events level ground of some kind, to start from. It is not so. A seedling pine springs up in some shallow chink or narrow crack in a mass of great stones; patiently it throws out long feeders on one side, which find their way down the steep faces of the rocks or run through any moist or narrow channels into the feeding ground beyond; it soon gathers strength enough to build a great trunk above the narrow chink from which it sprang, lapping its base over the close-embracing rocks much as a fungus would. I have seen trunks measuring 18 feet in circumference springing from masses of raised rocks, where one would not think a wiry juniper bush could live.

On looking at some compact brownish tufts of leaves, a few yellow Coronas are seen; these are somewhat "everlasting" in character, and have only faded with the snow-water. They belong to quite a distinct plant of the Buckwheat family—Eriogonum. The family we know is nearly all composed of weeds, and the genus, which has many members in America, is seldom in the least attractive; but this one is quite a gem of a rock-plant. Handsome umbels of primrose-yellow springing abundantly from dull brownish tufts of leaves two inches high, and making it as ornamental as it is distinct. Far away, on a bare, gravelly hill-side, vivid red tufts are seen; these prove to be another equally beautiful species of Eriogonum, the leaves of which assume a deep, shining blood colour. Above these are numerous Coronas with pretty yellow flowers.

Here and there the withered stems of Lilies may be seen; the Orange Lily creeps up high on the rocks of Piedmont; here Washington's Lily—a tall, noble, and fragrant kind—and several other Lilies occur abundantly. The stems of some which I found in little ravines were quite eight feet high, and I am told they grow even higher. The Soap-plant—a bulbous perennial—is abundant on all the lower mountains and on the coast hills. A morsel of it raises a lather immediately. Numerous bulbs of a high order of beauty occur on the mountains and plains of California.

Another very beautiful rock-shrub, quite distinct from anything we have in our European Alps, is the Bryanthus. After trudging for hours over snow and rock in quest of this, I had given it up, when a spray, with a withered truss of bloom, was seen, and soon I had dug a few score plants of it from beneath a couple of feet of snow. This Bryanthus may be roughly described as having the leaf of a heath, with handsome crimson flowers, like those of a small rhododendron, and forming bushes from four to ten inches high. It will prove one of our handsomest rock and border shrubs. Another rock-shrub, quite distinct from all others, is a creeping Ceanothus. The Ceanothuses are usually erect shrubs. This one runs along the ground as closely as the Periwinkle, or more so. On the lower hills, where it grows more freely, the shoots march in parallel lines over the ground, covering it with a rigid carpet of dark green leaves.

One of my objects in coming here was to see the Californian Pitcher-plant (*Darlingtonia*) in a wild state. This plant resembles the Sarracenias of the eastern side of the continent, the chief difference being that it has a cleft appendage to the margin of the orifice of the pitcher, each lobe being from 1 to 2 inches long. I came upon the Darlingtonia, greatly to my satisfaction, on the north side of a hill, at an elevation of about 4000 feet, growing among Ledum bushes, and here and there in sphagnum, and presenting at a little distance the appearance of a great number of Jargonelle pears, with their larger ends uppermost, at a distance of from 10 to 24 inches above the ground. This resulted from the pitchers being quite turned over at the top, so as to form a full rounded dome, and the uppermost part of the pitcher being of a decided ripe-pear yellow. The plants grow in small sloping bogs, resulting from springs on the hill-side; the soil is peat, resting on a quartz gravel. The plant is quite a strong grower. I found one large colony growing so vigorously among common rushes that Darlingtonia seemed to be quite beating the Juncus in the struggle for life. I was too late for seeds, but saw sundry stems 3 feet or more high, bearing empty seed vessels as large as large walnuts. All the pitchers have a spiral twist, which is much more marked towards the apex, and in the large specimens. But perhaps the most remarkable feature of the plant is its efficiency as a "fly-catcher." In the houses about here the pitchers are regularly used in summer for catching flies! Each of the developed pitchers that I cut off had from 3 to 5 inches of various forms of insect life, dead and closely packed in the lower part of its chamber. Pass a sharp knife through a lot of brown pitchers withering round an old plant, and the stumps resemble a number of tubes densely packed with the remains of insects. What attracts them is not so very clear, as the orifice is half hidden in the turned-over head, and by its two-lobed appendage. But, by raising the pitcher above the eye, and looking up into its dome, often 3 inches through in fair specimens, it seems a curvilinear roof of miniature panes set in a golden network. This is in consequence of the greater portion of the upper part of the pitcher being transparent in all the space between the veins, though no one transparent spot is more than a line or two across. Within the pitcher the surface is smooth for a little way down; then isolated hairs appear; and soon the chamber becomes densely lined with needle-like hairs, all pointing down, so decidedly indeed that they almost lie against the surface from which they spring. These hairs are very slender, transparent, and about a quarter of an inch long, but have a needle-like solidity, and are perfectly colourless. The poor flies, moths, ladybirds, etc., travel down these conveniently arranged stubbles, but none seem to turn back. The pitcher, which may be a couple of inches wide at the top, narrows very gradually, and at its base is about a line in diameter. Here, and for some little distance above this point, the vegetable needles of course all converge, and the unhappy fly goes on till he finds his head against the firm thick bottom of the cell, and his retreat cut off by myriads of bayonets; and in that position he dies. Very small creatures fill up the narrow base, and above them larger ones densely pack themselves to death. When held with the top upwards, sometimes a reddish juice, with an exceedingly offensive odour, drops from the pitchers. The plant throws out runners rather freely, by which means it increases. As to its culture, there can be no doubt about that—a soil of peat, or peat and chopped sphagnum, kept wet—not merely moist—the pots or pans to be placed on a moist bottom. Frame or cool house treatment is best in winter; warm greenhouse or temperate stove in summer. It would probably prove hardy in the south of England and Ireland.

PART II.

ALPINE FLOWERS.

A selection of the choicest Alpine Flowers alphabetically arranged, with instructions for the culture and position suited for each kind.

ALPINE FLOWERS.

PART II.

ACÆNA MICROPHYLLA.—*Rosy-spined Acæna.*

A MINUTE trailer from New Zealand, curiously beautiful from its small, close, round head of inconspicuous flowers being furnished with long crimson spines. The leaves are pinnate, the leaflets deeply incised, those at the apex of the leaf much the largest, the whole of a brownish green tint. The plant spreads into dense tufts, no taller than the Lawn Pearlwort, and in summer and autumn becomes thickly bestrewn with the showy and singular globes of spines. It is quite easily increased by division, is perfectly hardy, grows in ordinary soil, but thrives much the best in that of a fine sandy and somewhat moist character. Its home is on bare level parts of the rockwork, usually beneath the eye, and it is also good as a border or even an edging plant in soils where it thrives. Occasionally it may be used with a singularly good effect to form a carpet beneath larger plants not thickly placed.

ACANTHOLIMON GLUMACEUM.—*Prickly Thrift.*

A VERY compact and distinct little alpine plant, with dark-green pink-like leaves, with sharp spines at the points, and bearing one-sided spikes of pretty rose-coloured flowers—each a little more than half an inch across. It seems to thrive on almost any kind of soil, but is best suited for rockwork, on which it forms neat tufts from three to six inches high, wrapping itself round the stones, and blooming freely in summer. I have found it thrive perfectly well on slightly elevated rockwork far into London. It may be propagated by seed, cuttings, or division; but not very rapidly in the last way, and it should be divided very carefully. A native of Armenia, perfectly hardy everywhere in this country, at least when elevated on rockwork or banks. Synonyme *Statice Ararati.*

ACHILLEA ÆGYPTIACA.—*Egyptian Yarrow.*

A VERY silvery plant in all its parts, with finely cut leaves, and handsome heads of rich clear yellow flowers. It is distinct from any other kind, and, while quite equal to any of its relatives in beauty of flower, has something of the grace of an elegant fern in its leaves. A native of Egypt and Greece, and probably widely distributed in the East, it is not hardy in all soils and positions, but is quite so on well-drained sunny sides of rockwork, and I have observed it survive out of doors in borders. In a wild state it seldom grows more than about eight or ten inches high, but in rich light garden soil it reaches fifteen or eighteen inches. It is very suitable for the embellishment of rockwork among the taller plants, and may also be used in the mixed border or the summer flower garden. On the rockwork the best way to treat it would be to plant it in light loam mixed with brick rubbish, and in this it would grow compactly and survive many years. On chalky or very dry warm banks it would probably prove a hardy perennial. It flowers in summer and early autumn, and is very easily multiplied by division. When grown as a bedding plant, it is best kept over the winter in frames; and if the flowers are pinched off, it forms a dense mass of elegantly cut and very silvery leaves, and for this reason alone should prove very useful in the flower-garden.

ACHILLEA CLAVENÆ.—*White Alpine Yarrow.*

A DWARF and distinct sort, covered with a very short, silky down, which makes the plant almost of a silvery white. It seldom rises above six inches, and the corymbs of flowers, which appear in summer, are of a pure white; but the plant will probably be as much grown for its very silvery foliage as for its flowers. It likes light peaty soil or free loam, and should have a position on rockwork, where its white foliage and flowers would contrast well with the alpine plants that flower at the same season. Though cultivated nearly 200 years ago in the Edinburgh Botanic Garden, it is now very seldom seen in our gardens. A native of the Alps of Austria and Styria, increased by careful division of the root, and also by seed, though seed of it is not common.

Another white-flowered Achillea (*A. umbellata*) has lately been introduced to gardeners; it is smaller than the preceding, and useful as a silvery edging plant, but the flowers are not ornamental, and I am not certain of its hardiness.

ACHILLEA TOMENTOSA.—*Downy Yarrow.*

THIS is one of the little tufted plants that help to form the carpets of silver whereon large and handsome horned Violets and Gentians display their charms on the Alps, itself sending up in due time flat corymbs of bright yellow flowers. On elevated situations it is very dwarf and downy, but in rich soil in gardens it rises to six, nine, and twelve inches high. It is a good plant for the front margins of mixed borders, and also for the rockwork. A native of the European Alps, easily grown in ordinary soil, and readily increased by division.

ACIS AUTUMNALIS.—*Autumnal A.*

A VERY slender-leaved little bulb, with stems rising three or four inches high, and bearing a couple of flowers, that may be described as delicate pink snowdrops, drooping elegantly on short reddish footstalks, of a deep-red colour round the seed-vessel, and blooming in autumn before the leaves appear. It is a true gem for the rockwork, where it should be planted in a warm soil and sunny position, sheltered with a few stones, and on which it would look very well springing from a carpet of delicate, feeble-rooting *Sedum* or other dwarf plant. I have never seen it in nurseries except about Edinburgh, and first met with it in the late Mr. Borrer's garden in Sussex. Where the soil is of a fine sandy nature, it will thrive as a border plant, but is as yet so rare as to be worthy of the best position and care. A native of Spain and Southern Europe.

ADONIS VERNALIS.—*Vernal A.*

THIS, as regards size of flower, is the queen of all the Buttercup and Globeflower race. Early in May, its flowers, two to three and even four inches across on strong plants, spring from masses of light green finely cut leaves. Had Wordsworth seen a healthy plant of this in full blow, he would never have supposed the little Celandine had sat for its portrait to the artist who painted the "Rising Sun." It is a first-rate border plant, growing about a foot high, but it does very poorly in cold stiff soil, flourishing to great perfection where the air is somewhat moist and the soil light and good. Nothing can be finer where it grows healthfully; where it does not, it is not worth cultivating in the border, but on rockwork it will be easy to give it deep and light soil. It is a native of warm spots on the higher European Alps, flowering soon after the snow melts, and in our gardens in early summer. There is a variety called *A. sibirica*, which is said to have larger flowers, but is probably not in cultivation.

The Pyrenean Adonis (*A. pyrenaica*) is very like this plant, but has usually fewer, smaller, and more obtuse petals scarcely denticulated at the top, grows somewhat taller, and has radical leaves with long stalks, whereas those of *A. vernalis* are abortive or almost reduced to mere scales. It is not sufficiently removed from *A. vernalis* to merit culture except in large collections.

ÆTHIONEMA CORIDIFOLIUM.—*Lebanon Æ.*

A LITTLE glaucous half-shrubby plant, with an abundance of thin, wiry stems, bearing narrow grey leaves and a multitude of pretty rosy flowers, arranged at first in a compact head, which becomes elongated as the flowering season advances. This is one of the sweetest alpine plants in existence, and so hardy and free that it may be generally grown. I first met with it in M. Vilmorin's garden, near Paris, growing in quantity in a long bed of the sandy soil of the neighbourhood, the dense spray of leaves and wiry stems, about six inches high, thickly dotted with the delicate rose-coloured flowers. It had flourished in the same position for several years. Plants raised from seeds I brought from Paris have done quite as well in the neighbourhood of London. It succeeds perfectly well on the front margin of the mixed border; and though rockwork is not required for success with it, its presence will certainly be a gain to every rockwork where the highest beauty of alpine plants is sought. In consequence of the prostrate spreading habit of the stems, a pleasing result will be produced by planting it in one or two positions where the roots may descend into deep earth, and the stems fall over the face of rocks at about, or somewhat above, the level of the eye.

It is very readily raised from seed, is a native of Mount Lebanon, and will enjoy the sunny side of the rockery, though hardy enough for any position, if in a well-drained and sandy soil.

Æ. saxatile, parviflorum, and *membranaceum*, are also in cultivation, but I have observed none of them thrive so freely or look so well as this.

AJUGA GENEVENSIS.—*Erect Bugle.*

THIS has violet-blue flowers, springing thickly from the axil of every leaf and leaf-like bract, the stem being literally a cone of flowers for a length of four or five inches, or sometimes more. As the stems are produced almost as thick as they can stand, it is a very pleasing plant, and placed on the outer margins of shrubbery and mixed borders grows into round spreading tufts eight to ten inches high. It would also be suitable for rockwork, but where there are alpine plants rarer and more difficult of culture, it will hardly be wise to give it a place there, except in the roughest parts. It is probably the best of its family, and is easily increased by division. The true plant, widely distributed on the continent, is not found in Britain, but the variety with the floral leaves large and longer than the flowers, and having a dense leafy spike (*A. pyramidalis*), is found in Scotland, and is sometimes grown in gardens; it is not so ornamental as the typical form.

The common British Creeping Bugle (*A. reptans*) is grown in gardens under various names for the sake of its dark browny-purple leaves, and a variegated variety of it is sometimes grown in the spring garden, and in collections of hardy variegated plants.

ALYSSUM ALPESTRE.—*Alpine A.*

A PRETTY and bright little species, partaking of the brilliant colour and free-flowering properties of the well-known Rock Alyssum, and the neatness of habit and dwarfness of the Spiny or the Mountain A. It forms neat tufts of hoary entire leaves on stems woody at the base, the whole plant being covered with minute, shining, star-like hairs, and, so far as I have observed, not growing more than three inches high. It has, however, as yet been cultivated but very little in this country; and though recorded as in cultivation so long ago as 1777, it was lost to our gardens till recently re-introduced. A native of the Pyrenees and mountains of Switzerland, Italy, and Greece, its home with us is in sunny spots on rockwork, the soil to be of a light and poor, rather than of a rich, nature. Flowers in early summer, and is readily increased by seed or from cuttings. The Silvery A. (*A. argenteum*), a native of Corsica, is closely related to this species, but is taller and more robust, has small flowers, and is not so well worthy of culture.

ALYSSUM MONTANUM.—*Mountain A.*

A CHARMING and distinct species, spreading into compact tufts of slightly glaucous green, two or three inches high, and with oblong or obovate leaves. In April the flowers commence to open, and in May the plants are studded with yellow, alpine-wallflower-like blooms, sweet-scented, and produced abundantly on healthy specimens. The beautiful stellate hairs which are produced so freely by this family are large enough on this kind to be seen by the naked eye. It is a native of many mountainous parts of Europe, on hills and low mountain ranges, chiefly in sunny positions and on calcareous formations. I have grown it well on cold heavy soil, but it is almost certain to perish on such during winter. To succeed perfectly with it, it is desirable to place it on the rockwork in good sandy soil, or in some slightly elevated position, and so situated it will prove a beautiful ornament, especially when it grows into large cushions, on one side perhaps falling over the edge of a rock; readily increased by division, cuttings, or seeds, though it does not often seed freely with us.

ALYSSUM SAXATILE.—*Rock A.*

THE most valuable of the yellow flowers of spring. It is perfectly hardy in all parts of these islands, and the extreme brilliancy and profusion of its masses of bloom, combined with its capacity for growing in any soil or enduring any ill-treatment, have made it one of the most popular of garden plants. It is most frequently grown in half-shady places, under trees and shrubs, and where it has little chance of becoming fairly developed or showing its full flush of bloom; but it, like most rock plants, should be fully exposed. It is well fitted for the decoration of the garden of spring bedding plants, the mixed border, and rockwork, and also for association with the evergreen Candytufts, Aubrietias, &c., for fringing shrubberies, and for like purposes. On wet ground it is better to put a few plants in an elevated position and in poor soil : that is, if it be not grown on rockwork, as I have seen it perish in winter in heavy, rich clays, when on the level ground. Very easily raised from seed, or by cuttings. Comes from Podolia in Southern Russia, and flowers with us in April or May. There is a somewhat dwarfer variety, distinguished by the name of *A. saxatile compactum*, but it differs very little from the old plant.

ALYSSUM SPINOSUM.—*Spiny A.*

THE flowers of this are small and in no sense ornamental, but the plant forms such a distinct-looking, silvery, neat, and pretty little bush on any kind of soil, that I think it has quite as good a right to be named here as many others valued for their flowers alone. Small plants quickly become Lilliputian silvery bushes, three to six inches high ; when fully exposed, almost as compact as moss. The leaves are covered with small stellate hairs, and form interesting objects under the microscope. On established plants the old branches become transformed into spines : hence its specific name. It is entirely distinct in appearance from anything else in cultivation, and merits a place on some not over-valued spot on rockwork. It may also be used as a permanent edging plant, and should find a place in all collections of silvery-leaved plants. It is readily increased from cuttings, and comes from Southern Europe.

ANDROMEDA FASTIGIATA.—*Himalayan A.*

A REMARKABLY neat little shrub, with the branches closely overlapped along the stems, so as to make them square like those of *A. tetragona*, but distinguished from that plant by the leaves having a white, thin, chaffy margin terminating in a small point, and also a deep and broad keel. It is also larger in all its parts. The flowers, of a waxy white, produced at the top of each little branchlet, are turned down bell-fashion ; the reddish-brown calyx spreads half-way down the waxy flowers. This, one of the most rare and beautiful plants that we have obtained from the Himalayas, is, happily, not so difficult to grow as the mossy Andromeda, though it requires care. It has been successfully grown by Dr. Moore, in the Royal Botanic Gardens at Dublin, and should have a very sandy, moist peat soil. It is most likely to thrive in moist and elevated districts; but, safely planted on rockwork in deep, moist, but well-drained soil, and carefully guarded against drought during the warm season, it may be grown without difficulty; and doubtless, when it becomes sufficiently plentiful, it will be found, like the next species, to thrive very well in the natural soil of some districts.

ANDROMEDA HYPNOIDES.—*Mossy A.*

A MINUTE, spreading, moss-like shrub, one to four inches high, with wiry, much divided branches, densely clothed in all their parts with minute bright green leaves, and bearing small, waxy, white, five-cleft flowers, with reddish calyces. These flowers are freely produced, and are borne singly and drooping on slender reddish stems. It is one of the most interesting and beautiful of all alpine plants, and one of the most difficult to grow, being very rarely seen in a healthy state even in the choicest collections. Drought is fatal to it. It is a native both of Europe and America, either far north into the coldest regions of these countries, or on the summits of high mountains. It is such a delicate and fragile evergreen shrub that any impurity in the air is sure to injure it. In elevated and moist parts of these islands, it will succeed in very sandy or gritty moist but well-drained peat, freely exposed to the sun and air, and placed quite apart from more vigorous plants on rockwork. The chief difficulty would seem to be the procuring of healthy plants to begin with; once obtained, it would be desirable to carefully peg down the slender main branches, and to place a few stones round the neck of the plant, so as to prevent evaporation. It is a subject which the most skilful cultivator might be proud to succeed with, and worthy the best attention of those who delight in conquering difficulties.

ANDROMEDA TETRAGONA.—*Square-stemmed A.*

ONE of the neatest and prettiest of all the diminutive shrubs introduced to cultivation, seldom growing more than eight inches high. When in good health, the deep green branches are produced so densely that they form very compact and dressy tufts, pleasing at all times. The flowers are produced singly, but rather freely; of a waxy white, five-cleft, contracted near the mouth, and drooping. It is not likely to be confounded with any other plant except the much rarer *A. fastigiata*, from which it may be distinguished in a moment by the absence of the thin chaffy margin of the leaf. It is a native of Northern Europe and America, quite hardy, but requires a moist peat or very fine sandy soil for its perfect development. I have not elsewhere seen it so plentiful or so healthy as in the nurseries near Edinburgh, where it flourishes in common soil. It is a most fitting ornament for the rockwork, or for planting on the margin of beds of choice dwarf shrubs, in fine sandy peat, loves abundance of moisture in summer, and is easily increased by division, wherever it grows vigorously.

ANDROSACE CARNEA.—*Rose-coloured A.*

ONE of the prettiest and most distinct of its exquisite family, coming from the highest summits of the Alps and Pyrenees, where it flowers in summer, when the snow has at last yielded to the sun; opening in our gardens also perhaps among melting snow, but in early spring before any of its relatives. It is immediately known from any of the other cultivated kinds by its small pointed leaves, not, as in them, gathered in tiny rosettes, but more regularly clothing a somewhat elongated stem, so as to remind one distantly of a small twig of juniper, or of the juniper saxifrage. The flowers are of a lively pink or rose, with a yellow eye. It is not difficult to cultivate in a mixture of sandy loam and peat on rockwork—the spot to be exposed, and the soil at least a foot deep, so that its roots may descend, and be less liable to suffer from vicissitudes. Thorough watering should be given during the dry season, particularly when the plant is young, and before it has taken deep root. Treated thus it will form healthy tufts, and prove one of the most beautiful plants in the rock-garden in spring. Like most of the species, it may be easily raised from seed, which should be carefully sown in pans of sandy peat as soon as gathered, in the case of plants growing in gardens; but if gathered on the Alps late in summer, or early in autumn, it would, unless in the hands of a skilful propagator, be best kept over till early spring, when it ought to be sown in cold frames or pits.

ANDROSACE CHAMÆJASME.—*Rock Jasmine.*

THIS does not nestle into close moss-like cushions, like the Helvetian and other Androsaces, the foliage forming large rosettes of fringed leaves. The blooms are borne on stout little stems frequently not more than one inch high, but varying from that to five, according to the vigour of the plants and the position in which they grow. They are white at first, with a yellow eye, though this eventually changes to deep crimson, the outer part becoming a delicate rose. These changes may not be common to all the individuals of the species, but I have observed them in many specimens. When in good health, it flowers abundantly, is one of the most worthy of culture of all alpine plants, and one of the easiest to grow on an open spot on rockwork, in deep and well-drained rich light loam, the surface being nearly covered with small pieces of broken rock, to prevent evaporation and also to preserve the plant from injury. It should get abundance of water in summer, be exposed to the full sun, and be preserved from being overrun by weeds or grazed down by slugs. A native of the Tyrolese and Swiss Alps, where it flowers later than in our gardens. In England it blooms in April, May, and June, earlier or later according to the season, is propagated by division, and may be grown very well in pots along with the rarer Saxifrages, &c., plunged in sand or coal-ashes.

ANDROSACE HELVETICA.—*Swiss A.*

FORMS dense cushions, about half an inch high, of diminutive ciliated leaves, tightly packed in little rosettes. Each rosette rests on the summit of a little column of old and dead, but hidden half-dried and persistent leaves. A white flower, with a yellowish eye, rises from every tiny rosette, each flower being almost twice as large as the rosette of leaves from which it has arisen, and resting immediately on the little mass of glaucous green, the effect being quite charming. Looked at from the height of a man, the leaves are not distinctly seen, the flowers quite so; and thus the effect is somewhat as if you were looking from a considerable height down on some grey bush, with very large flowers and diminutive foliage almost indistinguishable in consequence of the distance. Requires considerable care in cultivation, perfect exposure to sun, and a thoroughly well-drained position on a well-constructed rockwork. It should be placed between and tightly pressed by stones about the size of the fist, which will guard it against danger from excessive moisture, and at the same time permit of the roots passing into the good loam and peat in the crevices of the larger rocks.

ANDROSACE IMBRICATA.—*Silvery A.*

THIS interesting species differs from the Pyrenean and Swiss Androsaces in having the rosettes of a beautiful silvery white colour. The pretty white flowers are without stalks, and rest so thickly on the rosettes as often to overlap each other. It will grow freely in rich loamy soil in narrow well-drained fissures of rockwork. A native of the Pyrenees, the Alps of Dauphiny, Switzerland, and North Italy. Flowers in summer, and is propagated by seeds and division. = *A. argentea.*

ANDROSACE LANUGINOSA.—*Himalayan A.*

THE European species of this diminutive family usually have their leaves in tufts more compact than the very mosses and lichens, and if they do in several instances throw out short runners it is in an underground and very careful sort of way. This kind is distinguished by its spreading and even sometimes, when in vigorous health, long stems, branched, and bearing umbels of flowers of a pleasing and delicate rose, with a small yellow eye; the leaves nearly an inch long, and covered with silky hairs. When growing freely, it is a lovely plant, but it is very rarely seen in good health. I have never seen it in such a perfect state as with Mr. John Bain, in the College Botanic Garden, at Dublin, where it grew on a narrow border, on the sunny side of a glass-house in very sandy deep soil. There it was perfectly hardy, and grew into luxuriant silvery tufts covered with flowers. It is very probable that many parts of the country are too cold for this plant, and that the southern and western counties, or warm and genial places near the sea, are those in which it may be grown with most success. It is, however, so distinct and pretty that in cold and dry places it will be well to preserve it over the winter in dry pits or frames, and plant it out in summer. The most suitable position for it is on the rockwork, planted in sandy peat or very sandy light loam, and so placed that its shoots may fall over the edge of a low rock. Where the soil is very free, and not too wet in winter, and the air moist and genial, it may be tried as a border plant. It is best propagated by cuttings, and flowers in summer and early autumn. A native of the Himalayas.

ANDROSACE OBTUSIFOLIA.—*Blunt-leaved A.*

THIS is said to be allied to *A. chamæjasme*, but has rather larger rosettes of leaves, lanceolate-oblong, somewhat spoon-shaped and obtuse, with stems clothed with short down, from one to four inches high, bearing sometimes one, but generally from two to five white or rose-coloured flowers, with yellow eyes. It seems to grow taller and more vigorously than *A. chamæjasme*, and in a native state is often gathered by handfuls, and placed in vases, with gentians and other alpine flowers. Widely distributed over the European Alps, occurring in France, Germany, and Switzerland, and usually flowering in midsummer; but in this country opening in spring. The treatment recommended for *A. chamæjasme* will be found equally suitable for this plant.

ANDROSACE PUBESCENS.—*Downy A.*

ALLIED to the Swiss and Pyrenean Androsaces in its rather large solitary white flowers, with pale yellow eyes, just rising above the densely packed, slightly hoary leaves, the surface of which is covered with stalked and star-like hairs. The unopened blooms look like small pearls set firmly in a tiny five-cleft cup, and are held on stems barely rising above the dwarf cushion formed by the plant. It may be distinguished from its fellows by a small swelling on the flower-stem close to the flower, and is an exquisite little plant, widely distributed over the Pyrenees, Alps, and other European ranges, generally flowering in July and August in its native state, and in our gardens in spring or early summer. It seems to grow without difficulty on sunny fissures in deep sandy and gritty peat.

ANDROSACE CILIATA.—*Fringed A.*

IS by some considered a variety of the preceding, with the flower-stems twice as long as the leaves, which are glabrous on the surface and ciliated at the margin, the old leaves not forming a column beneath each rosette. It is, however, sufficiently distinct for garden purposes. *Androsace cylindrica* is another variety with the stems rising to half an inch high, with persistent leaves which form columns on the stems. It is by some considered a species. bears pure white flowers in spring, and should be treated like *A. pubescens*.

ANDROSACE PYRENAICA.— *Pyrenean A.*

THIS spreads out into such a very dwarf, compact, and cushioned mass of tiny grey rosettes that one could almost use it for a pincushion. It is something like the Swiss Androsace, but the paper-white flowers with yellowish eyes are not quite so well formed as those of that kind, and the flower, instead of being seated or almost seated in the rosettes of leaves, rises on a stem from a quarter to half an inch high. The leaves are downy and have a keel at the back, and, like those of *A. helvetica*, the old leaves are persistent, and remain in little columns below the living rosette. This plant has been grown to great perfection by Mr. James Backhouse, of York, in fissures between large rocks, with, however, deep rifts of sandy peat and loam in them. In such a position it is more likely to be safe from the encroachments of rampant neighbours of the vegetable, or creeping things of the animal, kingdom, It will also grow on a level exposed spot, but in such a position should be surrounded by half-buried stones.

ANDROSACE VILLOSA.—*Shaggy A.*

A VERY pretty dwarf species, found on many parts of the Alps, Pyrenees, and mountains of Dauphiny, with leaves and stems thickly covered with soft white hair or down. The leaves are mostly covered with the silky hairs on the under side, united in a sub-globular rosette, and bear in umbels white or pale rosy flowers with purplish or yellowish eyes, on stems from two to four inches high. It is more inclined to spread than any of the nearly allied sorts, as it throws out runners, and is therefore suitable for planting so that one side of the specimen may fall down the face of a rock. It should be planted in loam and a mixture of peat, in a properly made fissure between limestone rocks or large stones; but it may also be grown on level spots on rockwork. In all cases it should have abundant moisture, is increased by seeds, and in Britain flowers about the beginning of May.

ANDROSACE VITALIANA.—*Yellow A.*

RARELY grows above an inch high, and produces, scarcely above the leaves, flowers large for so small a plant, and of the richest and most pleasing yellow. On the Alps it reminded me of a Lilliputian furze-bush, looked at through the wrong end of a telescope. It is lovely for association with the freer-growing Androsaces, dwarf Gentians, Primulas, &c. ; it may even be grown on a border in a not too dry district where the soil is free and sandy. A dry soil or a heavy one it does not like. On the rockwork it should be kept abundantly supplied with water during the dry months ; and when in suitable districts it is tried as a border plant on the level ground, it should be surrounded by stones, half plunged in the ground, to prevent evaporation, as well as to protect it from being trampled upon. It is abundant on the Alps in various parts of Europe, and is increased by careful division or by seeds. It is also known by the names of *Androsace lutea*, *Primula Vitaliana*, *Aretia Vitaliana*, and *Gregoria Vitaliana*.

Androsace Heerii, *Charpentieri*, *Wulfenii*, and *Haussmannii*, which are among the finest kinds, are not, I believe, yet introduced ; and one or two annual kinds in the country are not worthy of cultivation out of botanic gardens.

ANEMONE ALBA.—*White Windflower.*

THIS is best described as a dwarf and stout *Anemone sylvestris*, and is a native of Dauria, Russian Asia, the Crimea, and, doubtless, the Caucasus. I have not met with it in cultivation in England, but have seen it flowering very well in the open borders in the Jardin des Plantes, at Paris. The leaves were only a few inches high, and the handsome white flowers, somewhat like those of the fine large-flowered variety of *Clematis montana*, rose an inch or two above them. Till plentiful, it should be grown in deep fibry loam on the rockwork. Flowers in summer ; four to six inches high. Propagated by division and by seeds. The figure of this in the 'Botanical Magazine,' 47, 2167, does not do justice to the plant.

ANEMONE ALPINA.—*Alpine Windflower.*

THIS is almost too stately to be classed with the dwarf plants that we usually term alpines. But high on nearly every great mountain range in northern and temperate climes, it is one of the most frequent and well-marked plants. Cross a snowy range, and you will find it a few inches high and humbly holding up its velvety cups; in descending through the rich green meadows to reach again the roofs of men, you will brush against many of its stems nearly as tall as the knee, each bearing a large, soft, round head of silken-bearded seeds. It may be seen in every stage on the same day, and on the lower terraces of the great mountains and in the green slopes of the valleys it assumes somewhat about the same proportions as in our gardens. It is entirely distinct from most of its cultivated brethren in its large and much cut leaves, its size, and the very soft down on the exterior of its flowers. The interior of the flower is white, the outside being frequently tinted with pale purplish blue. It grows from four to eighteen inches and even two feet high. Being of a strong rooting and vigorous character, it should, if placed on rockwork, have a level spot with abundance of soil to grow in, and being also tall, it would be the better of close association with neat shrubs, plants of the stature of the vernal Adonis, *Primula cortusoides*, and the better kinds of Aquilegia. They would afford each other protection. Where the soil is good, it grows quite freely as a border plant. Flowers in its native country as the snow disappears, and in our gardens in the end of April or beginning of May. When plants are well established in good soil, they may be taken up and readily divided with advantage to themselves; it may also be raised from seed. Visitors to the Alps might bring home quantities of the seed, which ought to be sown as soon as possible after being gathered. Sometimes the flowers are yellow, in which state the plant is often known as *A. sulphurea.*

ANEMONE ANGULOSA.—*Great Hepatica.*

EVERY one who knows the charmingly bright flowers of the variously coloured varieties of the common Hepatica—the very bravest of our early spring flowers—will welcome this species, full twice the size of the common Hepatica in all its parts, with flowers of a fine sky-blue, as large as a crown piece, and distinguished from the common kind by its five-lobed and toothed leaves. It is a native of Transylvania, and hardy everywhere throughout these islands. Obviously the only thing to determine about such a valuable addition is how to best grow and enjoy it. It is naturally more an inhabitant of the elevated copse than the crest of the Alps; it is not able to flourish when thoroughly exposed to the fiercest blasts, like the little alpine plants that cushion down their stout, if diminutive, leaves shorter than the very moss, so that injury from the fiercest gale is out of the question. I have seen it in sandy soil in a thin shrubbery attain a height of more than a foot when not in flower, and the shelter and slight shade received from surrounding objects is decidedly favourable to its development. In all properly formed rockworks, or in their immediate vicinity, it will be possible to give it a suitable position; while in spaces between American plants and choice dwarf shrubs in beds it will succeed to perfection. When plentiful enough, it may be used as an edging to beds of choice spring-flowering shrubs, and for planting in wild open spots in shrubberies, or in open, rather bare, and unmown spots along the margins of wood walks. Let us hope that time will see it sport into several colours like its relative, our common Hepatica, one of the oldest, as well as brightest, inhabitants of English gardens.

ANEMONE APENNINA.—*Apennine Windflower.*

HAS erect flowers of a bright sky-blue—the blue of an Apennine, not a British, sky. These star-like flowers are larger in size than a half-crown piece, and are paler on the outside than within. The plants grow in dense tufts, so that, though there is but one flower to a stem, they are thickly scattered over the low cushion of soft green leaves. Although figured in most of our works on British plants, and naturalised at Wimbledon Park, Cullen in Banffshire, Tonbridge Castle in Kent, and various other places, it is not a true native of this island. But the hardiest of our native plants take not more kindly to our clime; and neither the Bluebell, the Forget-me-not, nor the Speedwell, surpasses its purity of colour. It is one of the sweetest of spring flowers, and among the many lovely plants that gem the alpine or Apennine pastures there is not one more worthy of being abundantly naturalised in the groves and shrubberies of all parts of these islands. It is welcome in the garden and on the rockwork; but it will be only when we see it scattered amongst and contrasting with the native Anemones in our woods, or making glorious mixtures of gold and blue with the Buttercup-like Windflower in open spots along shrubbery walks, or running wild among any other dwarf plants with which the woods or pleasure-grounds are graced, that we shall be able to realise the fact that this Italian beauty can add a new charm to the British spring. The Apennine Anemone flowers in March and April, is very readily increased by division, and grows about four to six inches in height.

ANEMONE BLANDA.—*Winter Windflower.*

This is a near relative of the Apennine Windflower, and a very lovely plant, deserving to be cultivated in every garden in the British Isles. It is of a fine deep sky-blue, like *A. apennina*, and has larger and more finely rayed flowers, dwarfer, harder, and smoother leaves, and blooms in the very dawn of earliest spring, during mild open winters, and in warm parts showing as early as Christmas, flowering continuously too, so that it may be seen in flower late in spring with its relative, *A. apennina*. It is perfectly hardy and vigorous, and, from the harder and smoother texture of the leaves, can stand exposure to cutting winds and sleets even better than the very hardy Apennine A. In a word, it combines every good quality of a hardy alpine plant; should be grown on every rockwork, planted on bare banks that catch the early sun in the pleasure-ground; should adorn the spring garden, and, when sufficiently plentiful, might be naturalised in half-wild places along with other free and hardy members of its charming family. It does not grow more than four inches high, and is multiplied easily by division. Botanically this is chiefly distinguished from *A. apennina* by its carpels being topped with a black-pointed style, and by the sepals being smooth on the outside. When visiting the York collection in the spring of 1868, this invaluable plant struck me as being distinct from the Apennine A., among batches of which, received from Greece, it was at first inadvertently distributed by Messrs. Backhouse, and I soon afterwards ascertained it to be *A. blanda*.

ANEMONE CORONARIA.—*Poppy A.*

A NATIVE of sub-humid pastures in the South of Europe, this plant has been one of the most popular in our gardens from the very earliest times. There are a great number of varieties, both single and double, all worthy of cultivation, and great ornaments of the spring garden. The single sorts may be readily grown from seeds, and they should be thus raised by those wishing a large stock of effective spring flowers. Infinitely varied as they are in colour, and possessing most vigorous constitutions, they deserve to be cultivated even more than many double varieties annually offered by our seedsmen. The plantation of these double varieties may be made in autumn or in spring, or at intervals all through the year to secure a continuity of flowers; but the best bloom is secured by September or October planting. The Poppy Anemone does best in a rich deep loam, but is not very fastidious. The roots of the more select kinds may be taken up when the leaves die down, but they are in few cases worth this special attention, simply because many splendid varieties may be grown as readily as any native herbaceous plant, and we had better cultivate new and distinct species of hardy plants rather than the numerous varieties of one kind. If the seed be sown in June, and the plants pricked out in autumn, they will flower very well the following spring, so that this fine old plant may be said to be almost as easily raised as an annual. Flowers in April and May, and often through the winter, red, white, and purple in variety. Height, six to nine inches. Propagated readily by seed or division.

ANEMONE FULGENS.—*Scarlet Windflower.*

THE white Lily is not more conspicuous for its purity among the border flowers of summer than this plant for its fiery brilliancy amidst the flowers of spring. It is perfectly hardy— vigorous too—the large scarlet flowers being boldly supported on stems about a foot high, springing from a dwarf mass of hard, deeply lobed and toothed leaves. A native of Greece and Southern Europe, it is by no means common in gardens, and is, indeed, unknown to the majority even of those who grow and care for spring flowers; but it will ere long become popular, being one of the noblest ornaments of spring, and, as a scarlet flower, almost unrivalled. It is admirably suited for culture as a border plant, indispensable for the rockwork and spring garden, and, when sufficiently abundant, may be tried amongst the other Anemones scattered about in half wild places. Flowers in April and May; vivid scarlet. Height, one foot. Propagated by division or by seeds.

ANEMONE HEPATICA.—*Common Hepatica.*

To add perfume to the Violet, paint the Lily, or gild the yellow Crocus, would seem to be no more wasteful excess than to praise this exquisite little flower. There is a cheerfulness and a courage about it on warm sunny borders in spring which no other flowers possess; they are hardy everywhere. are not fastidious as to soil, though they love a deep loam, and present a charming diversity. The principal varieties are the single blue, double blue, single white, single red, double red, single pink (*carnea*), single mauve purple (*Barlowi*), crimson (*splendens*), and *lilacina*. Every variety of the common Hepatica is worthy of care and culture. Is it possible to imagine a more beautiful feature than we may produce by planting a mixed edging of the various colours round say a bed of dwarf American plants, occupying space that perhaps would otherwise be naked? It is but one of many ways in which we may tastefully use them. The plant is a native of many hilly parts of Europe, usually found in half shady positions, which will be found to suit it best in a cultivated state also. It is readily increased by division or by seeds, the double kinds by division only.

ANEMONE NEMOROSA.—*Wood A.*

This hardy beauty, which not only embellishes the woods of these sea-girt isles in spring but also those of nearly all Europe and Russian Asia, is so abundant in the British Isles that there is little need to plead for its culture. It grows, or will grow, in every wood or copse, dotting its handsome flowers all over the ground, should other things not interfere, and seeming to invite us to plant other beautiful species of Anemone by its side. They tell us in the books that it grows in or near woods; and so it does in profusion, but I once met with it blooming sweetly on some of the very highest and almost inaccessible crags of Helvellyn, just under some cliffs where a peregrine falcon had built her nest, and very far away from either wood or copse. There are double varieties, and the colour of the flower is occasionally lilac, or reddish, or purplish. I have a single sky-blue variety, which has flowered densely in a fully exposed position, and produced the most exquisite cushions of cœrulean blue imaginable. One day it may become a popular rock-plant. Flowers from March to May; white, and reddish outside. Height, six inches. Propagated by division.

ANEMONE PALMATA.—*Cyclamen-leaved A.*

A VERY distinct kind, with leathery, kidney-shaped, slightly lobed leaves and large handsome flowers, of a glossy golden yellow, only opening to meet the sun. A native of North Africa, Spain, and other places on the shores of the Mediterranean, this charming flower requires and deserves a little more attention than most of its cultivated sisters. It is especially a rockwork gem, and should be planted thereon in deep turfy peat or light fibrous loam with leaf-mould. It should not be placed in positions on the face of rocks suited for Saxifrages and many other plants that are content with mere crevices, and drape the face of the rocks with the slightest encouragement, but rather on level spots, where it could root deeply and spread into firm tufts. Plants of very rapid growth or rambling habit should not be placed near it, as they might overrun and injure it. There is a double variety, *A. palmata fl. pl.*, and a white one, *A. palmata alba*, both now rare. Flowers in May and June; six to eight inches high, and is propagated by division or seeds.

ANEMONE PAVONINA.—*Peacock Windflower.*

THIS kind is very rarely seen in our gardens; though well worthy of being largely grown. The flowers are smaller than those of the common garden Anemone, but usually very double, from the great number of narrow pointed petals filling up the centre of each. These being of a gorgeous cinnamon-red, the effect is peculiarly rich when the flower opens well on fine days. Sometimes the central petals are green. The plant is a native of the South of Europe, should have a light warm well-drained soil, and is a charming ornament of the rockwork or border. In France I have seen it used with good effect as an edging plant for beds of spring and early summer flowers, but with us it is as yet too scarce to be employed thus. It is, however, rather abundantly grown in gardens in the South of France, and may be readily obtained by our nurserymen. Flowers in May and June; rich red. Height, six to eight inches. Propagated by division at the end of the summer growth or very early in spring.

ANEMONE PULSATILLA.—*Pasqueflower.*

Though sparsely distributed in Britain, this fine old border plant is a true native, and when it does occur on a bleak chalk down, it is generally freely dotted over the turf. The position is usually such as to suggest the aptness of the name Wind-flower for the family generally; and there are few sights more interesting to the lover of spring flowers than to see its purple blooms just showing through the hard grass of the blast-swept down on an early spring day. The plant is much smaller in a wild than in a cultivated state, usually devoting itself to the production of a solitary flower, which, while showing through the grass, seems careful not to rise above it. In the garden it forms rich healthy tufts, and flowers more abundantly and vigorously, the contrasts between the wild and cultivated states of the plant being very marked. There are several varieties, including red, lilac, and white kinds, but these are now rare. There is also a double variety. It prefers well-drained and light but deep soil. Flowers in March, April, May; purplish. Height, three to twelve inches. Propagated by division or by seeds.

ANEMONE RANUNCULOIDES.—*Yellow Wood A.*

Not unlike the Apennine and the common Wood Anemone in habit, this species is so very distinct in its clear golden flowers that it is well worthy of cultivation even by the side of the most admired kinds. Indeed, we may consider it an Apennine or a Wood Anemone done in gold! It is a South European species, and apparently is not so free on the generality of our soils as the blue A., but when grown into well-established tufts on a light or warm and well-drained soil, it displays qualities of which those who have merely seen isolated plants or figures of the plant can have no idea. I have not found it do well on clay soil, but on chalky soil it seems to grow as freely as the common

Crowfoot. It is quite charming for association with tufts of the Apennine or the Wood Anemone, the Pasqueflower, any of the varieties of *Anemone Hepatica*, the Aubrietias, and like plants. It comes in among the naturalised group of British plants, and grows in a semi-wild condition at Abbot's Langley in Herts, near Worksop in Notts, and it is also reported to occur in several other counties. It is one of the many beautiful hardy plants that may be freely naturalised in our woods and shrubberies. Flowers in the end of March and beginning of April. Height, four to six inches. Propagated readily by division, and also by seeds.

ANEMONE STELLATA.—*Starry Windflower*.

THIS native of Southern Germany, France, Italy, and Greece, if not so showy, is quite as beautiful, as the common garden A. The star-like flowers, ruby, rosy purple, rosy, or whitish, springing from the much dissected leaves, vary in a very charming way, and usually have a large white eye at the base, which contrasts agreeably with the gay or delicate coloration of the rest of the petals, and with the rich brownish violet of the stamens and styles that occupy the centre of the flower. It is not so vigorous in constitution as the Poppy A., and requires a little more care than that does, but this will only make it the more interesting to all who love variety in their collections of hardy plants. It likes a sheltered yet warm position, a light, sandy, well-drained soil, and seems to make little or no progress on heavy clay soils. It is suitable for association with the choicer kinds of Anemone on the rockwork, the mixed border, and the choice spring garden, and should be grown in every garden where spring flowers are appreciated. Flowers in May. Height, ten inches. Propagated by division or by seeds. = *A. hortensis*.

ANEMONE SYLVESTRIS.—*Snowdrop Windflower*.

A FREE-GROWING and handsome species, partaking somewhat of the size and vigour of the alpine or Japanese Anemone, and the neatness of habit and densely-blooming qualities of the dwarfer kinds. It grows vigorously on almost any soil; the handsome, pure white flowers, as large as a crown piece, being freely produced over a mass of fresh green leaves. A native of Siberia, North Italy, Germany, and France, it is perfectly at home in this country, should be grown wherever first-rate border flowers are appreciated, will associate well with the alpine Windflower, and plants of like size, about the lower and flatter parts of the rockwork, and, being naturally a native of the grove, will be found perfectly at home along our wood walks and half wild spots, in shrubberies, &c. The aspect of the drooping unopened buds has suggested its English name —the Snowdrop Anemone. Flowers in April and May; pure white. Height, one foot to fifteen inches. Propagated readily by division of root.

ANEMONE TRIFOLIA.—*Three-leaved Wood A.*

THIS is an interesting little species, much like the Wood A., but not so widely distributed. Although found in a wild state in pleasant groves on the woody hillsides of Piedmont or the Tyrol, it does perfectly well in our climate, and should be grown everywhere for variety's sake. It may be readily known from its relative, the Wood A., by its neatly toothed trifoliate leaves, and it seems to be a little smaller and dwarfer in habit. I have, however, never seen them under exactly like conditions. It is well suited for naturalisation along with *A. apennina* and others of the family, and is of course suitable for rockwork or borders. Flowers in April and May; white. Height, four to six inches. Propagated by division and by seeds.

ANEMONE VERNALIS.—*Shaggy Pasqueflower.*

ONE of the Pasqueflower division of the Anemone family, but very dwarf. The flowers are very large and shaggy, and covered with brownish silky hairs. It is a rare plant, and should be grown in some select spot on the rockwork, giving it good drainage and deep soil. A native of Norway, Sweden, and extreme northern countries, and also of very elevated positions on the Alps and Pyrenees, and is rarely seen in good condition in our gardens. It should as a rule be grown on a level spot on rockwork, in deep free soil, and be abundantly supplied with water in summer. Flowers early in spring; whitish inside. Height, four to eight inches. Propagated by division and by seeds.

Apart from the large and fine *Anemone japonica* and its varieties, the white one of which, known as Honorine Jobert, is the best, there are a few dwarf Anemones in the country unworthy of cultivation, insufficiently distinct, or difficult to obtain, and of the last class probably *A. Halleri*, *A. patens*, and *A. baldensis* are the best. But it is believed that the cream of the obtainable species is included in the foregoing.

ANTENNARIA DIOICA.—*Mountain Catsfoot.*

A LITTLE creeping perennial with leaves of a silvery grey tone, flower-stems from two to four or five inches high, bearing four to six flower-heads close together at the apex of the shoot. The flowers are usually whitish with pink florets in the centre, but in the variety best worth growing the flower-heads throughout are of a pleasing subdued rose or dull crimson colour. No alpine plant is more worthy of cultivation, whether for rockwork, pots, the front margin of the mixed border, or as an edging plant to nursery beds of bulbs or alpine flowers. In the last-named position or on rockwork it forms neat close-spreading tufts, dotted over with singularly pretty everlasting flowers in May, and seems to thrive in the low open border on good soil near London, as well as in more elevated and favourable parts. It is perfectly hardy, and may be increased to any extent by division. It is widely distributed over elevated and northern regions, and is abundant in many parts of Britain. *A. dioica minima* is a name given to a very small variety of the preceding; it is admirable for culture in pans, on open spots on rockwork, or in spreading tufts on the margin of the select mixed border, the contrast between the warmly toned flowers and the little carpet of grey leaves being very pleasing. *A. hyperborea* is a variety of *A. dioica*, with both sides of the leaves woolly. From this cause it is better adapted for edgings than *A. dioica*, but as we have better silvery edgings than either of them, it is not likely to be much employed except for variety's sake.

ANTENNARIA TOMENTOSA.—*Silvery Catsfoot.*

THIS is the best of all dwarf silvery-leaved plants for gardening purposes. It is very dwarf and spreading, scarcely rising above the ground, but forms a dense carpet of little flat spreading silvery leaves, and will prove a gem for those who wish for novelty and refinement in bedding out. It may be said to carpet the ground with silver ; and, as it is barely an inch high, it requires to be cut off from coarser plants by a line of some subject of moderate size, or by a bare space, and to be planted in a rather wide belt. It is somewhat like our own little mountain Antennarias in size and aspect, but whiter and brighter. I had it from Mr. Niven, of the Botanic Gardens, Hull, about five years ago, and have found it easy of propagation. For the following note with respect to its origin I am indebted to Mr. Niven :—"It is a native of the Rocky Mountains, from which it was sent along with another very pretty silvery-leaved species — even dwarfer, if that be possible — about the year 1848. We never had a specific name for either of them, and I find, on referring to my notes, that I could not identify either species with any described in De Candolle. So I was obliged to christen it myself." The flowers of this plant are not attractive, and whether on rockwork or in neat bijou arrangements in the flower-garden, it will be grown for the sake of its sheets of leaves ; it is best to remove the flowers when they appear. It is hardy on soils of ordinary warmth, but on low heavy clay ground I have noticed it perish in winter ; where grown for summer gardening, annual division and replanting will be desirable. On flat parts of rockwork exquisite effects might be produced by using it as a carpet, and then placing singly or in groups upon it plants with some length of stem, say, for example, the dwarf scarlet Lily, some graceful bulb like the autumnal Acis or the Atamasco Lily. Like combinations may be made in the flower-garden. One of the illustrations in Part I. shows how it has been used to give the effect of distant snow on the tops of miniature hills.

ANTHYLLIS MONTANUS.—*Mountain Kidney Vetch.*

A NEAR ally of our common Kidney Vetch or Lady's Fingers, this is a plant seldom seen in our gardens, but few hardy flowers are more worthy of general cultivation. It is very dwarf, about six inches high, the leaves being pinnate, and nearly white with down. On good light soils it grows larger. The pinkish flowers are produced in dense heads, rising little above the foliage, and forming with the hoary leaves pretty little tufts. There is a white variety, but I have not met with it in cultivation. The species is a most desirable one for every kind of rockwork, but chiefly valuable for its power of thriving on stiff, cold, and bad soils. I have never seen any dwarf alpine flower thrive better on the stiff clay of North London, which proves that the plant may be grown anywhere. Resisting any cold or moisture, it is peculiarly fitted for a position among the dwarf plants in the front rank of the mixed border, while it is of the first order of merit as a rockwork plant. A native of the Alps of Europe, propagated by division and by seeds.

ANTIRRHINUM MAJUS.—*Snapdragon.*

LIKE the Wallflower, this claims a place from the facility with which it may be grown on old walls and ruins, or even on the tops of walls far from old. Had we but the common variety, it would be well worthy of our attention from this habit, but when it is considered how many beautiful striped, and self-coloured, and flaked, and mottled, and delicately dotted kinds are now abundant in gardens, and raised from seed as easily as grass, few will doubt the desirability of naturalising them wherever there is an opportunity so to do. I speak of their merits apart altogether from ordinary garden culture, for which also they are so well adapted. In all but the rougher kind of rock-gardens, they would be out of place, though some very dwarf sorts recently raised, and now obtainable in nurseries, are small enough for association with subjects growing about half a foot high, or a little more.

AQUILEGIA ALPINA.—*Alpine Columbine.*

THIS plant, widely distributed over the higher parts of the Alps of Europe, is indispensable to the choice collection of alpines. The stems rise from less than one to more than two feet high, bearing showy blue flowers, and leaves deeply divided into linear lobes. There is a lovely variety with a white centre to the flower, which, in consequence of its exquisite tones of colour, is certain to be preferred, and many will say they have not got the "true" plant if they possess only the variety with blue flowers. It does not require any very particular care in culture, but should have a place among the taller ornaments of the rockwork, and be planted in a rather moist and sheltered but not shady spot in deep sandy loam or peat. It may be increased by seed or division. In moist districts, and in good free soil, it will prove a first-class border plant. Distinguished from *A. vulgaris* by the stamens being longer than the petals, and by its larger flowers.

AQUILEGIA CÆRULEA.—*Rocky Mountain Columbine.*

THIS native of the Rocky Mountains is as beautiful as it is distinct; and that it has the latter quality will be apparent when I state that the spurs of the flower are almost as slender as a thread, a couple of inches long, with a tendency to twist round each other, and with green tips. But it is in the blue and white erect flower that the beauty lies, the effect being even better than in the blue and white form of the alpine Columbine. It is a hardy herbaceous perennial, flowers rather early in summer, continuing a long time in flower. I have seen it flowering freely on very sandy soil in an exposed spot in Suffolk so late as September. It grows about from nine inches to fifteen inches high, and is worthy of the choicest position on the rockwork; is easier of culture than any of the other rare kinds, and is therefore suitable for the front margin of the choice mixed border, where the soil is sandy and deep, and not too wet in winter. Increased freely from seeds, and also by division. It was long lost to cultivation, but was reintroduced a few years since by Mr. Wm. Thompson, of Ipswich, whom we have to thank for distributing some of our most beautiful hardy plants.

AQUILEGIA GLANDULOSA.—*Glandular Columbine.*

A VERY beautiful species, with handsome blue and white flowers, and a tufted habit. Flowers in early summer—a fine blue, with the tips of the petals creamy-white, the spur curved backwards towards the stalk, the sepals dark blue, large, and nearly oval, with a long footstalk. Leaves much divided, the upper part of the stem covered with glandular hairs. A native of the Altai Mountains, and one of the most desirable kinds for the rock-garden, or the select border, in well-drained deep sandy soil. Increased by seed and by very careful division of the fleshy roots, when the plant is in full leaf. Mr. Wm. Jennings informs me that, if divided when it is at rest, the roots are almost certain to perish, at least on cold soils.

A. pyrenaica, much dwarfer than the preceding, and therefore suitable for rockwork, but not so attractive in point of colour;

the fine, but very tall *A. eximia (californica)*, the gaily coloured *A. canadensis*, *A. Skinneri*, and some of the more beautiful varieties of the common Columbine, are all worthy of culture, but for the most part too tall for the rock-garden.

ARABIS ALBIDA.—*White Rock Cress.*

THROUGH long years of neglect of all sorts of dwarf hardy plants this, the "white Arabis" of our gardens, has held its own, and is now seen in almost every garden in these islands, and in the barrow of every London flower-hawker in the spring. A native of the mountains of Greece, Southern Russia, and of many elevated parts in adjacent regions, it is as much at home in Britain as is the daisy, and will grow in any soil or situation, flourishing far into our cities as well as in the open country, where its profuse sheets of snowy bloom may expand unblemished under the earliest suns of spring. By seed, or division, or cuttings, it is as easily increased as a native weed, and is a valuable ornament of the mixed border, the spring garden, the rockwork, and for naturalisation in wild and bare rocky spots. On the rockwork it is peculiarly fitted for falling over the ledges of rocks; it may also be used as an edging to clumps of shrubs, though it is in better taste to associate it in such positions with groups of plants like the Aubrietias, the rock Alyssum, and other easily grown alpine flowers that bloom early in the year. *A. albida* is closely allied to the Alpine Rock Cress (*A. alpina*), so widely distributed on the Alps, and by some would be considered a sub-species of that plant, but it is sufficiently distinct, and by far the best kind. There is a variegated variety in cultivation, known by the name of *Arabis albida variegata*, which is useful as an edging-plant both in spring and summer flower-gardens. It is the dwarfest and whitest of the variegated rock cresses that are grown under the names of *A. albida variegata*. The yellower and stronger variety, frequently called *A. albida variegata*, and which is the best for general purposes, is a form of *Arabis crispata*, of which the ordinary green form is not worthy of cultivation.

ARABIS BLEPHAROPHYLLA.—*Rosy Rock Cress.*

THIS is not unlike the white Arabis in its habit, size, and leaves, but the flowers are of a deep rosy purple, and consequently make the plant very distinct from any flower of the same order in cultivation. It varies a good deal, but there is no difficulty in selecting a strain of the deepest and brightest rose. It is impossible to have anything more effective than healthy tufts of this plant in the month of April. Whether it prove sufficiently hardy to be a generally useful plant for the open air or not, it is certain to prove a very useful frame plant. It is best raised every year from seed, which, like most cruciferous plants, it yields freely. In all mild districts, and on light soils, plants should be tried out every winter, for there is no out-door flower which surpasses it in pleasing brilliancy during the month of April. The brighter forms are remarkably effective a considerable distance off, and therefore some plants should be placed in positions on the rockwork where they may strike the eye from afar. A native of North America, easily increased by seed.

Among other kinds of Arabis, *A. procurrens* a dwarf spreading kind, with shining leaves and small whitish flowers, is often grown, but is not worthy of culture. There is, however, a brilliantly variegated form of it (*A. procurrens variegata*) which is worthy of a place in a collection of silvery and variegated hardy plants. The prettiest of the variegated Rock Cresses is *A. lucida variegata*. It forms very neat and effective edgings in winter, spring, and summer flower gardens, from its striking and distinct character is effective on rockwork, and thrives best and is easiest to increase by division in open, sandy, and yet moist soil. The best time to divide it is early in autumn, April, or very early in May. It need scarcely be added that the flowers should be removed when they appear. I have grown the green form of the plant, but it is in no sense ornamental. *A. purpurea*, an interesting species for botanical, large, or curious collections, and bearing pale bluish and lilac flowers, is not worthy of general cultivation while we possess such brilliant plants as the purple Aubrietias. *A. arenosa*, from the South of Europe, is a pretty annual kind that may prove useful in the spring garden, and which might be naturalised on old ruins or dry bare banks. *A. petræa* (Northern Rock Cress) is a neat, sturdy little plant, with pure white flowers, a native of some of the higher Scotch mountains, and very rarely seen in cultivation, but when well developed on a moist yet well-exposed spot on rockwork is very pretty. There is a form of it with a purple tinge on the flowers. Other species are in cultivation which are unworthy of a position in any but a botanical collection, and there are various names erroneously applied to several of the kinds above enumerated.

ARCTOSTAPHYLOS UVA-URSI.—*Bearberry.*

A SMALL and prostrate but neatly creeping mountain shrub, somewhat resembling the Cowberry in aspect, but with the leaves more leathery, and their under side beautifully netted with prominent veins, and with the sepals at the base and not at the crown of the berry. The flowers are of a delicate rose-colour in clusters at the apex of the branches; the berries of a brilliant red, somewhat smaller than a currant. It is a native of rather dry heaths and barren places in hilly countries, and is much easier to cultivate than almost any other small mountain or bog shrub, thriving well in common garden soil, though it prefers a moist peaty one. I have noticed two forms in cultivation, one making compact tufts, the other much more rambling and somewhat larger and looser in all its parts. It is a useful plant in the large rock-garden, when its shining evergreen masses of leaves fall over the face of rocks, and also on the margins of beds of shrubs and on borders.

The Black Bearberry (*A. alpina*), a plant very rarely seen in cultivation, a native of high alpine or arctic regions, and of the northern Highlands of Scotland, distinguished from the preceding by its thin toothed leaves, which, unlike those of its relative, are not evergreen, but wither away at the end of the season, and by its bluish-black berries, is not so ornamental as the preceding, though it is a welcome plant in botanic gardens, or with cultivators of rare British species.

ARENARIA BALEARICA.—*Balearic Sandwort.*

THIS coats the face of rocks and stones with the dwarfest thyme-like verdure — clothes them with living beauty as the Ivy does the mouldering tower, and then scatters over the green mantle countless little starry flowers on slender stalks a little more than an inch long. I write this sitting on a rock to which its tiny carpet clings closer than the dwarfest moss. Beneath some rocks fall to the water; it has crept over the edge of these, and dropped its little mantle of green down to within eighteen inches of the water, but all the flowers look up from the shade to the light. Right and left there are boulders in all sorts of positions, on every face of which it may be seen, as every tiny joint roots against the face of the rocks, and the minute mat of leaves is so dense that enough of moisture is preserved to sustain the plant. To establish it on the stones, plant firmly in any common soil near the stones or rocks you wish it to cover, and it will soon approach and begin to clothe them. Flowers in spring and continuously, and is readily increased by division or seeds, and quite easy to grow on most soils. On cold ones it sometimes perishes in winter, but its true home is on the rockwork. It is easily known at any season by its dense tufted cushions of very small leaves. A native of Corsica.

ARENARIA MONTANA.—*Mountain Sandwort.*

A LITTLE grown but very ornamental plant, having the habit of a Cerastium, and fine pure white and large flowers—of sufficient substance to look waxy. It has slightly downy leaves, very narrow and ciliated, diffuse wiry stems, long, but, when well-grown, forming flat spreading tufts, on which the flowers appear so thickly in early summer as to obscure the foliage. It is the most ornamental of the Sandworts, and should be in every collection of herbaceous or alpine plants. On rockwork it would be well to plant it where its branches might fall over the face of a rock, giving it any kind of light soil. I have seen it thrive healthfully in borders in good sandy loam, and it is one of the most attractive early summer flowering plants for the front edge, succeeding the white, evergreen Candytufts and like flowers. Found wild in many parts of France, and is easily raised from seed.

ARENARIA PURPURASCENS.—*Purplish Sandwort.*

DISTINGUISHED from other cultivated kinds by its purplish flowers, produced in abundance on a dwarf densely tufted mass of smooth narrow-oval pointed leaves. It grows plentifully over all the Pyrenean chain, is perfectly hardy, and, like the other kinds, increased by seed or division. It should be associated on the rockwork with the smallest of its brethren, or with dwarf Saxifrages and other plants which, though very dwarf, are not slow growers.

ARENARIA TETRAQUETRA.—*Square-stemmed A.*

THIS is not in cultivation in Britain; but I saw strong tufts of it in M. Boissier's garden, in Switzerland, in 1868, and, as there is abundance of it in various parts of France, there can be no difficulty in procuring it. It forms very compact and singular-looking tufts in consequence of the leaves, each with a white cartilage along the margin, being disposed in four rows. The sepals are also margined—and these characters distinguish it at a glance from the others. I have not seen it in flower, but it is worth a place where the other small Sandworts are grown, if only for the peculiarity of its habit, and it is best fitted for the rockwork, on which it thrives without any particular attention.

ARENARIA VERNA.—*Vernal Sandwort.*

GROWS in dwarf grassy prostrate tufts, covered in April and May with multitudes of starry white flowers with green centres. It is useful for parts of rockwork where a very dwarf kind of vegetation is desired, but scarcely worth growing as a specimen alpine. In consequence of the prostrate habit of both shoots and flowers, the plant is seen to much greater advantage when placed on some little bank above the eye. The prettiest I have ever seen growing was on a little ledge about ten feet high, and receding about as much backwards, while tufts of the same plant at my feet looked comparatively insignificant. It is widely distributed over Europe, Asia, and America, is a native of the more northerly and elevated parts of Great Britain and Ireland, and is readily increased by seed or division.

Of other Arenarias in cultivation, the best and most interesting are *A. ciliata*, a rare British plant; *A. triflora*, a neat species in cultivation in some of our botanic gardens and curious collections; *A. laricifolia*, and *A. graminifolia*. These, however, are scarcely worth growing except in botanical collections.

ARMERIA CEPHALOTES.—*Great Thrift.*

THIS, compared to our British Thrift, is somewhat as the full-blown life-guardsman to his humble congener in the militia. From a dense mass of crowded leaves, four inches to six inches long, spring numerous stems fifteen inches to twenty inches high, each bearing a large, roundish, closely packed head of handsome satiny deep rose-coloured flowers. It is one of the finest and most distinct perennials in cultivation, and should be in every select mixed border, and on every rockwork among the taller and stronger plants. It comes from North Africa and Southern Europe, and, though hardy on free and well-drained soils, occasionally perishes during a very severe winter, especially on cold soils; it should therefore be placed in a warm position on rockwork, and in very well-drained, deep, and good sandy loam. It is known under various names—*Armeria formosa*, *A. latifolia*, *A. mauritanica*, *A. pseudo-armeria*, *Statice lusitanica*, and *Statice pseudo-armeria*. It is, fortunately, easily raised from seed; and, as it is not easily increased by division, it is a good plan to sow a little of it every year. Varies a little when raised from seed; but all the forms I have seen are worthy of cultivation.

ARMERIA VULGARIS.—*Common Thrift*.

THIS inhabitant of our sea-shores, and also of the tops of the Scotch mountains and the Alps of Europe, is very pretty with its soft lilac or white flowers springing from dense cushions of grass-like leaves; but it is the deep rosy form of it, which is rarely seen wild, that deserves universal cultivation in gardens. It is like the common Thrift in all respects but the colour of the flowers, which are of a deep and showy rose, and produced like those of the common form, in profusion. It is useful for the spring garden, for covering bare banks or borders in shrubberies, for making most attractive edgings, and for the rockwork. Easily propagated by division, and, as old and large plants do not bloom so long or so continuously as younger ones, occasional division (say every two or three years) and replanting are desirable.

ARUM ITALICUM.—*Italian Cuckoo Pint*.

A HANDSOME hardy plant, with shining green leaves, decidedly veined, and sometimes spotted, with white, which, beginning to push up in October when other plants are going to rest, are in perfection in mid-winter and early spring. When these die down in early summer, the attractions of the plant are not gone, for the brilliant scarlet fruits, packed in oblong masses at the head of erect stems about fifteen inches high, are among the most conspicuous objects in the garden in autumn; and just as the fruit is past its best, the new leaves begin to show. Like the common Cuckoo Pint, the whole plant is intensely acrid. Although perfectly hardy and free on soils of ordinary quality and warmth, it is desirable to place the Italian Cuckoo Pint in sheltered positions along the sunny fronts of shrubberies, amidst low-spreading evergreens, and in cosy spots about the flanks of rockworks and ferneries, to prevent its handsome foliage from being disfigured by cold wintry storms. It is a useful plant wherever winter or autumnal attractions are desired in the garden, and has been found wild in the Isle of Wight, but the form that occurs there is not so handsomely veined as the Italian one. It is one of those plants which, though unsuitable for the rockwork proper, or for intimate association with true alpine flowers, may yet be used with good effect in their immediate neighbourhood.

ASPERULA ODORATA.—*Sweet Woodruff.*

THIS little wood plant, abundant in Britain, is worthy of some attention in the garden and shrubbery, especially in localities where it does not occur wild. Many would like to cut and preserve its stems and leaves for the sake of the fragrant hay-like odour they give off when dried; and in May the pure white small flowers, profusely dotted over the tufts of whorled leaves, look very pretty. It may be seen covering the ground with its carpet of green frosted over with white, in some of the college gardens at Oxford, and it is one of the many plants that may be allowed to cover the earth in a shrubbery where the barbarous practice of annually digging and rooting up the borders is not resorted to. I have lately seen it used as edgings to the beds in cottage gardens, the odour filling the air. It is, however, as a wood or shrubbery plant—as a companion to the Wood Hyacinth, and the Wood Anemone—that it will be found most valuable. Readily increased by seed, or by division.

ASTER ALPINUS.—*Alpine A.*

THIS might be called the blue Daisy of the Alps, so diminutive is it when met with high up or even in rich green sub-alpine meadows. In a wild state it does not form the sturdy tufts which it does in gardens, but like the wild orange Lily is more beautiful when isolated in the grass. The flower is of a pleasing pale blue, with a tint of violet, and an orange-yellow eye, two inches across on plants cultivated in gardens, usually somewhat smaller in a wild state. It forms neat tufts eight to ten inches high, and is well suited for rockwork or border decoration. The leaves are roughish, three-nerved—slightly downy and sometimes velvety; it is, however, quite distinct from any other plant in cultivation. There is a white variety grown occasionally, like the ordinary one, in continental flower-gardens. Easily multiplied by division, thrives well in any sandy soil, and begins to flower in early summer.

Of the very large Aster or Michaelmas Daisy family, there are few dwarf enough to be associated with the preceding. The most ornamental of the dwarf species is that known as *bicolor* or *versicolor*, which, as it is somewhat prostrate, might be planted with very good effect on the lower parts of rough rock or root work. *A. altaicus* is also a dwarf species, with mauve-coloured flowers two inches across, and *A. Reevesii* is a dwarf neat species, occasionally seen, though not common. Where the embellishment of rough rocky ground is desired, some of the handsomest of the large-growing kinds might be used with good effect, the Pyrenean Aster, *A. pyrenæus*, for example.

ASTRAGALUS HYPOGLOTTIS.—*Purple Milk Vetch.*

A VERY dwarf, hairy perennial, with prostrate stems, and, for the size of the plant, large heads of bluish-purple flowers. In Britain it is found chiefly on the eastern side of the island from Essex and Herts to Aberdeen, and on dry, gravelly, and chalky pastures; in Ireland it is only found on the island of Arran, in Galway Bay. It forms a pretty object on level spots on the rockwork, and should always be associated with very dwarf subjects; and though it is not particular as to soil, it will be found to thrive best in open well-drained sandy loam, or in chalky soil. *Astragalus hypoglottis albus* is a very desirable variety—the paper-white heads of flowers sitting close upon the very dwarf carpet formed by the diminutive leaves. It looks showy for such a dwarf white plant, and, when closely examined, the flowers look singular from contrast with the short sooty or black hairs and points of the calyx. It well deserves a place on rockwork, and it is a good plan to plant it in company with several individuals of the usual purplish colour. It is so distinct from any other cultivated alpine plant in flower about the same period that it would be wise to form a little carpet of five or six plants of it in some level spot on the rockwork. It is not at all difficult to grow, and from its minuteness should always have a place on rockwork on a level part amidst very dwarf vegetation.

ASTRAGALUS MONSPESSULANUS.—*Montpellier A.*

A FINE vigorous species, with leaves a span long, the leaflets smooth on the upper surface, and with short whitish hairs thinly but almost quite regularly scattered over their under sides. The flowers are borne on stalks from six inches to a foot long, the racemes of bloom being from two to five inches long according to the strength of the plant. The closely set and unopened flowers at the head of the raceme are usually of a deep crimson, but as they open, they become of a pale rosy lilac, with bars of white on the standard or upper petals. The flowers, like those of the alpine Clover, are sometimes dull in colour. It is a valuable plant for the fronts of borders or the rougher portions of rockwork. The shoots, though vigorous, are prostrate, which causes it to be seen to greater advantage when drooping down the edge of rocks. It seems to grow well in any soil, and though it does not flourish so vigorously when in very gravelly or poor soil, yet in the very poor soil it will appear more floriferous in consequence of the small development of leaves. A native of the South of France, easily raised from seed. There are several varieties.

ASTRAGALUS ONOBRYCHIS.—*Saintfoin-like Milk Vetch.*

A VERY handsome species, in some varieties spreading, and in others growing about eighteen inches high, with pinnate leaves about four inches long, the leaflets smooth; handsome racemes of purplish-crimson flowers, supported on footstalks an inch or more longer than the leaves. As the individual flowers, when fully open, are a shade more than five-eighths of an inch long, and borne in clusters of from six to sixteen on each raceme, it is an attractive plant even among the many fine hardy flowers in this large family. It is a perfectly hardy, herbaceous perennial, will thrive well on any good loam, and is a capital subject for the mixed border, in the second rank from the front, and a suitable ornament for rockwork, on which it will want nothing but a sufficient depth of soil to root into. A native of various parts of Continental Europe and of Siberia, and flowers in June. There are several varieties enumerated, three of which, *alpinus*, *moldavicus*, and *microphyllus*, are prostrate in habit, and, if introduced, would probably prove valuable for rockwork, and one, *major*, which grows erect. The plant is particularly suited for the rougher parts of rockwork, and for positions where a rich effect rather than rare and minute beauty is sought. There are white forms of all the varieties.

ASTRAGALUS PANNOSUS.—*Shaggy Milk Vetch.*

A SINGULAR and attractive kind, from its very silvery and woolly pinnate leaves, which, growing in compact and luxuriant tufts about a span high, give the plant somewhat the appearance of a silvery fern. Attracted by this appearance, when I saw the plant in cultivation in Switzerland, I brought home some seeds, from which plants have been raised by Mr. J. Backhouse and Mr. W. Bull. I have not yet seen it in flower, nor do I know whence it comes, but from the beauty of its leaves alone it is likely to prove an excellent rock-garden plant, and probably a valuable bedding and edging one. It is easily increased by seeds.

AUBRIETIA DELTOIDEA.—*Purple A.*

A LITTLE alpine that will succeed on any soil, and never fails to flower abundantly, even should the cutting winds of spring shear all the verdure of the budding Weeping Willow. There is hardly a position selected for a rock-plant that may not be graced by this. Rockworks, ruins, stony places, sloping banks, and rootwork, will suit it perfectly; and no plant is so easily established in such places, nor will any other alpine plant so quickly clothe them with the desired kind of vegetation. It makes a neat edging, and may be used as such with good taste in any style of garden, geometrical or natural; though, as its chief period of flowering is the spring, it is not likely to be used as an edging in the summer garden, except around beds or clumps of neat shrubs, in which positions it would be highly appropriate. Growing in common soil, in the open border, or on any exposed spot, it thrives as luxuriantly as on the best-made rockwork, forming round spreading tufts; and on fine days in spring the blue flowers come out on these in such crowds as to completely hide the leaves, making, in fact, hillocks of colour. For covering bare ground beneath roses and shrubs, it might also be tastefully employed. It is quite easy to naturalise it in bare rocky places. London smut falls upon it without affecting its health in the least. It is easily propagated by seeds, cuttings, or division—the last mode the most facile. There are several so-called species very nearly allied to this plant, but I group them all under this name, believing them to be nothing more than marked varieties of the same species. Grown together, their affinity is clearly seen, and in these days of doubts about species few things may be more safely united under one specific name than the Aubrietias at present in cultivation.

Among the several varieties, *A. deltoidea grandifloro* and *A. Campbelli* are the most ornamental: *A. græca* is simply a variety. Aubrietias vary a good deal from seed, but their little differences make them all the more valuable as garden plants, and they all agree in carpeting the earth with dense cushions of compact rosettes of leaves, profusely clothed with beautiful purplish-blue flowers in spring, and, in the case of young plants in moist and rich soils, almost throughout the year. There are one or two pretty variegated varieties.

BEGONIA VEITCHII.—*Veitch's B.*

DR. HOOKER states this to be, "of all the species of Begonia known, the finest." This is surely a sufficient guarantee of its great beauty; but I doubt if any description can give an idea of what it presents to one who sees it for the first time in the open air on rockwork or border in the full gloss of its fine dark-green foliage, and brilliancy of its large vivid flowers. That any species of Begonia should flourish in the open air as this does, is interesting enough; but that the most magnificent of the genus that we know should do so, is surely welcome news to lovers of beautiful hardy plants. The foliage is of a rich, glossy green, and the flowers, which are considerably larger than a five-shilling piece, are of such a glorious colour, a "vivid vermilion cinnabar red," that the plant cannot be dispensed with in any collection of hardy or flower-garden subjects. It was gathered at an elevation of upwards of twelve thousand feet, near Cuzco, in Peru, by Messrs. Veitch's collector, and the plants grown in this country have already given proof of their hardiness by withstanding seven degrees of frost without the least injury. As to its position in the garden, a well-made rockwork with sunny exposure will certainly prove one of the best; and it may also be used with good effect in the more open parts of the hardy fernery. It is hardy in the South of England and Ireland, and may prove so everywhere in this country.

BELLIS PERENNIS.—*Daisy.*

DID we only find the Daisy in company with Androsaces on the high Alps, or even as far out of our way as the lowest Gentian, we would of course be enraptured with its neatness of habit, and delicate purity of tone; but the fact that it shows not a trace of the coarse raggedness of the great composite order to which it belongs, and combines the beauty of the plants of the glaciers with the constitution of lowland weeds, would not cause me to mention it here were it not for its varieties. The common Daisy is everywhere under our feet; its handsome double-flowered varieties are among the most effective and easily managed subjects in the system of spring gardening which has of late become so popular. Nobody would believe them capable of what they are, without seeing them tastefully arranged in such a garden as Cliveden, and in rough rockwork and borders they are quite as useful. There are various varieties, from the quilled white and the double red to the singular aucuba-leaved Daisy, with its leaves so richly stained and veined with yellow, and the quaint hen-and-chicken Daisy. A dozen or more varieties are in cultivation, and easily obtained, and all bloom gloriously from early spring to the end of May. The named and finer varieties may be increased very rapidly by dividing them into very small pieces early in April, and replanting them in rich ground, repeating the process several times during the summer, if the object be to increase any of the scarce varieties for the purpose of using them in quantity in the spring garden. They are also frequenly raised from seed, in which way, moreover, a variety of colours are obtained. When grown for the spring garden, they are removed to nursery beds in early summer, and again planted in the flower-garden in autumn. I probably should not have got thus far unless aided by the Latin name.

The Belliums, or "small Daisies," are nearly allied to the common Daisy. Three kinds are in cultivation : *B. bellidioides*, *crassifolium*, and *minutum*, none of which are so beautiful as the common Daisy, nor so hardy, and therefore scarcely worthy of cultivation, except in large and curious collections. Where grown without protection in winter, they should be planted in sandy warm soil, and in sunny spots on rockwork, on which I should certainly not be anxious to give them a place, considering the numbers of brilliant plants more fitted for the embellishment of the rock-garden.

BRYANTHUS ERECTUS.—*Hybrid B.*

A DWARF evergreen bush, from eight inches to a foot high, bearing pretty pinkish flowers. It is said to be a hybrid, and is in appearance somewhat intermediate between *Rhododendron Chamæcistus* and *Kalmia glauca*. In very fine sandy soil or in that usually prepared for American plants, it grows well, and is worthy of a place on rockwork or in collections of very dwarf alpine shrubs, whether planted in the rock-garden or in neat beds.

BULBOCODIUM VERNUM.—*Spring Meadow Saffron.*

GROWN in our gardens for generations, this very early bulb is generally seen in a state of single blessedness, probably in a pot in a musty old frame ; but if several tufts of it are put in good sandy soil on the rockwork or choice spring garden, it will prove one of the best as well as earliest of spring bulbs, sending up its fine large rosy-purple flower buds, distinct in colour from any other spring flower, earlier than *Crocus Susianus*—in fact, they often show for several weeks ere the snow takes leave of us. The flowers are tubular, nearly four inches long, and usually most ornamental when in the bud state, the colour being a sweet violet purple, the large buds appearing before the concave leaves, which attain vigorous proportions after the flowers are past. Associated with very early flowering plants like the Snowflake, Snowdrop, and *Anemone blanda*, it is very welcome indeed in the rock-garden, or in warm sunny borders. A native of the Alps of Europe, easily increased by dividing the bulbs, in July or August, replanting them at distances of four or six inches apart.

CALANDRINIA UMBELLATA.—*Brilliant C.*

A NATIVE of Chili, with reddish, much branched, little stems, half-shrubby at the base, and rarely growing more than three or four inches high. For vivid beauty and brilliancy of colour there is nothing to equal it in cultivation, the flowers being of a dazzling magenta crimson, attaining to an almost inconceivable glow, yet soft and refined. In the evenings and in cloudy weather it shuts up, and nothing is then seen but the tips of the flowers. It does very well in any fine sandy, peaty, or other open earth, is a hardy perennial on dry soils and in well-drained rockwork, and looks best in small beds, but may be used with advantage as a broad edging to large ones, and seems to live longest in chinks in well-made rockwork. It is as readily raised from seed as the common Wallflower, either in the open air in fine sandy soil, or in pots. As it does not like transplantation, except when done very carefully, the best way for those who wish to use it for very neat and bright beds in the summer flower-garden is to sow a few grains in each small pot in autumn, keep them in dry sunny pits or frames during the winter, and then turn the plants out without much disturbance into the beds in the end of April or beginning of May. As its beauty is concealed during dull or rainy weather, this may prove a drawback to its use in the flower-garden, but by employing it as a groundwork for some of the handsome Echeverias, and other neat succulents now beginning to be so extensively employed in good flower-gardens, this defect may not be so noticeable. When the plants are raised every year, they flower more continuously than old established specimens. It may also be treated as an annual, sown in frames very early in spring, but should in every case be associated with diminutive plants like itself.

CALLA PALUSTRIS.—*Bog Arum.*

MORE beauty than any native bog-plant affords results from planting in boggy places this small trailing Arad, which has pretty little spathes of the colour of those of its relative, the Ethiopian Lily. It is thoroughly hardy, and though often grown in water, likes a moist bog much better. In a bog, or muddy place, shaded by trees to some extent, it will grow larger in flower and leaf, though it is quite at home even when fully exposed. In a bog carpeted with the dwarf dark-green leaves of this plant, the effect is very pleasing, as its white flowers crop up here and there along each rhizome, just raised above the leaves. Those having natural bogs, &c. would find it a very interesting plant to introduce to them, while for moist spongy spots near the rock-garden, or by the side of a rill, it is one of the best things that can be used. A native of the North of Europe, and also abundant in cold bogs in North America, flowering in summer, and increasing continuously and rapidly by its running stem.

CAMPANULA ALPINA.—*Alpine Harebell.*

COVERED with stiff down, which gives it a slightly grey appearance, with longish leaves and erect, not spreading, habit, like the Garganica group, and with flowers of a fine dark blue, scattered in a pyramidal manner along the stems. It is a native of Transylvania and the Carpathian Mountains, hardier than the dwarf Italian Campanulas, and therefore valuable for the front margins of the mixed border, as well as for the rockwork. In cultivation it grows from five to ten inches high, and may be readily increased by division or seeds.

CAMPANULA BARBATA.—*Bearded Harebell.*

ONE of the sweet blue Harebells that abound in the rich green meadows of Alpine France, Switzerland, North Italy, and Austria, and readily known by the long beard at the mouth of its pretty pale sky-blue flowers, nearly an inch and a quarter long, nodding gracefully from the stems, which usually bear two to five or more flowers. Its rough, shaggy, lanceolate-oblong leaves distinguish it when not in flower. In elevated places in its native habitats, it sometimes grows no more than from four to ten inches high, but I have met with it nearly twice as high in the lower parts of the valleys in Piedmont. It is suitable for rockwork, or the front margin of the mixed border, though not a showy plant, is easily increased by seeds and also by division, and flowers in summer. There is a small few-flowered variety, and a white-flowered form, these, like the type, being well worthy of culture, and thriving freely in rather moist well-drained loam.

CAMPANULA CÆSPITOSA.—*Tufted Harebell.*

ONE of the most beautiful little gems in the alpine flora, abundantly distributed over the high ranges in the warmer and central parts of Europe, and thriving as well in all parts of the British Isles as in the pure cool air of its highland home. It grows only a few inches high, rivals the evergreen Candytufts in hardiness, and looks the same fresh, purely tinted, ever spreading, and bravely flowering little plant in margining a bed of roses in a British garden, as it does when seen mantling round the stones and crevices of rocks on the Simplon and other passes. There is a pure white variety as pretty and clear in tone as the blue, and both are admirable for the rockwork or mixed border and also as edging plants. They thrive best in rather moist peaty soil, and, when the plants are not too old, continue flowering from May till August, especially in moist ground, though they thrive in any soil. It is most easily increased by division and also by seed, but as a few tufts may be divided into small pieces, and quickly form a stock large enough for any garden, it is scarcely worth while raising it from seed, except where plants cannot be got. It is usually known as *C. pumila* in gardens, and under that name it was figured in the 'Botanical Magazine.' It is also known as *C. pusilla*, from which, however, it is distinct.

CAMPANULA CARPATICA.—*Carpathian Harebell.*

THIS, while bearing splendid cup-shaped flowers as large as those of the tall and vigorous peach-leaved Harebell, has the dwarf neat habit of the true alpine kinds, and is happily now spreading so rapidly in popular esteem that I need not plead for its culture. A native of the Carpathian Mountains and other parts of the same region, and fortunately easy of culture in all parts of these islands, growing from six inches to over a foot in height, according to the depth, warmth, and richness of the soil. It begins to flower in early summer, and often continues to bloom for a long time, especially if the plants are young, and the seed-vessels be picked off. There is a white variety, *C. c. alba;* a pale blue one, *pallida;* and a delicately toned white and blue kind, *bicolor*—names for the most noticeable variations raised from seed. It is quite easily raised in this way, or increased by division, and is a most valuable bedding and edging as well as rock and border plant. The white and white-and-blue varieties being somewhat slower in growth than the common one, are most worthy of a position on the rockwork.

CAMPANULA CENISIA.—*Mont Cenis Harebell.*

AN alpine plant growing at very high elevations. I have found it abundantly among the fine *Saxifraga biflora*, at the sides of glaciers on the high Alps, scarcely ever making much show above the ground, but, like the Gooseberry bush in Australia, very vigorous below, sending a great number of runners under the soil. Here and there they send up a compact rosette of light-green leaves. The flowers are solitary, blue, somewhat funnel-shaped, but open, and cut nearly to the base into five lobes. I have little experience of its culture in the open air in this country, and cannot say that it is so showy as some of the other species, but it is sufficiently interesting to merit a place on the rockwork. It should have a sandy or gritty and moist soil, and be somewhere near the eye if the rockwork be on a large scale. Easily increased by division, and perfectly hardy. A native of various parts of the Alps of Dauphiny, Provence, Savoy, and other Alps, as well as the particular mountain after which it is named.

CAMPANULA FRAGILIS.—*Brittle Harebell.*

THOSE who have increased this by division will probably bear witness to the unusual aptness of the specific name, as, in handling it, the leaves and stems break off almost as freely as if made of the slenderest ice. It is a desirable and useful kind, the root-leaves on long stalks heart-shaped in outline, and bluntly and shortly lobed, those of the stem more lance-shaped, the rather large pale blue open flowers somewhat bell-shaped, profusely borne on half prostrate stems, and making a pleasing show in summer. The whole plant rarely reaches six inches in height, and is perfectly smooth and rather fleshy. A native of the South of Italy; it is valuable for the rockwork in well-drained chinks into which it can root deeply without being too wet in winter. Probably on light soils it would not require this precaution. *C. fragilis hirsuta* is a variety quite covered with hair or stiff down in all its parts, so much so as to look almost woolly. It is of about equal value with the normal smooth form.

CAMPANULA GARGANICA.—*Gargano Harebell.*

A FINE showy species, with somewhat of the habit of the Carpathian Harebell, but smaller; the leaves that spring from the root are kidney-shaped, those from the stem heart-shaped, all toothed and downy. In summer the whole plant becomes a mass of brilliant bluish-purple starry flowers with white centres. It does not grow more than from three to six inches high, and is of a free prostrate habit, so that it is seen to great advantage when placed in interstices on the most vertical parts of rockwork, in warm and well-drained spots. The better and deeper the soil the finer and more prolonged the bloom will be. It is a native of Italy, flowers in summer, and is easily increased by cuttings, division, or seeds.

CAMPANULA HEDERACEA.—*Ivy C.*

THIS is not mentioned for its stature, being a weakly creeping thing, with almost thread-like branches bearing small delicate leaves, roundish or heart-shaped, with a few teeth; nor for the showiness of its flowers, which are of a faint bluish purple, less than half an inch long and drooping in the bud. However, as in the case of many other diminutive and delicate creatures, there is an interest and peculiar grace about it that we do not find in more robust members of the same family: besides it is a native of Britain, that creeps over bare spots by the sides of rills and on moist banks, and wherever there is a moist boggy spot near the rockwork, or by the side of a streamlet or in an artificial bog, it will be found worthy of a place as an interesting native. It occurs chiefly in Ireland and Western England, and less abundantly in the east. Increased by division.

CAMPANULA ISOPHYLLA.—*Ligurian Harebell.*

THIS is a very ornamental and profusely flowering Italian species, long known, but only recently introduced to cultivation by Mr. Traherne Moggridge. The leaves are roundish or heart-shaped, deeply toothed, and nearly all about the same size, and the flowers of a pale but very bright blue with whitish centre and protruding styles. It will make a charming ornament for every sort of rockwork, and should be placed in sunny positions in well-drained, rather dry fissures in sandy loam, and then it will repay the cultivator by a dense and brilliant bloom. It is one of many kinds of Campanula that might with great advantage be naturalised in rocky spots, the sunny walls of old quarries, chalk pits, and like places.

CAMPANULA PULLA.—*Violet Harebell.*

THIS is somewhat like the tufted Harebell in stature and appearance, but the stems only bear one flower, and that is of a fine deep bluish violet, larger than that of *C. cæspitosa*. It is also much rarer in consequence of being more delicate, and growing and increasing much less rapidly. It is, indeed, at present considered a rarity in botanic gardens, and those who wish to increase it should keep it in pots in a cold frame or pit, dividing it every spring till a sufficient stock is obtained. On the rockwork it should be placed on a level spot, free from other Harebells or rampant plants of any kind, and in sandy peat. It has a tendency to spread underground, and send up shoots in a scattered manner round the place where it is planted. It is in consequence better to give it an isolated spot, four or five inches across, and cover the surface with small bits of broken stones, say about the size of walnuts, which would serve to prevent weeds or other plants taking root, and allow this violet Harebell to send up its shoots at will. A native of the Tyrol and of other mountains in Central and Southern Europe; is increased by division or by seeds, and thrives very well in pans or pots.

CAMPANULA RAINERI.—*Rainer's Harebell.*

I DO not believe this species is in cultivation in England, but having seen it in M. Boissier's garden, Lausanne, in the summer of 1868, I depart from my usual rule of omitting species not in our gardens, as there can be no difficulty in some of our nurserymen obtaining it. It is one of the most beautiful, quite dwarf in habit, and distinct. The plants I saw had stems not more than three inches long (though it is said to reach twice that height), and, though small, were quite sturdy and firm; they were branched, and each little branch bore a large solitary somewhat funnel-shaped erect flower of a fine dark blue. They seemed of a vigorous nature, more so than several of the other dwarf Italian Harebells, and, when introduced, will no doubt form a valuable addition to choice collections of alpine plants. A native of high mountains in the North of Italy.

CAMPANULA TURBINATA.—*Vase Harebell.*

This combines the dwarfness of *C. cenisia* and the sturdiness of *C. nitida* with the large flowers of the Peach-leaved, and more than the neatness of habit of the Carpathian, Harebell. The leaves, rigid, of a greyish green, toothed, and pointed, with heart-shaped bases, form stiff tufts from two to three inches high, and an inch or so above them rise the cup-shaped flowers, of a deep purple, and nearly two inches across—an extraordinary size for such a very dwarf plant. It comes from the mountains of Transylvania, is perfectly hardy in this country, is not fastidious as to soil, and is one of the most valuable gems we possess for the embellishment of rockwork and also for the mixed border, on which, in deep light soil, the flowers sometimes reach a height of six or eight inches. As the great size of the individual flowers is the most pleasing character of this species, it is well in planting it to place it where they may be seen to advantage, in vertical spots on parts of rockwork that come near the eye.

Of other Harebells most likely to be attractive for the rock-garden, the following are among the best :—*C. muralis, nitida, modesta,* and *Barrelieri.*

CARDAMINE TRIFOLIA.—*Trefoil Ladies' Smock.*

"This small plant," says John Parkinson, "hath divers hard, dark round green leaves, somewhat uneven about the edges, always three set together on a blackish small footstalke, among which rise up small round blackish stalkes : half a foot high, with three small leaves at the joynts, where they branch forth ; at the tops whereof stand many flowers, consisting of four leaves a piece, of a whitish or blush colour very pale : after which come up small, thick and long pods, wherein is contained small round seed : the root is composed of many white threds, from the heads whereof run out small strings, of a dark purple colour, whereby it encreaseth. It was sent me by my especiall good friend John Tradescante, who brought it among other dainty plants from beyond the Seas, and imparted thereof a root to me." It is a hardy dwarf plant, a native of Northern Europe, not very ornamental as regards the flowers, but, being neat and compact in habit, and easily grown, merits a place in the full collection, and is suited for the rougher parts of rockwork, or for the margin of the mixed border.

CENTAUREA UNIFLORA.—*One-flowered Knapweed*.

THE flower-heads of this, previous to opening, look like withered balls, in consequence of each of the scales being terminated by a dark-brown feather-like point; and as these become developed, they lie down close upon the head, appearing to enclose it in a net. The stems rise six to fifteen inches high, each bearing a solitary flower two inches or more across, of a lilac rose. It is a distinct, curious, and rather ornamental kind, grows freely in well-drained and sandy soil, and merits a place on the rockwork or borders. The undivided leaves are irregular in form, sometimes lance-shaped in outline, and without a single tooth, sometimes sparsely and unequally toothed, and sometimes with one or two ears or short lobes at the base, and they have a silvery look from being covered with a cobweb-like close-lying down. I have met with the plant in a wild state amidst meadow grass on high mountains in North Italy, and expect it will prove more ornamental in cultivation.

CERASTIUM ALPINUM.—*Shaggy C.*

AN interesting British plant, found on Scotch mountains, and also more sparsely on those of England and Wales. Dwarf, tufted, and prostrate, spreading about rather freely, but seldom rising more than a couple of inches high, with leaves broad compared to those of any of the common weedy species, and densely clothed with a dewy-looking silky down, which gives the plant a singularly shaggy appearance. From these spring rather large white flowers in early summer. It is at all times a pretty and distinct-looking object on those parts of rockwork that come near the eye, and, being British, will be all the more interesting to the cultivator. It is not, like the common garden species, a plant fitted for forming edgings. Messrs. Backhouse say that it flourishes best under ledges that prevent the rain and snow falling on the foliage, but I have found it stand all sorts of weather, and winters in the open border in London. Readily increased by division, by cuttings, or seeds.

CERASTIUM BIEBERSTEINII.—*Bieberstein's C.*

A VERY silvery species, closely allied to the common *C. tomentosum*, from which, however, it is distinguished by its larger leaves, flowers, and fruit. Useful for the same purposes, and propagated and cultivated with the same facility, as *C. tomentosum*. It was once expected that it would surpass in utility the common kind, but this it has failed to do. A very good plant for borders or rough rock or root work, and, being seldomer seen than the common one, deserves a little more attention, and a better position in the mixed border or the lower and rougher parts of rockwork. A native of the higher mountains of Tauria, flowering with us in early summer.

CERASTIUM GRANDIFLORUM.—*Large-flowered C.*

ALLIED to *C. tomentosum*, but less downy and silvery. It is readily known from either *C. tomentosum* or *C. Biebersteinii* by having narrower and more acute leaves, and being less hoary, and it usually grows somewhat larger than either of the two very silvery kinds, rapidly forming strong tufts, and producing pure white flowers in great abundance. A fine plant for the front margin of the mixed border, or for the rougher parts of rockwork, but only in association with other strong and fast-growing things, as it spreads about so quickly that it would overrun and injure delicate and tiny plants if placed near them. It is not so likely to be appreciated for edgings as other kinds, though there is no reason why it should not be used for variety's sake as a bordering on a small scale. Like the other cultivated species, it is readily propagated by division or by cuttings inserted in the rudest way in the open ground, and is a native of Hungary and neighbouring countries, on dry hills and mountains. Flowers with us in early summer.

CERASTIUM TOMENTOSUM.—*Common Woolly C.*

THIS is now used in almost every garden for forming compact silvery edgings to flower-beds and borders. Its hardiness, power of bearing clipping and mutilation, and great facility of propagation, make it worthy of all the attention it receives. No plant has proved more useful in our great public gardens. It is also very useful as a border-plant, and for rootwork or rough rockwork, but is too common to be permitted a place on small or choice rockwork that might be devoted to some of the many rarely seen and beautiful alpine plants. A native of mountains in the South of Europe, flowering freely with us in early summer.

The preceding include all the kinds that are worth growing, except in botanical and unusually extensive collections. The other kinds enumerated in catalogues are :—*C. incanum, lanuginosum, ovalifolium, ovatum, tenuifolium, Wildenovii,* and *trigynum.*

CHEIRANTHUS CHEIRI.—*Wallflower.*

In a book advocating the culture of alpine plants on walls, we must not ignore the claims of the sweet old flower that has so long dwelt on walls and old ruins, hallowing their mouldering remains. It loves a wall better than any garden; while it grows coarsely in garden soil, it forms a dwarf enduring bush on an old wall, and grows even on walls that are quite new, planted in mortar. There is no variety of the Wallflower yet seen that is not worthy of cultivation; but the choice old double kinds— the double yellow, double purple, double orange, dark, &c.—are plants worthy of a place beside the finest rock-shrubs or border-plants, ornamental in a high degree, and endeared to us by many associations. These are the varieties most worthy of a place on dry stony banks near the rockwork, and also on old ruins, on which the common kind is likely to find a home for itself. The fine mixed " German " kinds, that are so easily raised from seed, would also be worthy of introduction to ruins and stony places. A packet of the seed strewn in such places would be all that is necessary. Writing of these to the ' Field,' Mr. Henry Kingsley says :—

"A dim haze of colour. Not gaudy, bright, vulgar, positive colour, but a nearly innumerable number of half-tints, ranging from dull yellow, through brown, into purples; worth, to an educated eye, all the fantastic barbarisms made out of bad scarlet, bad yellow, and (of all colours in heaven and earth) bright blue (lobelia). There is not more contrast between gaudy yellow and red beds and one of wallflowers (or stocks later on) than there is between Miss Braddon and George Elliot. Some like one, some another. I confess, on my own part, that I like the tint of George Elliot best; he never allows you to lose one colour until you have found another. On the other hand, Miss Braddon beds out rather too gaudily. I should say that ' Silas Marner ' was very like a bed of wallflowers myself—that is to say, to an eye educated to colour, one of the most beautiful things ever seen."

CHIMAPHILA MACULATA.—*Spotted Wintergreen.*

A LOW wood plant of North America, having leathery, shining, slightly toothed leaves, the upper surface of which is pleasingly variegated with white, and bearing whitish umbellate flowers— one to five—on rather long stems. The plant attains a height of only from three to six inches, and is a very suitable subject for a half shady and mossy, but not wet, place in the rock-garden, associating well with such plants as the dwarf Andromedas and the Pyrolas, and succeeding best in very sandy decomposed leaf-soil.

C. umbellata, with glossy unspotted leaves, and somewhat larger reddish flowers, is suited for like positions. Both are rare in cultivation and very seldom seen well grown. They flower in summer, and are increased by careful division.

COLCHICUM VARIEGATUM.—*Chequered Meadow Saffron.*

THIS is the prettiest of the easily procured *Colchicums*, and is often grown under the name of, and mixed with, the common meadow Saffron, *C. autumnale*, but is distinguished by its rosy flowers being distinctly and regularly mottled over with purple spots, and being more open, and its leaves undulated. Like the common species, it flowers abundantly in autumn, grows well in ordinary soil, and may be tastefully associated with the autumn-flowering Crocuses on rockwork, borders, or edgings to beds of dwarf shrubs, &c. There are several varieties, both double and single, of the common meadow Saffron worthy of cultivation.

The large-leaved *C. byzantinum* and the small *C. alpinum* are also worthy of a place. Most of the species grow so readily in almost any soil that they are excellent subjects for naturalisation in grassy places in the pleasure-ground and woodland glades.

CONVALLARIA MAJALIS.—*Lily-of-the-valley.*

So long have we been accustomed to this in our gardens that we can scarcely think of it as an alpine plant. But as the traveller ascends the flanks of many a great alp, he sees it sweetly blooming low among the Hazels and other mountain shrubs; and it is widely distributed over Europe and Russian Asia, from the Mediterranean to the Arctic Circle. It is needless to say anything of its appearance or culture, but it would be seen to much greater advantage if people would transplant a few tufts of it from the matted and often exhausted beds in which it is usually grown in the kitchen-garden to half shady and half wild spots near their gardens, alongside of wood walks, where it would become naturalised among low shrubs on the fringes of the rock-garden or hardy fernery. It might also be planted in tufts among American plants or other choice shrubs, as in any of these positions its graceful beauty will be more appreciated than when it is seen in a formal mass, grown as prosaically as kitchen Thyme or Spearmint, while its delicious odour will be spread over the place. Of course, planting it in this way need not prevent its being grown in quantity for cutting, though, if sufficiently plentiful in the positions mentioned, there would be little occasion to grow it in any other way. Of late it has been deservedly much forced in hothouses in early spring. There is a variety with double flowers, one with single rose flowers, one with double rose flowers, one with the leaves margined with a silvery white, and one richly striped with yellow—all of which are worthy of cultivation. Although growing in almost any soil, it prefers a deep sandy loam enriched with leaf-mould, and loves a partial shade, notwithstanding that it thrives well in the full sun.

CONVOLVULUS LINEATUS.—*Dwarf Silvery Bindweed.*

This, so far from having any of the free-climbing tendencies of our common wild Convolvulus, is quite a pigmy compared to the dwarf and popular *C. minor*—the whole plant often showing nothing but a tuft of small silky rather narrow and pointed leaves above the ground. Among these appear in summer delicate flesh-coloured flowers more than an inch across, and in full perfection at less than three inches high, though in warmer soils and districts than those on which I have observed the plant it sometimes grows an inch or two higher. Few subjects are hardier or more suitable for embellishing some arid part of the rockwork near, and somewhat under, the eye, as its beauty is not of a telling, though of a high, order. A native of the Mediterranean region; easily increased by dividing the root.

CONVOLVULUS MAURITANICUS.—*Blue Rock Bindweed.*

A BEAUTIFULLY coloured and graceful plant, without the rampant tendencies of many of its race, but withal throwing up a number of elegant shoots, which bear numbers of clear light-blue flowers, about one inch across, and with a white centre. It is quite distinct from any other Convolvulus in cultivation, and, happily, is hardy in sunny chinks of rockwork, and also on raised borders. It is seen to the best advantage in a somewhat raised position, so that its free-flowering shoots may fall freely down, though it may also be used with good effect on the level ground in the flower-garden, or as a vase plant. A native of the North of Africa; readily increased by cuttings.

CONVOLVULUS SOLDANELLA.—*Sea Bindweed.*

THIS is at once recognised from its fellows by its leathery, round, or kidney-shaped leaves, and by its stems being short, heavy, and without the twining tendency so common in the family. The flowers are large, handsome, of a light pink colour, and freely produced. I have observed it thrive and flower freely in ordinary soil far away from the seaside, and therefore recommend it with confidence as worthy of a place among the trailers of the rock-garden. A native of maritime sands, in many parts of the world; not uncommon on our own coasts, and flowering in summer.

CORNUS CANADENSIS.—*Dwarf Cornel.*

A VERY pretty but neglected miniature shrub, of which each little shoot is tipped with white bracts, pointed with a tint of rose. I know nothing prettier than this Cornus when well established, and it is not at all fastidious, but, being very dwarf, rarely comes in for a proper situation. It is lost among coarse herbaceous plants, and totally obscured by ordinary shrubs, and should therefore be planted among alpines on a rockwork, or round or near the edge of a bed of very dwarf Heaths or American plants. Many know and appreciate the singular beauty of the *Mussænda frondosa* of the hothouse, with its white bracts tipping the clear green leaves; this little Cornus may be described to those who do not know it as a diminutive hardy plant of equal beauty. It grows about the size of the Partridge Berry, or somewhat larger.

Wherever placed, rather damp sandy soil will be found to suit it best. A native of North America, in damp cold woods.

Growers of British plants may like to possess *Cornus suecica*, but I have not seen it either well grown under cultivation, nor is it so ornamental as the preceding.

CORONILLA MINIMA.—*Dwarfest C.*

A DIMINUTIVE evergreen, generally prostrate, rarely rising more than a few inches above the ground, and of a glaucous green. The small rich yellow flowers are freely produced, six to twelve in each crown, in April and May. It is a plant of easy culture, and well worthy of a warm spot on rockwork, where its tiny shoots may lap over the stones. A native of France and Southern Europe ; not particular as to soil, and readily increased by seeds or division. Deep light soil in sunny fissures will suit it best, and in such places its diffuse little stems will be seen to greatest advantage.

CORONILLA VARIA.—*Rosy C.*

A VERY handsome, free, and graceful plant, with a profusion of pretty rose-coloured flowers, widely distributed on the Continent, and found on many of the railway banks in France and Northern Italy. It forms low dense tufts, sheeted with rosy pink, which attract the traveller's eye, their beauty and dressy appearance marking them among the weeds which inhabit such places. It ought to be grown in every garden as a border-flower, or for naturalising in semi-wild spots. Perhaps, however, the most graceful use that could be made of it would be to plant it on some tall bare rock, and allow its vigorous shoots and bright little coronets to teem over and form a lovely curtain down the face of the stone. It is also admirable for chalky banks, or for running about among low trailing shrubs like the common Cotoneaster. When in good soil, the shoots grow as much as five feet long, and it thrives on almost any sort of soil. It is readily increased by seeds, which are frequently offered in our seed catalogues. I have seen a fine deep rose-coloured variety of this in the Jardin des Plantes which would be well worthy of culture. It may occur in other collections, and when the plant is much raised from seed, highly coloured varieties may be selected.

Coronilla montana, somewhat larger than the preceding

species, attaining a height of fifteen to eighteen inches, and bearing numerous yellow flowers, is somewhat too large for association with small alpine plants, but, being an ornamental species, is excellent for the rougher parts of the rock-garden or for the mixed border.

CORTUSA MATTHIOLI.—*Alpine Sanicle.*

SOMEWHAT like the tender *Primula mollis*, with large seven- or nine-lobed leaves, the leaf-stalks thickly and the leaves sparsely covered with colourless short hairs. A wiry thread of vascular matter runs through the stem leaves, and may be drawn through the blades as well as footstalk of the leaves without breaking. The flowers, borne on stems about fifteen inches high, very downy on the lower half and smooth above, are pendulous, and of a peculiarly rich and deep purplish crimson, with a white ring at the base of the cup, six to twelve being borne on a stem. It does well on rockwork, especially in the angle formed by two rocks where its leaves cannot be torn with the wind. Flowers in early summer, and comes from the Alps of Piedmont and Germany; increased by careful division of the root, or more abundantly by seed sown as soon as possible after being gathered.

CORYDALIS LUTEA.—*Yellow Fumitory.*

THIS well-known plant is not so much esteemed as it deserves, for not only are its graceful masses of much cut, delicate pale-green leaves profusely dotted over with spurred yellow flowers, very pleasing in borders, but it grows to perfection on walls, which renders it doubly valuable. I have seen it in the most unlikely spots on walls in hot as well as in cold and moist countries, and know nothing to surpass it for garnishing ruins, walls, stony places, and poor bare banks, the tufts often looking as full of flower and vigorous when emerging from some old chink where a drop of rain never falls upon them, as when planted in fertile soil. It also makes a handsome border-plant, and is well suited for the rougher kind of rock and root work. A naturalised plant in England, and widely spread over Continental Europe. Readily increased by division or by seeds; in any stony position it spreads about with weed-like rapidity.

CORYDALIS NOBILIS.—*Noble Fumitory.*

A WELL-NAMED plant, for it is truly ornamental, compared to the other members of the genus that are known and cultivated. The leaves are much divided, and the plant strong and neat, growing, when in flower, ten inches or a foot high; the flower-stems are stout and leafy to the top, and bear a massive head of flowers, composed of many individual blooms in various stages. They are arranged in a short and close spiral whorl, the unopened ones at the apex of the flower forming a light green rosette, and contrasting pleasingly with the fully opened flowers lower down. The open flowers are of a rich golden yellow, with a small protuberance in the centre of each, of a reddish chocolate colour; and the effect of this, with the yellow and the green rosette when the bloom is young, makes the plant very ornamental. It is quite easy of culture in borders, but is rather slow of increase, and, where it does not thrive as a border-plant, should be planted in deep light and rich soil on the lower flanks of rockwork, associated with plants of the vigour and stature of the Vernal Adonis, the American Cowslip, and the Rocky Mountain Columbine. A native of Siberia; increased by division, and flowering in early summer.

CORYDALIS SOLIDA.—*Bulbous Fumitory.*

A COMPACT tuberous-rooted species, from four to six or seven inches in height, and freely producing dull purplish flowers. It has a solid bulbous root, is quite hardy, and of easy culture in almost any soil. A pretty little plant for borders and the rougher portion of rockwork, for naturalising in open spots in woods, and also for use in the spring garden. It is naturalised in several parts of England, but is not a true native, its home being the warmer parts of Europe; easily increased by division, flowers in April, and is often known as *Fumaria bulbosa* or *F. solida*.

CROCUS LUTEUS.—*Yellow C.*

ONE of the commonest and most vigorous of all our garden Crocuses, a native of Eastern Europe, and, it need hardly be added, at home everywhere in Britain. " It is observable that all the wild specimens of this species seem to have grown with the bulbs five inches or more under ground. Depth is very necessary to their preservation, for mice, which I have found usually to meddle with no other species, will scratch very deep in quest of them. The fine common large yellow *Crocus luteus* of the gardens differs sufficiently from the varieties to make it pretty evident that it is a natural local variation of the species, and not a garden variety; but we know not whence it was derived. All the varieties of this species seem to prefer a very light soil upon a clay subsoil."—Herbert, in 'Trans. Hort. Soc.' Grown in nearly every garden, it is needless to speak of the positions for which it is most fitting, but it might be naturalised with great advantage in the rougher parts of the pleasure-ground, or in open sunny spots or banks along wood-walks.

CROCUS NUDIFLORUS.—*Pale Purple Autumn C.*

A BEAUTIFUL pale bright purple Crocus, flowering in autumn after the leaves of the year are withered, thriving freely in any sandy or light soil, and naturalised abundantly in meadows about Nottingham, Derby, Halifax, and Warrington. The corm flowers when about the size of a pea, sending out stolons in spring, the thickened apices of which afterwards form new corms. The leaves appear in very early spring, and are very slender, with a narrow white line in the centre. Flower with the tube from three inches to ten inches, and the segments one inch and a half to two inches long; stigmas reddish-orange, cut into an elegant fringe. A native of South-Western Europe, and well worthy of general cultivation as a rockwork or border plant; particularly suitable for forming edgings or clumps round beds of autumn flowers. It somewhat resembles the Meadow Saffron at first sight, but is easily distinguished from it by having three, not six, stamens.

CROCUS ORPHANIDIS.—*Orphanides' C.*

A LOVELY Crocus, with soft lilac-blue flowers, having yellow throats, two inches and a half in diameter, and opening in autumn. The bulbs are unusually large, nearly two inches long, "closely covered with a bright chestnut-brown tissue." The leaves appear with the flowers, exceeding them in length, and getting much longer afterwards. This has recently been sent to the Royal Gardens at Kew by Professor Orphanides, of Athens, and named after him by Dr. Hooker, who describes it as most lovely, and very distinct. A native of Greece, and, till plentiful, should be exclusively planted on warm slopes of the rock-garden.

CROCUS RETICULATUS.—*Cloth of Gold C.*

This is the little rich golden Crocus with the exterior of its flowers of a brownish black. It is the earliest of the commonly cultivated spring Crocuses, and a native of the Crimea and South-Eastern Europe. There are several varieties, and among them a lilac and a white, but these I have never seen in cultivation. Suitable for association with the earliest and dwarfest flowers of the dawn of spring, thriving in ordinary soil. It is generally known as *C. susianus*.

CROCUS SATIVUS.—*Saffron C.*

This species was formerly cultivated in England for the production of saffron, which is made from the fringed and rich orange style. Its native country is not known with certainty, but it is probably from the shores of the Mediterranean. It blooms in autumn from the end of September to the beginning of November according to position and soil. The flowers are of a pale violet, with deeper-coloured veins, the tube of the flower long; the erect anthers nearly three quarters of an inch long, of a rich clear yellow, on small pale lilac filaments; the stigmas of a rich orange red, divided into three branches, each about an inch and a quarter long, becoming thicker towards the apex, hollow, and so heavy that they droop out of the flower—contrasting singularly with the erect yellow anthers. The flowers have a sweet and delicate odour. The sharp-pointed, very narrow, leaves appear about the same time as the blooms, remain in rigid bundles throughout the winter, and acquire their full development in spring. The bulbs of the Saffron Crocus should be planted from four to six inches under the surface, and it loves a deep good sandy loam and a sunny warm position. Where permanently planted, six inches apart will be near enough to place the bulbs, and in soils where the plant thrives, it will be necessary every second or third year to raise and divide them. Where the natural soil is too cold for the full development of this plant, it will be easy to give it a suitable position on sunny parts of the rockwork, on which, or on sunny borders, it may be associated with the Meadow Saffron or other autumn Crocuses.

CROCUS SIEBERI.—*Sieber's C.*

A SMALL species, from the mountains of Greece, which has only been introduced within the past few years, as if to remind us that, though in its native country little now remains of the race of heroes that once populated it but "the rifled urn, the violated mound," nature there is fresh and fair as when the "isles that crown the Ægean deep" saw their happiest and most glorious days.

> "Eternal summer gilds them yet,
> But all except their sun is set!"

We have Crocuses that flower in spring, and Crocuses that flower in autumn; but this hardy mountaineer flowers in winter and earliest spring, anticipating all the others. Very dwarf, with pale violet flowers; is not at all difficult to cultivate, and should be placed on some little sunny ledge on rockwork, or other spot where it may be safe from being overrun or forgotten.

CROCUS SPECIOSUS.—*Showy Autumn C.*

THIS is the finest of the autumn-flowering Crocuses, and comes into beautiful bloom when the wet gusts begin to play with the fallen leaves, at the end of September or beginning of October; the flowers bluish violet, striped internally with deep purple lines, smooth at the throat, the divisions most deeply veined near their base; the stigmas, of a fine orange colour, cut so as to appear as if fringed; the leaves appearing about the same time as the flowers, but not attaining their full development till the following spring. It is particularly suitable for forming edgings and tufts near beds or groups of autumn flowers, and also for rockwork or borders. It is not particular as to soils; but, being scarce, it would be much better to encourage it in good light loam or peaty soil, in which it thrives to perfection. As the leaves grow vigorously in summer, unaccompanied by flowers, it may be necessary to caution workmen against the barbarous practice of cutting them off, as is frequently done with the common kinds. It seeds freely in this country, and may be readily increased in that way, and also by division. In a wild state it is said to prefer dry rich soil on table-land near trees. A native of East Transylvania, the Crimea, and neighbouring regions.

CROCUS VERNUS.—*Purple C.*

THE parent of most of the blue, white, and striped kinds generally cultivated in our gardens. It has sported into a great number of varieties under cultivation, every one of them beautiful and worthy of culture, and is naturalised abundantly in meadows near Nottingham, at Hornsey, and some other places. " It is one of the most widely extended Croci, and of the easiest culture, producing seeds abundantly, which, as neither the birds nor the mice seem to eat them, become almost a nuisance, from the multitude of self-sown seedlings which come up spontaneously. It is the Crocus of the Alps, but its flower is small there, promiscuously purple and white or whitish, generally with the throat purple on the outside, but always white and hairy within. It reaches Cevennes; and I am told it is to be found, though rare, on the Pyrenees. It extends, with white acute flowers, into Carinthia, and is found white, with very blunt obovate flowers, on the Bavarian Alps, sometimes assuming a blush of purple. I believe it is found only in particular spots on the Pyrenees, affecting the oolite or jurassic limestone. On the Alps it reaches above 5000 feet of altitude. I have seen it both white and purple from the Tyrol. The finer Neapolitan variety inhabits the loftiest mountains of Carinthia and Lucania, not descending lower than 5000 feet. On the Wengern Alp its flowers actually pierce the remaining snow in June. The Odessa variety, which grows on part of the Steppes, is much finer, and from that stock the finest garden varieties seem to be derived. The segments of the flower are so rounded and concave that the half expanded flower is nearly spherical. They are white, sometimes beautifully striped in the inside, or deep purple."—Herbert, in 'Trans. Roy. Hort. Soc.'

CROCUS VERSICOLOR.—*Striped C.*

THIS is a pretty and distinct spring-flowering kind, which has spread into a good many varieties, and is abundantly grown in Holland. The ground colour of the flower is white, but richly striped with purple, the throat sometimes white, sometimes yellow, the inside being smooth, by which it can be readily distinguished from *Crocus vernus*, which has the inside of the throat hairy. Dean Herbert says this "likes to have its corm deep in the ground. If its seed is sown in a three-inch pot plunged in a sand-bed, and left there, by the time the seedlings are two or three years old, the bulbs will be found crowded and flattened against the bottom of the pot; and, if the hole in the pot is large enough to allow their escape, some of them will be found growing in the sand under the pot." It, however, thrives in any ordinary garden soil.

In addition to the preceding, for the most part easily obtained and very distinct, *C. Boryanus, Imperati, medius, biflorus*, and several others, are mentioned in catalogues, and are for the most part scarce, except the last, commonly known as the Scotch Crocus.

CYCLAMEN COUM.—*Round-leaved C.*

TUBER round, depressed, smooth, fibres issuing from one point on under side only. Leaves of a plain dark green, cordate, slightly indented; these, with the flowers, generally spring from a short stem rising from the centre of the tuber. Corolla short, constricted at the mouth; reddish purple, darker at the mouth, where there is a white circle; inside striped red. Flowers from December to March, and is a native of the Greek Archipelago. This, with the others of the same section—viz. *vernum* of Sweet (*coum zonale*), *ibericum, Atkinsii*, and the numerous hybrids from it—though perfectly hardy, and frequently in bloom in the open ground before the Snowdrop, yet, to preserve the flowers from the effects of unfavourable weather, will be the better for slight protection, or a pit or frame devoted to them in which to plant them out. I grow many in this way, and during the early spring, from January to the middle of March, they are one sheet of bloom. When so cultivated, it is best to take out the soil, say one foot and a half to two feet deep, place a layer of rough stones nine to twelve inches deep at the bottom, covering them with inverted turf to keep the soil from washing down and injuring the drainage; then fill up with soil composed of about one-third of good free loam, one-third of well-decayed leaf-mould, and one-third of thoroughly decomposed cow manure. Plant one inch and a half to two inches deep, and every year, soon after the leaves die down, take off the surface as far as the top of the tubers, and fresh surface them with the same compost, or in alternate years they may only have a dressing on the surface of well-decayed leaves or cow manure. During summer, or indeed after April, the glass is removed, and they are slightly shaded with larch-fir boughs (cut before the leaves expand) laid over them, to shelter from the extreme heat of the sun. As soon as they begin to appear in the autumn, gradually take these off, and do not use the glass until severe weather sets in—at all times, both day and night, admitting air at both back and front, and in fine weather draw the lights off, remembering that the plants are perfectly hardy, and soon injured if kept too close. They do not like frequent removal. *C. coum album* is a variety raised by Mr. Atkins, of Painswick, which received a first-class certificate from the Royal Hort. Soc. 1868. It has the dark plain foliage of *coum*, with flowers white, and dark mouth; hardy; same treatment as *coum*. It is a very distinct and interesting variety, well worthy of culture. *C. Atkinsii*, a hybrid of the *coum* section, raised by Mr. Atkins, has larger flowers, white, with dark mouth, and nearly round or ovoid leaves, variously marked.

C. vernum of Sweet is considered by many as only a variety of *coum*, and for it I would suggest the name of *C. coum*, var. *zonale* (from its marked foliage). I was for a long time unwilling to give it up as a distinct species, but now doubt there being sufficient permanent specific distinction to warrant its being retained as such, especially after seeing the many forms and hues the leaves of other species of this genus assume. Though this, as well as *C. coum*, retains its peculiarities as to markings very correctly from seed, so do some undoubted varieties of other species of Cyclamen. In Loddiges' 'Bot. Cab.' t. 108, some years previous to Sweet's publication, it is well figured as *C. coum*. There are specimens in various herbariums of this form under the name of *C. vernum* (Sweet), mostly from Iberia and Tiflis.

CYCLAMEN IBERICUM.—*Iberian C.*

THIS also belongs to the *coum* section. I fear the original type of the species as first imported into this country is lost; the greater portion now sold as such are hybrids of the *Atkinsii* group. There is some obscurity respecting the authority for this species and its native country; but there are specimens of it in the Kew and Oxford Herbariums marked "ex Iberia." Leaves very various. Flowers: corolla rather longer than in *coum;* mouth constricted, not toothed; colour various, from deep red-purple to rose, lilac, and white, with intensely dark mouth; produced more abundantly than by *coum.*

CYCLAMEN EUROPÆUM.—*European C.*

TUBER of medium size and very irregular form, sometimes roundish or depressed and knotted, at other times elongated. The rind is thin, smooth, yellowish, sometimes "scabby." The underground stem or rhizome is often of considerable length and size, sometimes even more than a foot in length. The leaves and flowers originate from stalks or branches, which emerge from all parts of the tuber. The root fibrils spring from the lower surface of the tuber as freely as from the upper, but are never so numerous as in *C. hederæfolium;* and there are usually two or three stems springing from different parts, and growing in different directions, from which the leaves and flowers arise. When these stems are much elongated and irregular, the plant becomes the *C. radice-anemone,* or *C. anemonoides* of some old authors. The leaves in this, as well as in most of the other species, vary much in outline as well as extent of the markings on the upper surface and colour beneath. Those from the more northern habitats are coarser and more decidedly dentate than those from some localities south of the Alps, where they assume in a measure the finer texture, rounder form, and more delicate markings, of *C. persicum.* The leaves appear before and with the flowers, and remain during the greater part of the year. Flowers from June to November, or, with slight protection, until the end of the year. The petals rather short, stiff, and of a reddish-purple colour. The base or mouth of the corolla pentagonal, not dentate. Some of the southern varieties, by attention to cultivation under glass, may even assume a perpetual flowering character. The varieties *Clusii, littorale,* and *Peakeanum* are of this section. In these varieties the flowers become much longer, of a more delicate colour, often approaching peach colour, and are almost the size of those of *C. persicum.* Pure white are rare, but pale ones are not uncommon. They are very fragrant. Thrives freely in various parts of the country in light loamy well-drained soil, as a choice border and rockwork plant. Where it does not do well in the ordinary soil, it should be tried in a deep bed of light loam, mingled with pieces of broken stone. In all cases it is best to cover the ground with cocoa-fibre. It is a very desirable species, on account of its delightful fragrance and long succession of flowers. I have often seen them luxuriate in the débris of old walls, and on the mountain-side, with a very sparing quantity of vegetable earth to grow in.

CYCLAMEN HEDERÆFOLIUM.—*Ivy-leaved C.*

A NATIVE of Switzerland, South Europe, Italy, Greece and its isles, and the north coast of Africa. Tuber not unfrequently a foot in diameter when full-grown; its shape somewhat spheroidal, depressed on the upper surface, rounded beneath. It is covered with a brownish rough rind, which cracks irregularly, so as to form little scales. The root fibres emerge from the whole of the upper surface of the tuber, but principally from the rim; few or none issue from the lower surface. The leaves and flowers generally spring direct from the tuber without the intervention of any stem (a small stem, however, is sometimes produced, especially if the tuber be planted deep); at first they spread horizontally, but ultimately become erect. The leaves are variously marked, and the greater portion of them appear after the flowers, continuing in great beauty the whole winter and early spring, when they are one of the greatest ornaments of our borders and rockeries, if well grown. I have had them as much as six inches long, five inches and a half in diameter, and 100 to 150 leaves springing from one tuber. They are admirably adapted for table decoration during winter. The flowers begin to appear at the end of August, continuing until October. Mouth or base of the corolla ten-toothed, pentagonal, purplish red, frequently with a stripe of lighter colour, or white, down each segment of the corolla. There is a pure white variety, and also a white one with pink base or mouth of corolla, which reproduce themselves tolerably true from seeds. Strong tubers will produce from 200 to 300 flowers each. I have had as many as 150 from one plant blooming at the same time. The varieties from Corfu and other Greek isles are very distinct and valuable additions; there do not appear to be sufficient permanent characters for specific distinction. They generally flower later, and continue longer in bloom. Their leaves rise with or before the majority of the flowers, both being stronger and larger than the ordinary type, with more decided difference of outline and markings on the upper surface of the leaves, the under surface being frequently of a beautiful purple. Texture thick, shining, and wax-like. Some of them are delightfully fragrant. They are quite hardy, but are worthy of a little protection to preserve the late blooms, which often continue to spring up till the end of the year.

This species is so perfectly hardy as to make it very desirable not only for the rock, but also for the open borders. It will grow in almost any soil and situation, though best (and it well deserves it) in a well-drained rich border or rockery. It does not like frequent removal. It has been naturalised successfully on the mossy floor of a thin wood, on a very sandy, poor soil, and it may be naturalised with perfect success almost everywhere in these islands. It would be peculiarly attractive when seen in a semi-wild state in pleasure-grounds and by wood walks. It is very frequently sent out by English nurseries and bulb dealers as *C. europæum*, though perfectly distinct from that species. It is well figured in Baxter's 'British Flowering Plants,' p. 505, and is the so-called British species; but it is doubtful whether it is a true native plant.

C. græcum is a very near ally, if more than a variety, of *C. hederæfolium*; it requires the same treatment. The foliage is more after the *C. persicum*, or the southern var. of *C. europæum*, type than most of the *hederæfolium* section; the shape of corolla and toothing of the mouth the same. *C. africanum (algeriense macrophyllum)*, much larger in all its parts than *C. hederæfolium*, otherwise very nearly allied, is hardy in warm sheltered situations.

CYCLAMEN VERNUM.—*Spring C.*

TUBER round, depressed, somewhat rough or russety on outer surface; fibres issue from one point on the under side only; under cultivation it has little or no stem, but leaves and flowers proceed direct from the upper centre of the tuber, bending under the surface of the soil horizontally before rising to the surface. Corolla long, segments somewhat twisted, mouth round, not toothed; colour from a delicate peach to deep red purple, very seldom white; deliciously fragrant. Flowers from April to end of May.

Native of South Italy, the Mediterranean and Greek isles, and about Capouladoux, near Montpellier. Leaves rise before the flowers in the spring; they are generally marked more or less with white on upper surface, and often of a purplish cast beneath; fleshy; semi-transparent whilst young. For many years I believed this species to vary in the outline and colouring of the foliage less than any other, but I have now received imported tubers from Greece, with much variety in both particulars, some of the leaves quite plain and dark green, others dashed all over with spots of white, others with an irregular circle of white varying much in outline. Among these every intermediate form occurs, up to that figured and described by Sibthorp and Sweet. The latter variety is the one more generally met with, and is reproduced from seed very true and unvarying. This, though one of the most interesting species and perfectly hardy, is seldom met with cultivated successfully in the open borders or rockery; it is very impatient of wet standing about the tubers, and likes a light soil, in a nook rather shady and well sheltered from winds, its tender fleshy leaves being soon injured. The tubers should also be planted deep, say not less than two inches to two inches and a half beneath the surface. I have grown them for many years in a border and on rocks without any other protection than a few larch-fir boughs lightly placed over them, to break the force of the wind and afford a slight shelter from the scorching sun. Some authorities give *C. repandum* as a distinct species, but I consider them identical, the only difference being in the shape and markings of the leaves, which are very variable. It is generally cultivated in England under the name of *repandum*, but most of the best continental botanists adopt the name of *vernum* for it, and it is, no doubt, the original *C. vernum* of L'Obel.

GENERAL CULTURE.—Perfect drainage at the roots is indispensable for the successful culture of all Cyclamens, growing as they often do in their native habitats amongst stones, rock, and débris of the mountains, mixed with an accumulation of vegetable soil—the tubers being thereby often covered to a considerable depth, and not exposed to the action of the atmosphere, as is too often the case under culture by placing them on the surface of the soil. This practice is in most instances very injurious, drying up and destroying the incipient young leaf and flower buds when the tubers are apparently at rest; for I find in most species that, though leafless, the fibres and young buds for the ensuing year are still making slow but healthy progress under favourable circumstances. Collectors from abroad should be specially careful in this particular. We seldom find tubers of some of the species that have been much dried or exposed to the air vegetate freely, or sometimes at all. I have now by me some roots imported nearly six years since (I believe from the Greek isles), that were thus exposed, and though the tubers have remained sound and sent out tolerably healthy fibres, they have not until this season produced healthy leaves. They have made two or three abortive attempts before, but always failed. Now, having recovered vigour of foliage, I fully expect them to bloom next autumn. In *C. hederæfolium* and its varieties the greater portion of their fibres issue from the upper surface and sides of the tuber, indicating without doubt the necessity of their being beneath the soil. The habit in *C. coum*, *C. vernum*, and their allies, of the leaf and flower stalks, when in a vigorous state, running beneath the soil, often to a considerable distance from the tuber, before rising to the surface, points in the same direction. Though Cyclamens require perfect drainage at the root, they like plenty of moisture above when in full vigour of growth.

Cyclamens generally like a rich soil, composed of good friable loam, well-decayed vegetable matter, and cow manure, reduced to the state of mould, and rendered sweet by exposure to the atmosphere before use. *C. hederæfolium* and its varieties require a stiffer loam and stronger manure than the others. They are all admirably adapted for rockwork; they enjoy warm nooks, partial shade from mid-day sun, and shelter from the effects of drying, cutting winds. Neither of these can they bear with impunity. An eastern or south-eastern aspect is best, screened from cutting winds, as affording the requisite protection against heat; but a northern one will do well. They love an open yet sheltered spot; pure air is their delight. I have a northward piece of rockwork covered with them, which from the end of August, when they begin to bloom, up to the end of March, when the leaves begin to die down, is much admired both for the flowers and also for the beauty of the wax-like foliage. During the dead period of winter it is in full perfection; and few things are more ornamental.

Cyclamens are best propagated by seed sown as soon as it is ripe, in well-drained pots of light soil. I generally cover the surface of the soil after sowing with a little moss, to ensure uniform dampness, and place them in a sheltered spot out of doors. As soon as the plants begin to appear, which may be in a month or six weeks, the moss should be gradually removed. As soon as the first leaf is tolerably developed, they should be transplanted about an inch apart in seed pans of rich light earth, and encouraged to grow as long as possible, being sheltered in a cold frame, with abundance of air at all times. When the leaves have perished the following summer, the tubers may be planted out or potted, according to their strength.

From the earliest times there appears to have been great difficulty felt by our best botanists in clearly defining the species of Cyclamen, from the great variation in shape and colouring of the leaves both above and below. Too much dependence on these characters has been the cause of much confusion and an undue multiplication of species. Some of the varieties of this genus become so fixed, and reproduce themselves so truly from seed, as to be regarded as species by some cultivators. The following are some of the more important synonyms—*æstivum (europæum); anemonoides* or *radice-anemone (europæum); autumnale (hederæfolium); Clusii (europæum); hyemale (coum); littorale (europæum); neapolitanum (hederæfolium); odoratum (europæum); Peakeanum (europæum); Poli (hederæfolium); repandum (vernum); vernum* of Sweet *(coum,* var. *zonale); zonale (vernum* of Sweet*). Anemonoides, Clusii,* and *littorale,* are southern varieties of *C. europæum,* quite distinct from the northern type.

CYPRIPEDIUM ACAULE.—*Rosy Lady's Slipper.*

A VERY handsome and perfectly hardy dwarf orchid, with a fine large purplish-rose flower, nearly two inches long, with a deep fissure in front which immediately distinguishes it from the other cultivated kinds. It is very common in North America, usually growing in woods under evergreens, and the best position for it in cultivation is in some nicely sheltered and half shaded spot on the lower flanks of rockwork, or among shrubs planted near it in sandy loam, with an abundance of leaf-mould. It also succeeds in sheltered and somewhat shaded and well-drained spots on rockwork, and, being so highly ornamental and distinct, deserves universal cultivation. It is occasionally found with pale and, more rarely, with white flowers. Flowers in summer, and may be propagated by division, but the plants in the country, at present, are too small and puny to bear this.

CYPRIPEDIUM CALCEOLUS.—*English Lady's Slipper.*

THE largest, and, when well grown, the handsomest of our native orchids, and therefore an object of much interest to cultivators of hardy plants, as well as to botanists. When grown under tolerably favourable conditions, the stem rises to a height of from sixteen to twenty inches, with large pointed leaves, and bearing large flowers; the lip yellow, variegated with purple; the long sepals and petals of a brownish purple. Although reputed to be extinct in Britain, it is known to exist yet in a wild state with us, but in very few places, and let us hope the last remaining plants may long remain undisturbed; it is abundantly distributed over Continental Europe, and should not be difficult to obtain. I have never seen this fine plant nearly so well grown as by Mr. James Backhouse, of York. He plants it on an eastern shaded aspect of his rockwork, in rich, deep, fibrous loam, in narrow, well-drained fissures, between limestone rocks. The condition in which this and other orchises are obtained has a great influence on their well-being. The roots are often dried up, and nearly or quite dead when obtained; and in this condition they would have but a poor chance of surviving, even if planted in the wilds most favourable to their natural development. Given good sound roots, there will not be the least difficulty in establishing plants in deep loam, in any well-drained, half shady spot, with some shelter afforded by low bushes and plants to prevent the leafy growth of the plant from being destroyed or injured by wind. It is propagated by division of the root, but should not be disturbed for that purpose till the plants are well established, and have begun to spread about.

CYPRIPEDIUM SPECTABILE.—*Noble C.*

A NOBLE hardy orchid; a native of meadows and peat bogs, in the Northern, and on mountains in the Southern, United States, and easily known from all its fellows by its large, much inflated, rosy lip. When well developed in the open air, I know of no hardy plant to surpass this in boldness of chiselling, and delicate purity of colour. The plant is as hardy as the common Rhubarb. It is a strong deep-rooting thing when in a congenial soil and position, and therefore to attempt its culture in shallow pans, orchid fashion, as I have seen some do, is quite useless. Doubtless a few good plants may be highly desirable in pots, if these are tolerably large and deep, and afford good room for root development; but the truest and most satisfactory way is to establish it as a free-growing hardy perennial in the open garden. Good strong roots should be procured; hardy orchids are often dried to death, or just prepared for rotting when put into the ground. I have grown the plant very well in free sandy soil mulched with cocoa-fibre, and in a partially shaded spot; but it does not usually thrive near London if not accommodated with a shady nook and deep moist soil. Doubtless in the north and west, and in most moist neighbourhoods, it will be found to succeed with pretty full exposure, or in fact as a border-plant; but to grow it well I would not recommend that course for gardeners generally. There are few gardens that do not afford some shady nook near the houses, where a deep hole might be dug, and filled with rich peat or spongy free loam, mixed with plenty of decayed vegetable matter. In such a position it would luxuriate, and also in any shady place where a deep and somewhat unctuous soil exists. The best plants I have ever seen were at Glasnevin, behind one of the ranges of plant-houses there, planted close against the wall in deep rich soil—a mixture of free, rich, very moist loam and peat. Wherever there is any kind of a bold or diversified rockwork, there should be no difficulty in succeeding with this fine plant. It should be placed on the lower flanks, and in different positions and aspects, mostly sheltered ones; and if it does not in all cases attain the stature of the Glasnevin plants, it will command admiration as the finest of hardy orchids.

C. pubescens is also in cultivation, but rare in this country. It is, however, not sufficiently distinct in aspect from the English Lady's Slipper to be of much interest for the garden. There are other hardy kinds, but none of the obtainable ones equal to the foregoing kinds.

DAPHNE CNEORUM.—*Garland-flower.*

A LITTLE trailing but compact shrub, growing from six to ten inches high, and bearing a multitude of rosy-lilac flowers, the unopened buds being crimson. The flowers, like the plant, are very compact, arranged in neat terminal umbels, and so deliciously sweet that, where it is much grown, the air often seems charged with its fragrance. It is a native of most of the great mountain chains of Europe, and is one of the most suitable of all plants for rockwork. It is also a beautiful object in the front margin of the mixed border, or for forming edges round beds of choice low shrubs. Where it thrives, the margins of the shrubberies should here and there be finished off at the grass-line by its round low-spreading tufts. It seems to delight in peat and very free, moist, sandy soils, but in some very dry and stiff soils usually proves a failure. Wherever the soil is favourable, it should be much used, and is usually increased by layers, but it is hardly worth while to propagate it thus, as it is very easily procured in most of our great nurseries.

DIANTHUS ALPINUS.—*Alpine Pink.*

A RARE, beautiful, and distinct plant, recognised at a glance from any other cultivated Pink by its dense, shining green oblong, and obtuse leaves, not pointed or ascending like those of most of the other species. Each stem bears a solitary flower, of a circular form, deep rose spotted with crimson, and, when the plant is in good health, so freely produced as to hide the leaves. In poor, moist, and very sandy loam on rockwork, it thrives, and forms a dwarf carpet, though the flower-stems may rise little more than an inch in height: both leaves and stems are much taller and more vigorous in deep, moist, peaty soil. The finest specimens I have seen of it were at Glasnevin, where it grew in peat soil on the level ground. Wireworms, rather than unsuitable soil, often cause its death. It should be placed in an exposed position, and carefully guarded against drought, especially when recently planted; comes true from seed, and is not difficult to increase in that way, or by division where it grows freely. A native of the Alps of Austria, flowering in summer.

DIANTHUS BARBATUS.—*Sweet William.*

FOR ages deservedly one of the most admired of our borderflowers, and though not so popular now-a-days in "great" gardens, happily still very much so in cottage gardens. It is to be hoped that no excuse is needed for introducing it here. More than two hundred years ago there were various varieties of it cultivated in English gardens, and in the present day many beautiful kinds, both single and double, are easily raised from seed. Many of these are valuable for borders and the rougher parts of rockwork. The plant often perishes in winter in cold soils, and although a true perennial in a wild state, and on dry and raised soils, it is one of those plants of which seedlings should be raised every year. A native of gravelly or mountain fields, in various parts of Central and Southern Europe; the flowers white or rose with white dots. Parkinson, alluding to the varieties of Sweet Williams, says: "We have them in our gardens, where they are cherished for their beautiful variety;" and speaks of a pretty "speckled kinde, termed by our English gentlemen, for the most part, London Pride."

DIANTHUS CÆSIUS.—*Cheddar Pink.*

ONE of the neatest and most attractive of the dwarf Pinks with which rocks and rocky places are studded over so great a portion of the northern and temperate regions of the earth. The short leaves are very glaucous, and the large, fragrant, rosy flowers supported on stems six inches in height, and sometimes a few inches more if the plant be grown in rich soil. It requires peculiar treatment, as in winter it perishes in the ordinary border, but flourishes freely and flowers abundantly on an old wall. I have seen many dwarf compact cushions of it on walls at Oxford, and should advise anybody who wants to succeed perfectly with it to try it in a like position. It is a native plant, and grows on the rocks at Cheddar, in Somersetshire, but is found in many parts of Europe besides. To establish it on the top or any part of an old wall, the best way would be to sow the seeds on the wall in a little cushion of moss, if such existed, or, if not, to place a little earth with the seed in a chink. It may also be grown upon a rockwork in firm, calcareous, sandy, or gritty earth, and, if possible, placed in a chink between two small rocks. Flowers in May and June, and is readily propagated by seeds.

DIANTHUS CARYOPHYLLUS.—*Carnation*

THE parent of all the races of Carnations, Picotees, and Clove Pinks, so variously and beautifully coloured and laced, so deliciously fragrant, neat in habit, and profuse in flower, as to make them, perhaps, on the whole, the most valuable of our hardy border-flowers. The plant occurs in a wild state on old castles and city walls in various parts of England, and more abundantly in similar places in the West of France, the flowers of the wild form being usually purple or white. Cultivated from time immemorial, it has given rise to the various races above named, and to innumerable varieties of these, of almost every colour, blue excepted. The varieties of Carnations alone numbered as many as 400 more than 150 years ago, and numbers have been raised since in this country, where it is not, and never was, so popular as in Italy and Germany. The Carnation is divided by florists into classes, according to the markings of the flower. Thus we have scarlet, crimson, pink, and purple bizarres when the flowers are irregularly marked with two colours on a white ground, one colour, however, predominating; purple, scarlet, and rose flakes, striped distinctly and largely with one colour on a white ground; and Picotees, with serrated edges, and usually a beautiful margin of colour. The fine old Clove Carnation is more deliciously sweet than any of the others, and it should be in every garden, as should the white Clove, of which there are now several varieties.

The Carnation and Picotee are best propagated by layers in the month of August. The florists do not consider two-year-old plants good enough to furnish what are called exhibition blooms, but for ornamental purposes they are better than young plants; and even old tufts in a suitable spot on rockwork, &c. will furnish good flowers for years. But, generally speaking, it is when the plants are about two years of age that they are most valuable for general garden decoration. The masses of flowers they then furnish are most pleasing to look at, and very useful for cutting from. In fact, wherever many flowers are required for in-door decoration, the various kinds of Pinks are worth growing to a considerable extent, merely for the sake of cutting their flowers, even if they are not desired as flower-garden ornaments. The special beds where the florists' kinds may receive attention should be in the kitchen-garden or some by-spot, but the miscellaneous kinds may be judiciously planted in the mixed border, and will there prove highly ornamental. From the borders, when they get a little scraggy, as they are wont to do when a few years old, they must be removed, and the stock kept up with young plants. Therefore it is desirable to propagate a few score Pinks and Carnations every year. Almost every kind is worth growing, from the common white Pink, so much sold in London in early summer, to Anna Boleyn, so greatly prized in some gardens in the olden days, and still grown extensively in many places, being now as worthy of it as ever, and the grand old Clove Carnation. They are generally left in the hands of a few enthusiastic florists, whereas they ought to be in every garden, small or great.

In specially cultivating the better kinds in beds, it is usual to cover the surface with an inch or more of fine rotten manure that has been passed through a sieve, and plenty of water is also given in dry weather; but as many will not care about paying more attention than is necessary, I may state that neither water nor top-dressing is usually required in ordinarily good garden soil, and the result will be quite as valuable from an ornamental point of view. But when grown in special little beds, as before suggested, in some warm border in the kitchen-garden, a top-dressing, composed of one barrow of mould to three of decayed manure, could be given in a very short time, and if the weather or soil were very dry, an occasional heavy watering would of course improve matters. To this it is only necessary to add that Picotees enjoy a stronger soil than Carnations, the latter having a tendency to "run," or lose their admired regularity of colouring under such conditions.

DIANTHUS DELTOIDES.—*Maiden Pink.*

THIS true native of Britain differs from its cultivated neighbours in its close spreading tufts of smooth, green, pointless leaves, and bright pink-spotted or white flowers, rather freely produced on stems from six to twelve inches long. Although the flower is little more than half an inch across, there is a bright and cheerful look about it which makes it indispensable to the collection of dwarf hardy flowers. It has a good constitution, and will grow almost anywhere, on border or on rockwork, not appearing to suffer from wireworm, as most other Pinks do. It is rather abundant in some parts of Britain, but wanting in many counties. It frequently flowers several times during the summer, may be readily raised from seed, or easily increased by division. A native of many parts of Europe and Asia, as well as Britain, but not of Ireland.

DIANTHUS DENTOSUS.—*Toothed Pink.*

A DISTINCT and singularly pretty species ; dwarf, with violet-lilac flowers, more than an inch across, the margins toothed at the edge, the base of each petal having a regular dark-violet spot, which produces a dark eye nearly half an inch across in the centre of the flower. I first saw this plant in cultivation in Madame Vilmorin's garden, near Paris, and was much struck with its beauty, the large, glaucous leaves spreading into broad tufts, and being quite covered with flowers, on stems not more than five or six inches high, and not unlike a purplish form of the alpine Pink. It comes readily from seed, is a native of Southern Russia, flowering in May and June, continuing till autumn, and thriving well in sandy soil, in borders, or on rockwork.

DIANTHUS NEGLECTUS.—*Glacier Pink.*

IT is impossible to exaggerate the beauty of this plant. It forms, very close to the ground, tufts resembling short wiry grass, of slightly glaucous leaves, concave, pointed, and, except in vigorous specimens, from half an inch to an inch long, the lower leaves on the stems being longer, the flowers on stems from one inch to three inches high, according to the vigour of the plants. The petals are quite level and firm-looking, with the outer margins slightly notched, and the flower about an inch across, in vigorous specimens an inch and a quarter ; the colour is of the purest, deepest, and most brilliant rose. It is so dwarf in habit, and has flowers so large, that tufts of it might at first sight be taken for the alpine Pink ; but it is immediately distinguished from that by its short, narrow, pointed, grass-like leaves. In a wild state, and in poor earth on rockwork, specimens of it may be seen in perfect bloom at one inch and a half high, and even less ; but when cultivated in rich, deep, sandy loams, it attains greater dimensions, at some slight loss of neatness and compactness. It is surpassed by no alpine plant in purity or vividness of colouring, and is, happily, very easily grown, not appearing to have any of the fastidiousness characteristic of *D. alpinus* or *D. cæsius* in some soils and positions. It grows with freedom in very sandy loam, either in pots or on rockwork, rooting through the bottoms of the pots into the sand as freely as any weed, and is perfectly hardy and easily grown in all parts of these islands. A native of the highest summits of the Alps of Dauphiny, and the Pyrenees, Switzerland, and North of Italy; easily increased by division, and also from seed. Sometimes known as *D. glacialis*.

DIANTHUS PETRÆUS.—*Rock Pink*.

A CHARMING species, with very short sharp-pointed leaves, forming hard tufts an inch or two high, from which spring numerous flower-stems, each bearing a solitary fine rose-coloured flower. I once grew a group of about fifty plants of this, which formed a sort of turf, and flowered so freely as to be conspicuous at a great distance. It grew along with other species, and nearly all the plants raised from seed varied, so that it ought to be increased by division. It seemed to escape the attacks of wireworm when nearly every other species was destroyed. A native of Hungary, flowers in summer, and is well worthy of a position on rockwork, where it ought to be planted in very sandy and rather poor moist loam.

DIANTHUS PLUMARIUS.—*Pink*.

THIS plant is considered the parent from which our numerous varieties of Pinks have sprung, and, as the progenitor of such a stock, is entitled to some consideration, even if it does not come up to the popular standard so well as many of its race. It has single purple flowers, rather deeply cut at the margin, and is naturalised on old walls in various parts of England, though not a true native. It is rather handsome when grown into healthy tufts, but on the level ground it is apt to perish or get shabby, which points to the desirability of establishing it on old walls. But the almost innumerable beautiful and fragrant double varieties command the highest admiration, and there is no reason whatever why these should not be cultivated as rock-plants, particularly as they live much longer and thrive better on such elevations. They are such prettily shaped flowers, so compact, hardy, and fragrant, that one would think a few words in praise of them quite unnecessary; but the fact is, with many other charming plants, they have been driven into comparative obscurity by the widely-spread taste for common bedding plants. They have for many years been amongst the most favourite "florists'" flowers in European countries. In cultivating the finer florists' varieties, it is desirable to give them a special little bed in some convenient part of the garden—one of the beds in a neat little nursery border in the kitchen-garden would do best; and if the natural soil be not a good sandy loam, it should be made so by additions from a neighbouring pasture, and well rotted manure should be added in any case. The bed should be slightly raised above the alleys, say from six to ten inches, according to the soil; of course, the wetter and stiffer it is, the more acceptable will a slight elevation prove, while the beds should, generally speaking, be made in a dry and warm part of the garden, the spot to be well drained and the soil deep. The best growers usually plant out about the end of September, but, like most hardy evergreen or herbaceous herbs, they may be safely transplanted at any time, either in spring or autumn. The propagating of the Pink is always very much easier of accomplishment in the beginning, middle, or end of June, say about the 20th, than at any other time. It is effected by pipings—another word for cuttings, the difference consisting in the way they are made—pulled off from the shoot, the little stem parting readily, and coming out from its embracing leaves with a slender shank. There is no necessity for doing it thus: on the contrary, the better way is to cut off some of the side shoots, trim them up a little way with a sharp knife, as you do most cuttings, and place them firmly in pots or in a gentle hotbed, where they will soon root, and should then be hardened off. If put in pots, they should be placed in a gently heated frame till struck, but the best plan is to put four or five inches of sandy soil over a shallow, gentle hotbed, surface the soil with a little fine sand, and then put in your cuttings, covering them with common hand-lights, which must of course be shaded at first; then the lights may be taken off gradually as the plants become rooted, and finally removed altogether. Thus treated, the young plants will be in nice condition for planting out in autumn.

It is not desirable to enumerate what are considered the finest kinds in a book of this character, inasmuch as they are liable to continual change; tastes vary, and occasionally varieties with no names at all are superior to the florists' kinds for general garden use. It should be generally known that a race of perpetual-flowering Pinks, the result of a cross between that old favourite Anna Boleyn and some of the florists' varieties, is now in existence, and well worthy of culture. Garibaldi, Most Welcome, and Tennison Pink are among the best of these. They are for the most part of a fine rosy-crimson tone.

It is to be regretted that good single as well as double varieties are not sought after; they flower so very profusely, are more showy, and have better constitutions. One of the best I know of is the *D. plumarius annulatus*, a very fragrant variety with a large dark ring in the centre, and it would not be difficult to select other finely coloured single varieties valuable as rock, border, or bedding plants.

DIANTHUS SUPERBUS.—*Fringed Pink.*

A HANDSOME and very fragrant species, easily known by its petals being cut into lines or strips for more than half their length, which gives the plant a singular and not ungraceful appearance, quite different to that of any other kind in cultivation. It inhabits many parts of Europe from the shores of Norway to the Pyrenees, and is a true perennial, though it perishes so often in our gardens, when very young, that many regard it as a biennial. It is more apt to perish in winter on rich and moist soil than on that which is somewhat poor, light, and well-drained, and it should be planted in fibry loam, well mixed with sand or grit, where it is desired to establish it as a perennial. It is, however, very free to grow on nearly every description of soil; and by raising it every year from seed an abundant stock may be kept up even where it perishes in winter. Unlike some of the other kinds, it comes quite true from seed, generally grows more than a foot high, flowers in summer or early autumn, and is perhaps more suited for mixed beds and borders than for the rockwork. On this it should not get a choice position, as it is one of the easiest kinds to grow; besides, it is somewhat too large for association with the jewel-like flowers which form the true ornaments of the rock-garden.

DIAPENSIA LAPPONICA.—*Lapland D.*

A STURDY and very dwarf little evergreen alpine shrub, very rarely seen even in botanic gardens, and usually considered impossible to cultivate, but which may be grown very well on fully exposed rockwork in deep sandy and stony peat, kept well moistened during the warm season. It grows in very dense rounded tufts, with narrow closely packed spoon-shaped leaves, and solitary white flowers with yellow stamens, about half an inch across, the whole plant being often under two inches high. A native of Northern Europe and North America, on high mountains or in arctic latitudes, flowering in summer, and probably most easily increased from seed, though as yet the plant has been so little grown that much cannot be said on this subject.

DIELYTRA EXIMIA.—*Plumy D.*

AN elegant plant, much longer in cultivation than *D. spectabilis*, but far less popular, though well deserving of extensive cultivation; its leaves, much divided into oblong strips, form graceful, somewhat fern-like tufts, in good soil, usually a foot or more high. It freely bears pretty reddish-purple flowers in racemes in early summer, and continuously if the soil be generous, and, like *Thalictrum minus*, it may be used in borders or on rockworks with a similar object to that we have in view when placing fine-foliaged plants in a conservatory, or a spray of Maidenhair fern in a bouquet. A native of North America, quite hardy, and easily increased by careful division.

Dielytra formosa, also of North America, is said to differ in being smaller, with the lobes of the flower longer; but, while many forms of a species, which botanists would not even dignify with the name of varieties, are, in appearance and for garden purposes, really distinct, these two, or the plants which pass for them in this country, present no such differences, and the cultivator who possesses *D. eximia* will not find a new beauty in *D. formosa*.

DIELYTRA SPECTABILIS.—*Mountain D.*

Now too well known to need description or recommendation, nearly every garden in the country being embellished with its singularly beautiful flowers, opening in early summer, gracefully suspended in strings of a dozen or more on slender stalks, and resembling rosy-crimson hearts. It is a native of China and Siberia, is perfectly hardy, and unquestionably one of the handsomest and most useful plants ever introduced into this country. It usually grows too large for association with the subjects to which this book is devoted, but it is of such remarkable beauty and grace that it may be used with the best effect near the lower flanks of, or in bushy places near, rockwork, or on low parts where the stone or "rock" is suggested rather than exposed. It need hardly be added that it is well worthy of naturalisation by wood walks, &c., especially on light rich soils. There is a "white" variety, by no means so ornamental as the common one, though worth growing for variety's sake.

DIOTIS MARITIMA.—*Sea Cottonweed*.

A VERY distinct-looking plant, found native on the sea-shore sands of the southern half of Great Britain, and also recorded from Ireland, but most abundant in St. Owen's Bay, Jersey. It is readily known by both sides of its oblong leaves being densely covered with a very white cottony-looking felt. The yellow flower-heads are not ornamental, and except in the botanic garden, the plant is most likely to be grown for the singular appearance of its stems and leaves. It forms a suitable ornament on rockwork, and I have also seen it employed with some effect as an edging plant in the flower-garden, though it is apt to grow rather straggling, and should be kept neatly pegged down and cut in well to prevent this tendency, and should have a very sandy and deep soil. Increased by cuttings, as it seldom seeds in gardens. There is only one species of the genus.

DODECATHEON MEADIA.—*American Cowslip*.

THE American Cowslip, bright, graceful, and perfectly hardy, is second to none of our old border-flowers. Its blooms should be seen in early summer in every spot worthy of the name of a garden. They are supported in umbels on straight slender stems from ten to sixteen inches high, each flower drooping elegantly, the purplish petals springing up vertically from the pointed centre of the flower, much as those of the common greenhouse Cyclamen do, and this gives the bloom such a gay and singular appearance that one can understand the natives of the Western United States calling it, as they do, "Shooting Star." It inhabits rich woods in North America, from Maryland and Pennsylvania, in the North, to North Carolina and Tennessee, in the South, and far westward, loves a rich light loam, and is one of the most suitable plants for the rock-garden or well-arranged mixed borders, the fringes of beds of American plants, &c. In many deep light loams, the plant flourishes without any preparation, but where a place is prepared for it as is often necessary, it is very desirable to add plenty of leaf-mould. In a somewhat shaded and sheltered position, it attains its greatest size and beauty, though it often thrives in exposed borders, and is best increased by division when the plants die down in autumn; when seed is sown, it should be soon after being gathered.

There are several varieties of the American Cowslip : "*album*," "*elegans*," "*splendidum*," "*giganteum*," "*lilacinum*," &c., which occur in the catalogues. These are beautiful, and well worthy of cultivation, though there are more names than distinct varieties. Some consider the three kinds here described as varieties of one species, but they are sufficiently distinct for gardening purposes.

DODECATHEON INTEGRIFOLIUM.—*Small American C.*

A LOVELY and gaily-coloured flower, deep rosy crimson, the base of each petal white, springing from a yellow and dark orange cup, and appearing in May on stems from four to six inches high. The leaves are much smaller than those of *D. Meadia*, oval, and quite entire. A native of the Rocky Mountains, a gem for the rock-garden, planted in sandy peat or sandy loam with leaf-mould, and increased by careful division of the root and by seed, which it ripens freely in this country. It is easily grown in pots, plunged, in the open air, in some sheltered and half shady spot during summer, and kept in shallow cold frames during winter. Where alpine plants are grown in pots for exhibition, it should not be omitted.

DODECATHEON JEFFREYANUM.—*Great American C.*

A NOBLE kind, which I have grown as high as two feet in very favourable circumstances, and have known to grow much larger even in London gardens than the old American Cowslip. It has much larger and thicker leaves, of a darker green, and with very strong and conspicuous reddish midribs, the flower being like that of the old kind, except that it is somewhat larger and darker in colour. It is a thoroughly hardy and first-class plant, flourishing freely in light rich and deep loam, and thriving best in a warm and sheltered spot, where its great leaves may not be broken by high winds. Spots suited for the handsome *Cypripedium spectabile*, in the hollows and in the fringes of the rock-garden, will suit it to perfection, and, when sufficiently plentiful, it will no doubt prove one of our most valuable border-flowers. A native of North America.

DRABA AIZOIDES.—*Seagreen Whitlow-Grass.*

THIS may be taken as the typical species of the Golden Drabas; it is indigenous to Britain, but only found in one locality in South Wales, where, let us hope, its inaccessible position in many instances will protect it from the hand of the destroyer. In growth it does not exceed three inches in height, and when planted on the slope of a sunny border, in sandy soil, which it loves, it forms a dense golden carpet in the early part of March. It does not ripen seed so freely as the following kind, but increases readily by division, and in these two respects we have a very marked and tangible distinction from the following otherwise closely allied species. A very neat plant for rockwork, and also an attractive subject for naturalising on moist old walls, mossy ruins, &c.

DRABA AIZOON.—*Evergreen Whitlow-Grass.*

A NATIVE of the mountains of Carinthia, closely allied to the previous species, but a much more vigorous grower; the leaves are broader and of a darker green, and arranged so as to form a most complete rosette, not unlike the Sempervivums. From the centre of this rosette it sends up a stem five or six inches long, bearing numbers of bright-yellow flowers, and ripens its seeds freely. *Draba bœotica* I am disposed to consider a narrow-leaved form of the above. In the cultivation of both it must be borne in mind that, unlike *D. aizoides*, the old stems will never throw out roots, consequently they cannot be classed as spreading plants. They increase freely from seed, some of which it would be interesting to sow on old walls and ruins, with a view to naturalising them in these positions. They are most effective when grown in small pots, in which they might, for early spring use, be plunged in a close line, say round the margin of a raised "pin-cushion" bed, with admirable effect.

DRABA ALPINA.—*Alpine Whitlow-Grass.*

AN arctic plant, with dark-green, smooth, somewhat ovate leaves, growing about two inches high, and producing bright golden flowers. It is rather a delicate subject, and best adapted for pot culture, or well-drained chinks in rockwork. The true species is somewhat scarce in cultivation. It, like *D. tridentata*, is liable to suffer from slugs, and both should be carefully guarded against their attacks, especially during the winter months. Allied to this is *Draba aurea*, a Danish plant, with flowers produced in a dense corymb, on a leafy stem some eight or nine inches high; the habit is not neat, otherwise it is a very distinct, well-defined species.

DRABA CINEREA.—*Grey Whitlow-Grass.*

THIS native of Siberia, frequently called *D. borealis*, is in my opinion the most effective of the white-flowering Drabas. Of dwarf habit, producing an abundance of clear white flowers in the earliest spring, well relieved by the dark-green leaves, and of a free-growing and permanent character. It should be in every collection. Seeds abundantly, and by that means, as well as by root division, it may readily be increased.

DRABA CUSPIDATA.—*Pointed Whitlow-Grass.*

A NATIVE of the highest mountains in Spain, closely allied to *D. ciliaris*. They both possess many of the characteristics of *D. Aizoon*, but are more compact in growth, as well as more diminutive. *D. cuspidata* has the points of each of the ciliated leaves, of which the dense little rosettes are formed, somewhat incurved, and for close examination it is the gem of the yellow Drabas, forming a comparatively thick woody stem. It is only to be increased by means of seed, which it produces but sparingly. My experience in raising the seeds of this plant leads to the conclusion, from the varied forms produced in the offspring, that *ciliaris* is only a slight variety of *cuspidata*, or vice versâ. I have, however, not yet succeeded in getting seed from *D. ciliaris*.

Draba lapponica, a native, as the name indicates, of the arctic regions, though bearing the aspect of *D. rupestris*, is dwarfer in habit, and devoid of the ciliated hairs on the leaves; it forms dense tufts, and flowers freely in early spring, producing an almost equally abundant bloom in the autumn; it also seeds freely.

Draba rupestris, frigida, and *Chamæjasme,* are three very dwarf, compact-growing plants, closely allied, in fact so much so that they may be considered as mere varieties of the typical species, *D. rupestris*. The flowers in each case are small, but are produced abundantly. Considering the neat habit of the plants, every collection should possess at least one of them.

Draba nivalis, a native of the Swiss Alps, is the most diminutive of the genus. The leaves are of a whitish green, owing to the presence of minute stellate hairs. The plant, when in flower, is not over two inches high, of nice compact habit, but rather a shy grower, and consequently is rarely met with in cultivation.

DRABA TRIDENTATA.—*Three-toothed Whitlow-G.*

THOUGH classed amongst the yellow Drabas, this is quite a distinct plant from the preceding species—in fact, in general contour it is unallied to any other species that we know; it is a native of the mountains of Southern Russia, and forms a dark-green, branching plant about four inches high, with a very delicate root stock; the leaves being tolerably broad, and, as the name indicates, tridentate in character. The flowers are produced abundantly, and are of a most intense golden yellow. It seeds pretty freely, and ought to be planted on a rockwork, so that the seeds may vegetate at once round the parent plant, which, by the way, must be looked upon as little better than a biennial.

Amongst the spring-flowering alpines. the genus Draba must always take an important position. In addition to the brilliant golden colour of the flowers of one section of the genus, the plants are characterised by a dwarf compact habit, and by much neatness in the arrangement of the bristly ciliated hairs, which not unfrequently become bifurcate; thus the attractive appearance in the matter of colour is enhanced on a closer inspection by the beauty of form and detail. In another section, we find white to be the predominant colour, and though in many cases the flowers are small, still, in the mass, filling up a nook or crevice in a rockwork, and contrasted with the dark-green leaves, they become very effective. They should be placed in the sunniest aspect on a rockery; the more effectually the plants are matured by the autumn sun the more freely will they return these favours by an abundant bloom in early spring. The third section, which includes plants of a purple and violet tint of colour, is chiefly, if not altogether, confined to the representatives of the genus that grow abundantly in the high mountain lands of South America. Of these we have but one in cultivation, *Draba violacea*, and of so recent introduction that it may be considered rash to pass any opinion on it beyond the fact that it is a remarkably beautiful plant, of doubtful hardiness. We may here observe that the sections I have adopted must be considered as more strictly chromatic than botanical.

DRACOCEPHALUM AUSTRIACUM.—*Austrian D.*

A SHOWY species, with blue flowers more than an inch and a half long, in whorled spikes, the plant of rather a woody texture, spreading into masses about a foot high, the floral leaves velvety, trifid, and with long fine spines, the leaves three- or five-cleft, with narrow segments. A native of nearly all the great mountain chains of Europe, flourishing on rockwork in light soil, and increased by seed or division. Quite free to grow in most ordinary garden soils, but, like many other mountain plants, only attaining perfect ripeness of texture on rockwork, unless in very well-drained, warm, and sandy soils.

DRACOCEPHALUM GRANDIFLORUM.—*Betony-leaved D.*

A PLANT rarely seen in our gardens, distinct in appearance from its relatives, not diffuse or procumbent, but in habit more like a dwarf Betony; the flowers, however, are handsome, blue, in whorled oblong spikes two to three inches long; the leaves oblong, obtuse, heart-shaped at the base, and crenated; the whole plant little more than half a foot high, though it varies from two inches to a foot high. A native of Siberia, frequent in the Altaic Alps, and thriving best on somewhat elevated sandy borders, or low spots on rockwork in good sandy and thoroughly drained loam. It should be guarded as far as possible against slugs, which are fond of it, and may quickly destroy young and small plants. Flowers in early summer, and is increased by division.

DRACOCEPHALUM RUYSCHIANUM.—*Ruysch's D.*

HAS flowers smaller than those of *D. austriacum*, produced in rather close spikes at the summit of the stem; the leaves smooth, narrow, entire, and opposite; the floral leaves or bracts also entire. A pleasing border or rock perennial, flowering rather late in the summer, and thriving best on slightly elevated rockwork, for which it is also well fitted by its spreading, somewhat prostrate habit, forming tufts about a foot high. Increased by division or seed.

DRYAS OCTOPETALA.—*Mountain Avens.*

FEW have travelled in alpine or arctic regions without seeing how abundantly the mountains are clothed with the creeping stems and large creamy-white yellow-stamened flowers of this plant. The leaves are shining above and white and downy beneath, and the fruit has a feathery appendage above an inch long. It is an evergreen, and, being neat in habit as well as handsome in bloom, ought to be grown in every collection of rock-plants. Widely distributed through the mountain region of Europe, Asia, and North America, and very abundant in Scotland. It is easy of culture in moist peat soil, in which it grows so freely about Edinburgh that I have observed it forming dressy edgings to beds in some of the nurseries there. Propagated from seed, or by cuttings and division where it grows freely.

Dryas Drummondi, a species very like the preceding, but with yellow flowers, is also in cultivation, but far from common. It would probably succeed under the conditions that suit the other, but I have not seen it out of frames.

ECHEVERIA SECUNDA.—*Silvery E.*

A MEXICAN plant, with somewhat of the appearance of a large European Houseleek, but forming more open rosettes, from three to six inches in diameter, and of a very pleasing silvery glaucous tone. The flowers are reddish, freely produced in long racemes drooping at the top, making an attractive object in the greenhouse. It is, however, chiefly grown for the effect of its rosettes in the open garden, for which purpose it is kept over the winter in frames or greenhouses, and put out in early summer; but it is hardy enough to survive the winter in some situations, and in dry places on rockwork may be tried in the open air. It is almost indispensable for association with dwarf succulents in geometrical arrangements, and is valuable also for the rockwork. Easily increased by offsets, which root readily.

The large-leaved *Echeveria metallica*, almost like a "foliage-plant" from the aspect of its great rosettes of metallic-looking leaves, is most valuable for producing a striking effect among dwarf bedding and edging plants. It should always be placed singly, forming centres to masses or rings of its dwarf relative *E. secunda*, or Sedums, Sempervivums, and Saxifrages. Increased by seeds or by the leaves; both those of the flower-stem and of the rosette soon strike root in a temperate house. It requires a warm greenhouse in winter, and is only mentioned among "alpine flowers" in consequence of being occasionally associated with them in flower-gardens.

EPIGÆA REPENS.—*Ground Laurel.*

A PROSTRATE, trailing evergreen found in sandy or rocky soil especially in the shade of pines, common in many parts of North America, and remarkable for its delicate rose-coloured flowers in small clusters, exhaling a rich aromatic odour, and appearing in early spring. The leaves are rounded-heart-shaped, covered with russety hairs. It is a plant very seldom met with in good health in this country, though occasionally seen flourishing in heath soil. In planting it, it would be well to bear in mind that its natural habitat is under trees, and plant a few specimens in the shade of pines or shrubs. I have seen it thrive planted out in a shady cold frame in leaf-mould and peat. In New England it is known as the Mayflower.

ERANTHIS HYEMALIS.—*Winter Aconite.*

A SMALL plant, with yellow flowers, surrounded by a whorl of shining-green divided leaves, with a short, blackish, underground stem resembling a tuber; the flowers, an inch or more across, being thrown up on stems from three to eight inches high. It is naturalised in woods and copses in various parts of the country, but has probably escaped from cultivation, and is not considered a native, its true home being shady and humid places on southern continental mountains. It is pretty well known, being frequently sold by our bulb merchants, and is too common a plant for the choice rock-garden. I only introduce it here to say that its best use is for naturalisation in shady spots under trees and shrubs. Where the branches of specimen trees are allowed to rest on the turf of lawn or pleasure-ground, a few roots of these scattered over the surface will soon form a dense carpet, glowing into sheets of yellow in the very dawn of spring. It will also cover any bare place under trees, and thus we may enjoy it without giving it positions suited for rarer and more fastidious plants, or taking any trouble whatever about it.

ERICA CARNEA.—*Spring Heath.*

ONE of the most valuable plants of the spring Flora of our gardens, forming dense, neat, dwarf tufts of shoots; these, in the very dawn of spring, become covered with rosy-red flowers, which, in the bud state throughout the winter, seem to await the coming of the first fine and sunny days to fully blush into masses of colour. It thrives best in peat, but often also in ordinary garden soil. It is becoming very popular, and is much grown with American shrubs in nurseries, and often used as an edging to beds of these plants. It should be grown in every garden, either in isolated tufts, in borders, or around shrubberies. On rockwork masses of it cushioning over the edges of rocks on the sunny side look charming in spring. A native of the Alps, and nearly allied to *E. mediterranea.* Indeed, some consider them varieties of the same species, but for garden purposes they are sufficiently distinct, as I have seen the last attain a height of five feet, in gardens where *E. carnea* spread about dwarfer than the common Thyme. It is also much hardier and more ornamental. The plant found in Ireland (*E. hibernica*—Syme), and which is sometimes united with *E. carnea*, is also very distinct, being much larger, less hardy, and not an early-flowering species.

The varieties of our common British Heaths afford exquisite beauty of colour. There are a number of forms of the common Ling (*Calluna vulgaris*), very pretty and dwarf; then there is the showy and beautiful Scotch Heather (*Erica cinerea*), always attractive in a wild state, but particularly so in its variety *coccinea;* and there are the following varieties in cultivation—*alba, atropurpurea, coccinea, rosea,* and *rubra. Erica Tetralix,* the large-flowered *E. ciliaris,* the white variety of the Irish

Heath, *E. hibernica alba* (*mediterranea alba*), and Mackie's Heath, *E. Mackieana*, are well suited for the rougher parts of the rock-garden.

ERIGERON ROYLEI.—*Royle's E.*

A VERY rigid plant, two to four inches high when in bloom, the centre of the flower almost black when young, but changing to a mixture of gold and black when fully open, as if gold dust were sprinkled over black velvet. The flower is two inches across, sometimes more, and the rays of a dark lavender-blue are barely one-third the width of those of *Aster alpinus*, to which the plant bears at first sight some resemblance. It is, however, easily distinguished from the alpine Aster by the narrow, sharply pointed, and violet-brown scales of its flower-head, whereas those of *A. alpinus* are blunt, green, and recurved. A native of the Himalayas. It is sometimes known as *Erigeron speciosus*, but is quite distinct from *Stenactis* (*Erigeron*) *speciosa*, a handsome plant with yellow flowers and long, pale-purple rays, very desirable as a border-plant, often attaining a height of nearly two feet, and therefore too tall for intimate association with the gems of the rock-garden.

ERINUS ALPINUS.—*Wall E.*

A NEAT and distinct little plant, with violet-purple flowers in short pubescent racemes, abundantly produced over very dwarf tufts of downy, oblong, and toothed leaves, obtuse at the apex. A native of the Alps of Switzerland, the Tyrol, and the Pyrenees, perishing in winter on the level ground in most gardens, but quite permanent and producing masses of flowers when allowed to run wild on old walls or ruins. I have seen brick garden walls with every chink between the bricks filled with this plant, so as to look at a distance as if covered with moss in winter, and in summer becoming covered with masses of lovely colour. It is easily established on old ruins or walls by sowing the seeds in mossy or earthy chinks, and is of course well suited for rockwork, growing thereon in any position, often flowering bravely on earthless mossy rocks and stones. *E. hirsutus* is a variety covered with long and whitish pubescence. A pure white variety was raised by Mr. Atkins, of Painswick, and it is a very desirable addition.

ERODIUM MACRADENIUM.—*Spotted Heronsbill.*

ALLIED to the rock Heron's Bill, but immediately distinguished from it by the two upper petals being marked with a large blackish spot, the lower petals being larger and of a delicate flesh-colour, veined with purplish rose, two to six flowers being borne on stalks from two to six inches high. The leaves are twice divided, and form graceful little tufts. The flowers are very beautiful, and the entire plant has a peculiar and agreeably aromatic fragrance. It comes from the Pyrenees, and is easily grown in chinks and thoroughly drained spots on the sunny side of rockwork, in dry and warm rather than rich soil, and is increased with the greatest facility from seeds, and also by division.

ERODIUM MANESCAVI.—*Noble Heronsbill.*

A VIGOROUS and showy species, with numerous long, much-divided leaves, from which spring many stout flower-stems, each bearing an umbel of from five to fifteen purplish flowers, each more than an inch across and very handsome. It is quite distinct from any other kind, and deserves a place in every collection, flourishing healthfully on the level ground as well as on rockwork, on which, being a vigorous grower, it should be associated with the strongest plants only. A native of the Pyrenees, flowering in summer, and, when the plants are young and in rich soil, for a long time in succession. Easily raised from seed, and in cultivation grows from ten inches to two feet and a half high.

ERODIUM PETRÆUM.—*Rock Heronsbill.*

A NEAT species, with much divided, usually somewhat velvety leaves, and rather large, lively rose, or white-and-veined, but not spotted, flowers, attaining a height of from three to six inches, and well suited for the embellishment of warm and dry chinks or nooks on the sunny sides of rockwork. It and its ally, *E. macradenium*, are just the plants to try on old walls or ruins; on the level ground, or in moist spots on rockwork, they are not so attractive, as the leaves become developed at the expense of the flowers, and the softness of tissue resulting from the same cause predisposes them to perish in winter. There is a smooth variety, *E. lucidum*, and one with more curled and downy leaves,

E. crispum; all are natives of dry rocky places in the Pyrenees and Southern Europe, and are increased by seed or division. The two last-mentioned varieties are probably not in cultivation.

ERODIUM REICHARDI.—*Reichard's Heronsbill.*

A TUFTED stemless plant, a native of Majorca, and so minute as to seem fitted for nestling under the deep snow with the *Androsaces*, and with the highest traces of vegetable life on the Alps. The heart-shaped little leaves rest upon the ground, and the flower-stems attain a height of two or three inches, each bearing a solitary white flower, faintly veined with pink. It flowers pretty freely, and usually from spring or early summer till autumn; is quite easy of culture in moist sandy peat or loam, and is worthy of a position on every rockwork, either on flat bare exposed spots or in chinks. Where alpine plants are grown in pots or pans, its neatness of habit and dwarfness, added to the fact that it is so easily grown, make it worthy of a place.

ERODIUM ROMANUM.—*Roman Heronsbill.*

A PRETTY species, with gracefully cut leaves like those of the British *Erodium cicutarium*, to which it is allied; but it differs in having larger flowers, with equal-sized petals, in being stemless and a perennial. The flowers are purplish, and freely produced in the end of March or beginning of April. It is easily grown, and comes up thickly from self-sown seeds, at least in light and chalky soils; would thrive on old walls and ruins, and is a suitable ornament for the less important spots on rockwork. It was cultivated in this country as far back as one hundred and fifty years ago, but was probably long lost till reintroduced by the Rev. Harper Crewe, in whose interesting garden at Drayton-Beauchamp I first saw it. A native of France and Italy.

It is probable that we are as yet but imperfectly acquainted with the species of Erodium most worthy of culture. Among those I have not as yet observed in cultivation in this country is *E. trichomanefolium,* the Maidenhair Cranesbill from the East, with leaves so deeply cut as to appear, as the name indicates, like a finely-cut fern. I saw it in the botanic garden at Geneva, and believe it would prove hardy on rockwork in our gardens, and well suited for the positions recommended for the Rock and Spotted Heronsbills.

ERPETION RENIFORME.—*New Holland Violet.*

THIS mantles the ground with a mass of small, kidney-shaped leaves, has numerous slender, creeping, and rooting stems, and bears blue and white flowers of exquisite beauty, rising not more than a couple of inches from the ground, and produced continuously throughout the summer. A violet it is indeed, but a violet of the southern hemisphere, one at home under a Port Jackson sun, but without the vigour and depth of colour of our northern sweet Violet, which braves and bestows its sweets on the " hard, hard, hard north-eastern breeze that breeds hard Englishmen," yet having a simple loveliness that prevents its omission here, even though it is not hardy enough to stand our winters. It is peculiarly fitted for planting out over the surface of a bed of peat or very light earth, in which some handsome plants would be put out during the summer in a scattered or isolated manner, and the little herb allowed to crawl rapidly over the surface. For example, that handsome succulent, *Echeveria metallica*, has been found to grow admirably in the open air in England for several summers past, and in consequence of its bold habit it is necessary to place it, say, a couple of feet from plant to plant. Our little Australian friend is one of the very best things to fill up the surface; and, as the practice of placing plants of some character in flower beds is likely in due time to get the preference which it deserves over the massing system pure and simple, dwarf plants to form a carpet around and beneath the taller ones will be requisite. This should form one for the choicest positions. Being very small and delicate as well as pretty, it should not be used under or around coarse subjects. It must of course be treated like an ordinary tender bedding plant—taken up or propagated in autumn, and put out in May or June. In every place where alpine plants are grown in pots, it should find a home; and in mild parts of these islands, say the south and west coast, it would probably maintain its ground in sunny spots without perishing during winter.

ERYSIMUM OCHROLEUCUM.—*Alpine Wallflower*.

THIS handsome and distinct plant forms, when under cultivation, very neat, rich green tufts, six to twelve inches high, and is in spring covered with a dense profusion of beautiful sulphur-coloured flowers. Rockwork will be found to offer the most congenial home for it; it does very well on good level ground, but is apt to get somewhat naked about the base, and will perhaps perish on heavy soils during an unusually severe winter. I have found it thrive best when rather frequently divided. It is propagated by division and by cuttings. Most probably this plant would find the conditions that best suit it on old walls, ruins, &c.; but I have never tried it in such positions. It is capital as a dwarf border-plant on light soils, the flowers bear some resemblance to those of *Vesicaria utriculata*, and in spring, tufts of it may be seen, covered with clear yellow bloom, in the barrows of the London costermongers. A native of the Alps and Pyrenees, flowering in spring and early summer. There are several varieties. It is readily known from the following species by its much greater size. = *Cheiranthus alpinus*.

ERYSIMUM PUMILUM.—*Lilliputian Wallflower*.

A REMARKABLE little plant, very rare in cultivation, resembling in the size and colour of its flowers the alpine Wallflower, but without the vigorous and rich green foliage of that species; producing flowers very large for the size of the plant, often only an inch high, above a few narrow, sparsely toothed, leaves barely rising above the ground. I have seen specimens of it in full bloom with the flowers nearly as large as those on healthy tufts of the alpine Wallflower, and yet the whole plant, flowers and all, could be almost covered by a thimble. In richer soil and less exposed spots it is larger; but the specimens above alluded to were grown in England. A native of high and bare places in the Alps and Pyrenees, requiring to be grown on rockwork in an exposed spot in very sandy or gritty loam, surrounded by a few small stones to guard it from excessive drought and accident, and associated with the choicest and most minute alpine plants. It is very nearly related to the alpine Wallflower, *E. ochroleucum*, but is at once separated from that plant by its minuteness and the dull greyish-green colour of its leaves.

ERYTHRONIUM DENS CANIS.—*Dog's-tooth Violet.*

ONE of the loveliest of all our old garden-flowers, now seldom seen, though it should be in every place where spring flowers are welcome—its handsome oval leaves, rounded below and pointed above, being so marked with patches of reddish brown as to make it worthy of being grown as a diminutive foliage-plant, even if its fine flowers never appeared. These are borne singly on stems four to six inches high, drooping gracefully, and are cut into six rosy purple or lilac divisions. There is a variety with white, one with rose-coloured, and one with flesh-coloured flowers. The plant is said to like shade, but I have seen it attain its highest beauty in moist, sandy, peaty soil, in positions fully exposed to the sun. It is one of the most valuable subjects for the spring or rock garden, or border of choice hardy bulbs, and, where sufficiently plentiful, for edgings to American plants in peat soil. The bulbs are white and oblong; hence its common name; and it is increased by dividing them every two or three years, replanting rather deeply. A native of the great continental mountain-chains.

FICARIA GRANDIFLORA.—*Large Pilewort.*

A LARGE-FLOWERED kind, a near relative of our very common Pilewort or lesser Celandine, *F. ranunculoides*, but about twice as large in all its parts, the flower being nearly or quite two inches across; the bases of the leaves meet, whereas they are divergent in the common one. I have little experience of this plant in cultivation, and it is as yet very rare. I brought a few plants from France in 1868, and hope soon to see it generally grown; it will no doubt prove a desirable addition. It is a native of Southern Europe and Northern Africa, and I believe it to be well worthy of a place on the rockwork in sandy loam, in a warm and well-drained spot. When plentiful, it may be tried as a border-plant.

Our British and very common *Ficaria ranunculoides*, or lesser Celandine, would be well deserving of culture were it not so very plentiful. Its white and double varieties, *F. ranunculoides alba*, and *F. r. fl. pl.*, may, however, have better claims to a place in a collection.

GALANTHUS NIVALIS.—*Snowdrop*.

OUR old friend, the Snowdrop, has long asserted its claim to be included among our most favourite "alpine flowers," by doing annually in our gardens what many plants do on the high Alps— piercing the snow with its flowers. It is almost needless to describe its appearance or speak of its culture, as it grows as freely as any weed, and is happily yet to be seen in many gardens, though the neglect of hardy plants in favour of mixed borders had much reduced its numbers before the recent taste for spring gardening had commenced. In only one point need a remark be made concerning its uses—it is seen to much greater advantage dotted over the grass in pleasure-grounds than in borders or as edgings. The leaves perform their functions so early in the year that it may be planted in grass that is repeatedly mown, as well as on banks in pleasure-ground or half wild places. The bulbs may be inserted a couple of inches into the turf, and the spot afterwards made firm and level, especially if it be on a trimly kept lawn.

The as yet comparatively scarce Crimean Snowdrop, *Galanthus plicatus*, much larger in foliage than the preceding (the leaves being sometimes an inch wide), and also distinguished by a longitudinal fold on both sides of the leaf near the margin, is a fitting subject for sandy soil among the rare bulbs in the rock-garden till sufficiently plentiful to be spared for the fringes of shrubberies and the sides of shady walks, associated with other early spring flowers.

GAULTHERIA PROCUMBENS.—*Creeping Wintergreen*.

THIS plant barely rises above the ground, on which it forms dense tufts of shining oval leaves, with small drooping white flowers in June, which are succeeded by a multitude of bright-red berries about the size of peas, formed by the fleshy calyx of the flower. The neat little shrub is of itself pretty, but the berries give it quite a charm through the autumn and winter months, when it is, or rather ought to be, one of the most attractive objects on every well-made rockwork. A native of North America, in sandy places and cool damp woods, often in the shade of evergreens, from Canada to Virginia; and as the leaves, when properly dried, make an excellent tea, it is also known by the name of the "mountain tea." Loudon says it is difficult to keep alive, except in a peat soil kept moist; but I have never seen it prettier or so full of berries as on clayey loam. The plant was thoroughly exposed, and the only advantage it had corresponding to those usually mentioned as necessary was that the soil was moist. It does well in moist peat, and forms capital edgings to beds in some places where the natural soil is of that quality, or a very light loam. Easily increased by division or by seeds, and suitable for the rockwork, the front margins of borders, and occasionally as an edging to beds of choice and dwarf American plants.

GENISTA SAGITTALIS.—*Winged G.*

A VERY handsome and singular plant, known at once and at all times by its branchlets being winged (by the stem expanding into two or three green membranes), more like those of a miniature Epiphyllum than a Genista, and producing an abundance of rich yellow flowers in summer; the shoots are usually prostrate, and the plant is rarely more than six inches high. It is met with growing abundantly in the grass in the mountain pastures of many parts of Europe. In cultivation it is a valuable plant, hardy and vigorous in the wettest and coldest soil, forming neat profusely-flowering tufts when fully exposed, and excellent for either rockwork or borders. Easily raised from seed.

GENISTA TINCTORIA.—*Dyer's G.*

A NEAT native shrub, with numerous slender branches, usually forming compact tufts from a foot to a foot and a half high, and becoming quite a mass of pretty yellow flowers in early summer, over the usually smooth and shining stalkless leaves. The flowers are in dense racemes, each bloom springing from the axil of a small leaf or bract. There is, however, no fear of confounding the plant with any other. It is grown in many of our nurseries, and merits a place among rock-shrubs on the rougher parts of rockwork, or on the margin of shrubberies. There is a double variety rather common in cultivation. Not unfrequent in many parts of England, but rare in Scotland and Ireland.

GENTIANA ACAULIS.—*Gentianella.*

A WELL-KNOWN old inhabitant of our gardens—sometimes erroneously called *G. verna,* than which it is a very much larger and stronger plant in all its parts. Its large solitary flowers, two inches long, of the deepest and most lustrous blue, with dotted throat, barely elevated above the low earth-mantling spread of dense leathery leaves, quite distinguish it from any other species worthy of general cultivation. Sometimes the flowers are so abundantly produced that, when fully opened to the sun, they cover the whole plant. The form with the points of the corolla distinctly tipped with white is a very lovely one. The plant is too well known to require further description, and happily, while among the most beautiful of the Gentians, it is very easily cultivated, except on very dry soils. In some places edgings are made of it, and where the plant does well, it should be used in every garden to some extent, as, when in flower, edgings of it are of the most exquisite beauty, and when out of bloom, the masses of little leaves, gathered into compact rosettes, form a very dwarf and firm edging, peculiarly appropriate for margining beds, or a small garden devoted to interesting or rare alpine or herbaceous plants. It is at home on a rockwork, where there are good masses of moist loam into which it can root, and it is particularly well suited for those spots of depressed rockwork where the stone is suggested here and there rather than exposed. It may be successfully grown in pots, and that would be worth the trouble where the plant would not grow in the open air from a very dry soil or any other cause. It is sometimes sold in Covent Garden in pots when in flower in spring, and is readily propagated by division, and also by seeds ; but these are so small and so slow in germination that attempting its propagation in this way is never worth the attention of amateurs. It is abundant in many parts of the Alps and Pyrenees. With us the flowers open in April and May, but in its native region their opening is regulated by position, somewhat like those of the vernal Gentian. The traveller leaving England in early June, who has seen the Gentianella in flower in British gardens in April, may meet with it not yet open in descending one of the mountains of Savoy, though lower down he will find it passed out of flower and in fruit. No garden should be without such an easily grown plant, so attractive from its associations as well as its great beauty.

G. alpina is a marked variety of the preceding plant, with very small broad leaves, and there are several other varieties.

GENTIANA ANDREWSII.—*Closed Gentian.*

THE kinds of Gentian which attract so much attention for their beauty on European mountains open their flowers wide when the sun shines. This does not do so, but forms apparently closed tubes about an inch long, in clusters, and of a deep dark blue. Then, instead of spreading low and mantling the ground with rosettes of leaves like *G. verna* and *G. acaulis*, the shoots grow erect and a foot or more high. It is handsome, seems to grow perfectly freely in a sandy loam, but has been hitherto so little grown that experiences of its likes and dislikes are not yet obtainable. It is, so far as I have observed, far more beautiful and worthy of culture than the Soapwort Gentian (*G. saponaria*), a perennial more frequently met with in our gardens. The flowers are closely set in clusters near the tops of the shoots, the leaves ovate, lanceolate, acute, narrowed at the base. A native of moist rich soil in North America, flowering in autumn, and increased by division and by seed. Suited for association with the larger alpine plants in moist deep soil in the rock-garden.

GENTIANA ASCLEPIADEA.—*Asclepias-like Gentian.*

A TRUE herbaceous plant—*i.e.* dying down every year, and thus keeping out of danger in winter time. This, of course, helps to explain the fact that it is easily cultivated in almost any soil. It grows erect, with shoots almost willow-like in their freedom, and from fifteen inches to two feet high, according to the nature of the soil; bearing numerous large purplish-blue flowers, arranged in handsome spikes. Little need be said of its culture, as it is not fastidious, but in a deep sandy loam or peat it will grow twice as large as in a stiff clay, and as in a wild state it likes sheltered valleys, a little shelter in the garden will save its long shoots from being injured by winds, which could not affect the dwarf evergreen Gentians, no matter how much exposed. In consequence of its tall habit, this species is best adapted for the lower parts of rockwork, or in the borders near at hand. A native of European mountains, and readily propagated by division of the root.

GENTIANA BAVARICA.—*Bavarian Gentian.*

IN size and flower this species resembles the vernal Gentian, but is very readily known from that by its smaller box-like leaves of a yellowish-green tone, and by all its tiny stems being thickly clothed with foliage, forming close, dense little tufts, from which spring flowers of the deepest and most brilliant blue, which, as the observer gazes at it in admiration, seems occasionally flushed with a slight tinge of deep, rich, purplish crimson; but he cannot define the hue—it is too subtle for description. The flower is even a shade more lovely than that of *G. verna*. The plant is a native of the high Alps of Europe, and in 1868 I saw it in great abundance near the monastery of the Simplon. *G. verna* occurs in the same place abundantly; but, while it is found on dry ground, or ground not overflowed by water, *G. bavarica* is seen in perfection in spongy boggy spots, where some diminutive rill has left its course and spread out over the grass, not covering it, but saturating it so that, when you walk upon it, the water bubbles up around. There can be no doubt that we must imitate these conditions as far as possible if we desire to succeed with the plant in England. The best thing to do with it is to plant it near the margin of a rill that falls from a rockwork, taking care to let no Carices, Couch-grass, Cotton-grass, or other strong-growing subjects get near the spot, or they would soon cover and destroy the plant. It may also be grown in pots, plunged in coal ashes or sand during the summer; sandy loam to be the soil used; the plants to have repeated and abundant waterings from early spring till the heavy autumnal rains set in, or to be placed standing half plunged in water. In all cases it must have free exposure to light. To try to establish it in such positions as it is found in naturally will prove an interesting experiment for those having opportunities of doing so.

GENTIANA PNEUMONANTHE.—*Marsh Gentian.*

A BRITISH perennial, scarcely less beautiful than any alpine Gentian, with tubular flowers, an inch and a half or more long, of a beautiful blue within, with five greenish belts without, the lobes of the mouth short and spreading; the flowers arranged in opposite pairs in the axils of the upper leaves, and on stems six inches to a foot high; the upper leaves nearly linear, the lower ones shorter and broader, all obtuse and rather thick. A native of boggy heaths and moist pastures, and in cultivation requiring moist peat or boggy soil. It is not recorded from Scotland or Ireland, though not difficult to obtain in some parts of England. Few plants are more worthy of a place on the rockwork or in the artificial bog. Increased by careful division.

GENTIANA PYRENAICA.—*Pyrenean Gentian.*

SOMEWHAT like the vernal Gentian in stature and size, but with imbricated narrow and sharp-pointed leaves and dark-violet almost stalkless flowers, the flat portion or limb of the flower being formed of five oval lobes, with a triangular appendage between each nearly as long as the lobes. It requires much the same treatment as *G. verna;* flowers in early summer. It is well worthy of a place in the choice rock-garden, though not of quite such a vivid hue as *G. bavarica* or *G. verna.*

GENTIANA SEPTEMFIDA.—*Crested Gentian.*

A LOVELY plant, bearing on stems six to twelve inches high flowers in clusters, cylindrical, widening towards the mouth, of a beautiful blue and white inside, greenish brown outside, having between each of the larger segments of the flowers one smaller and finely cut. A native of the Caucasus, and one of the most desirable species for cultivation on the rockwork, thriving best in moist sandy peat, and increased by division.

GENTIANA VERNA.—*Vernal Gentian.*

VERY rarely seen in good health in gardens, but known to many as the type of all that is beautiful in alpine vegetation. It covers the ground with rosettes of small leathery leaves, often spreading into tufts from three to five inches in diameter, and producing in spring flowers that even the botanist calls "beautiful bright blue," though botanical books are usually above taking any notice of colour at all. Sometimes the blooms barely rise above the leaves, and at other times are borne on stems two or three inches high. A few things are essential to success in its cultivation, and far from difficult to secure. They are good, deep, sandy loam on a level spot on rockwork, perfect drainage, abundance of water during the warm and dry months, and perfect exposure to the sun. Grit or broken limestone may be advantageously mingled with the soil, but if there be plenty of sand, they are not essential ; a few pieces half buried on the surface of the ground will tend to prevent evaporation and guard the plant till it has taken root and begun to spread about. It is so dwarf that, if weeds be allowed to grow around, they soon injure it. In moist districts, where there is a good deep sandy loam, it may be grown on the front edge of a border carefully surrounded by half plunged stones. It may also be grown in pots of loam with plenty of rough sand, well drained and plunged in beds of coal ashes or sand, thoroughly exposed to the sun, and well watered from the first dry days of March onwards till the moist autumn days return. In all cases good well-rooted specimens should be secured to begin with, as failure often occurs from imperfectly rooted, half-dead plants that would have little chance of surviving even if favoured with the air of their native wilds. In a wild state this plant is abundantly distributed over mountain pastures on the Alps of Southern and Central Europe, and those of like latitudes in Asia.

In addition to the preceding kinds, there are various other Gentians in cultivation : *G. caucasica, adscendens, gelida, cruciata, Fortunei, lutea, punctata,* and *purpurea.* Of these, the four first mentioned are those most likely to be attractive, especially *G. gelida.* Some of the last are scarcely worthy of cultivation, and certainly not in choice collections. Most of the Gentians may be raised from seed, but it is a very slow process, and, except in the hands of careful propagators, a very uncertain one.

GERANIUM ARGENTEUM.—*Silvery Cranesbill.*

A LOVELY alpine Geranium, with leaves of a silvery white, and large pale rose-coloured flowers, on stems seldom more than two inches high, and usually nearly prostrate. It is nearly allied to the grey Cranesbill (*G. cinereum*), but is known from that plant at first sight by its more silvery, somewhat more deeply divided, leaves, and it is of smaller stature. It comes from the Alps of Dauphiny and the Pyrenees, is perfectly hardy, flowers in early summer, and is a gem for association with the choicest plants on rockwork. It loves a firm, sandy, and well-drained soil, and should, as a rule, be placed near and somewhat below the eye, if the rockwork be an extensive one, as, though the plant is of a high, it is not of a conspicuous, order of beauty. Increased freely by seeds.

GERANIUM CINEREUM.—*Grey Cranesbill.*

A BEAUTIFUL dwarf plant, with five- or seven-parted leaves, clothed with a slightly glaucous pubescence, and bearing very large and handsome pale pinkish flowers, veined with red. A native of the Pyrenees, two to five or six inches high, grows freely on rockwork, and is easily propagated by seeds; but it is as yet comparatively rare in our gardens. Grown into strong tufts on good sandy soils, it forms a very attractive ornament for the front of the mixed border. On rockwork it is peculiarly at home, and is fitted for association with the choicest kinds. Where alpine plants are grown for exhibition, it and the Silvery Geranium are among the best plants that can be used, both growing freely in pots or pans. It flowers and seeds abundantly, and may be easily raised from seed.

GERANIUM MACRORHIZUM.—*Long-rooted G.*

A DWARF and distinct species, with large thick stems, and leaves in five divisions, each being deeply and irregularly lobed. The surface of the leaves is sparsely clothed with very short colourless hairs, and the margins on both sides are of a reddish-brown colour; calyx of a dull red and almost round; the flowers of a bright purple, freely produced when the plant is grown on light soil. It is suitable as a border-plant on light soil, seldom grows more than a foot high, is easily increased by seeds or division, flowers from May to July, and comes from Italy and Southern Europe. As it flowers very freely when in warm, rather poor, and very sandy soil, it is worthy of a place among the stronger and more robust plants on the lower and rougher flanks of rockwork.

GERANIUM SANGUINEUM.—*Bloody Cranesbill.*

A NATIVE species, forming very neat and somewhat spreading close tufts from one to two feet high, the leaves cut into five or seven segments, which are again cut into narrow lobes. The flowers are large, nearly or quite one inch and a half across, of a deep crimson purple, produced singly, and very profusely when the plant is grown on very sandy soil. Its close firmly-seated habit instantly distinguishes this plant from any other cultivated species, and the flowers being more showy and beautiful than those of any other, it deserves to have a place in every garden border, and also among the larger and more easily grown plants on rockwork. It grows on any soil, is readily propagated by division or seeds, and occurs in a wild state in some parts of Britain, though not a generally distributed or common plant.

There are two forms or varieties of the Blood Geranium. One, the common or "true" species, with ascending stems matting into vigorous but compact tufts; the other more hairy, less vigorous in its growth, and usually prostrate in habit. This last form usually occurs on sandy sea-shores. A form of this variety, with pale pink flowers veined with red, was found at Walney Island, in Lancashire, and has been distinguished as a species under the name of *G. lancastriense*, but it has no right to rank as such, merely differing in colour from the sea-shore variety. Both these forms, being smaller and less vigorous than the common one, are more adapted for rockwork, though where they do well, they make suitable ornaments for the front margins of the mixed border. I have noticed that in the heavy clay to the north of London, where *G. sanguineum* thrives vigorously, the pale-flowered sea-shore form was with great difficulty cultivated.

GERANIUM STRIATUM.—*Striped Cranesbill.*

A VERY old and charming border-plant, still to be seen in many cottage gardens, and worthy of a place in every collection. "This beautiful Cranes-bill," says Parkinson, writing nearly 250 years ago, "hath many broad yellowish green leaves arising from the root, divided into five or six parts, but not unto the middle as the first kinds are: each of these leaves hath a blackish spot at the bottom corners of the divisions: from among these leaves spring up sundry stalks a foot high and better, joynted and knobbed here and there, bearing at the tops two or three small white flowers, consisting of five leaves apeece, so thickly and variably striped with fine small red veins that no green leafe that is of that bigness can show so many veins in it, nor so thick running as every leaf of this flower doth." It is a native of Southern Europe, growing very freely in warm sandy soils, and is easily increased by seed or division. Being of a low spreading habit, it is suitable for the rougher parts of rockwork as well as for borders.

GLOBULARIA NANA.—*Dwarf G.*

A MOST dense trailing shrub, forming a firm mass of thyme-like verdure, about half an inch high, and dotted over with compact heads of bluish-white flowers, with stamens of a deeper blue or mauve. The flower heads are not half an inch across, and barely rise above the foliage. It should be placed on rockwork in very sandy or gritty soil, and so that it may crawl some little way over the face of the surrounding rocks or stones, and in a very open sunny spot in such a position it will not be so liable to be overrun by coarse plants. A native of the Pyrenees, and increased by division or seeds.

There are several other Globularias in cultivation : *G. nudicaulis, trichosantha,* and *cordifolia,* but these are somewhat overrated and scarcely worthy of culture except in large collections. The most desirable of them for the rockwork is the neat *G. cordifolia,* which is a little prostrate trailing shrub, with bluish flowers.

GYPSOPHILA PROSTRATA.—*Spreading G.*

NOT a brilliant plant, but valuable from its dwarf spreading habit, its multitudes of pink or white flowers, veined with rose, on thread-like stems, and its adaptability for rockwork or stony ground. The leaves are glaucous and spread into dwarf tufts, and the plant is well suited for old walls, dry banks, or any poor stony soil, growing, however, in the best as well as in the worst of soils, though scarcely worthy of a position where there is only a small extent of rockwork. It is sometimes grown as *G. repens,* and is a native of the Pyrenees and Alps, flowering in summer, and growing six or eight inches high.

I am by no means certain that the preceding is the best worth growing of the perennial kinds. Some of the annual kinds, *G. muralis,* of France, and *G. elegans,* from the Caucasus, for example, are even more beautiful and profuse in bloom, and well merit being naturalised on old ruins and bare rocky places, particularly the former, which is very dwarf and compact in habit, and produces myriads of pale rosy flowers. *G. paniculata* is a fine large kind for borders, banks, &c. All the kinds are easily grown in any soil.

HEDYSARUM OBSCURUM.—*Creeping-rooted H.*

A HANDSOME, creeping, vetch-like plant, with beautiful large purplish-violet flowers in long spikes, and leaves composed of seven to nine pairs of leaflets, the stipules united and opposite the leaves. From six to twelve inches high and sometimes more in rich soil. Readily increased by division or seeds, grows freely in ordinary garden soil on level ground, and is a valuable rock or border plant. A native of the Alps of Dauphiny and the Tyrol. Cultivated in our gardens more than two hundred years ago, but now rarely seen there, though not difficult to obtain.

HELIANTHEMUM FORMOSUM.—*Beautiful Sunrose.*

A SHRUBBY kind, but sufficiently dwarf for cultivation on the warmer and rougher slopes of rockwork, with downy and hoary leaves and shoots, and large handsome yellow flowers with a dark spot in the lower part of each petal, produced in summer. It is somewhat tender if planted in wet ground, but will flourish in calcareous or dry soil in thoroughly drained fissures in dry and sheltered parts of rockwork, where its distinct, abundant, and beautiful inflorescence will well repay the tasteful planter. A native of Portugal; increased by seed or cuttings. = *Cistus formosus.*

HELIANTHEMUM OCYMOIDES.—*Basil-like Sunrose.*

A NATIVE of dry rocky hills in Spain and Portugal, with bright yellow purple-eyed flowers nearly an inch and a half across, and hoary opposite leaves an inch to an inch and a half long, narrow, and pointed. Like *H. formosum*, this will be found very useful on the warmer and drier parts of rockwork, among the stronger alpine shrubs and choice herbaceous plants. Increased by seed or cuttings. = *Cistus algarvensis.*

HELIANTHEMUM TUBERARIA.—*Truffle Sunrose.*

A DISTINCT and beautiful rock-plant, from the shores of the Mediterranean, bearing flowers like those of a single yellow rose, two inches across, and with dark centres, drooping when in bud, and on stems about nine inches high. It is quite removed from all the other cultivated Sunroses in not producing woody stems, but sending up large hairy leaves, somewhat like plantain-leaves, from the root, and scarcely looking like a Sunrose at all. It flowers in summer, and continuously if in good health and in good soil. It is said to grow abundantly where truffles abound, and is well worthy of a position in a well-drained spot, or dry fissure on the sunny side of rockwork.

HELIANTHEMUM VULGARE.—*Common Sunrose.*

A WELL-KNOWN British under-shrub, growing in dry pastures and heaths, and producing an abundance of bright yellow flowers on stems from a few inches to nearly a foot long. In a cultivated state this plant varies a good deal in colour, and numerous plants passing under different names in our gardens are really forms of this species. Some of the most attractively coloured are well worthy of cultivation. While thriving in almost any soil, they attain ripest health, and flower most profusely, on chalky and warm ones, and on soils of this description they may be used with good effect on the margins of shrubberies, especially the copper-coloured and red varieties. They are only suited for the rougher parts of rockwork, except where less common and more beautiful plants cannot be obtained. The best way to obtain varieties of different colours is by seed, which is offered in most of the catalogues. Many beautiful members of this family are lost to cultivation, or have not yet been introduced, and not a few passing as species in some of our botanical collections are mere varieties, but often very showy, and useful in gardens.

The pretty annual spotted Sunrose (*H. guttatum*), found in the Channel Islands, on the Holyhead Mountain, in Anglesea, and widely distributed on the Continent, deserves a place in the curious collection, and indeed has beauty enough to recommend it to the general cultivator. It is quite easily grown, but is best raised in pots in spring, and then planted out in May. Once established, it sows itself annually. The hoary Sunrose, *H. canum*, a native of limestone rocks in Britain, but somewhat rare, is much dwarfer than the common kind, and produces in great abundance very small pale yellow flowers. The whole plant does not grow more than three inches high, and is likely to possess attractions for cultivators of interesting British plants.

HELICHRYSUM ARENARIUM.—*Yellow Everlasting.*

THIS is the beautiful little plant which affords the "everlasting flowers" so much used for Immortelles and ornaments. The grey leaves are closely covered with long down, and the flower-stems, ascending from four to ten inches, are clothed all the way up with narrow hoary leaves, having their edges turned backwards, and support a number of flowers, of a bright, glistening, golden yellow. To preserve the flowers, they should be gathered when fresh and newly blown, as, if allowed to become matured, they are apt to fall away. A native of sandy and sunny places in Central and Southern Europe, and only succeeding perfectly in this country on warm, sandy, and thoroughly drained soils. Increased by division, and worthy of a place in every rock-garden.

HELLEBORUS NIGER.—*Christmas Rose.*

ALTHOUGH this hardy and familiar old plant is too vigorous for association with the often minute and brilliant gems to which this book is chiefly devoted, yet its fine evergreen foliage and handsome large flowers form ornaments in the rougher parts of rockwork, or banks near it, or in the hardy fernery. Although hardy enough to grow almost anywhere, yet, as it flowers at the dreariest season, when low ground is often saturated with cold rain, it always repays for being planted in slightly elevated positions, and where it may enjoy as often as possible the faint wintry sun, by producing clearer and larger flowers, and finer foliage. *H. n. maximus*, the very large and noble variety grown about Aberdeen and other places in Scotland, is the best, and flowers a month or so earlier than the common kind; and *H. n. minor*, a smaller and much scarcer variety, is well suited for rockwork and bare banks.

Besides the Christmas Rose, there are other species of Helleborus well worthy of cultivation; and among the best is *H. atrorubens*, with flowers of a dark purple. The colour, though somewhat dull, by turning up the usually pendent flower, is seen to greater advantage, being then contrasted with the yellow stamens. It has the quality of throwing its flowers well above ground to a height of nine to twelve inches, and is a free grower, but rather scarce, requiring, as all the Hellebores do, a considerable time to establish itself after being disturbed. *H. olympicus*, with large rose-coloured flowers, and good habit, is very similar, if not identical, with one grown as *H. abchasicus*, though the true plant is really quite distinct from it. *H. argutifolius* is remarkable in the genus for its beautiful, whitish, trifoliate leaves, each secondary vein being terminated by a well-defined point. Its flowers are a fresh lively green, and produced about the month of March.

The small *Helleborus trifolius* of Linnæus, now generally known as *Coptis trifolia*, is a very diminutive and interesting kind, with white flowers. It is quite easy of culture, but in many gardens flowers seldom or sparsely; in others abundantly, especially when on rockwork in moist peat soil. It is is also very desirable for planting on the margins of beds of Rhododendrons and hardy Azaleas, in peat soil, associated with *Rhexia virginica* and other dwarf plants.

HIPPOCREPIS COMOSA.—*Horseshoe Vetch.*

A SMALL prostrate British plant, with pretty little deep-yellow flowers, in coronilla-like crowns, the upper petal faintly veined with brown, the pinnate leaves small, and leaflets smooth. It is a capital little plant for the upper ledges of rocks in dry positions, as in such places the shoots will fall down some eighteen or twenty inches; easily raised from seed; partial to chalky soils; rather common in the South of England, but not a native of Ireland or Scotland.

HOTEIA JAPONICA.—*Japanese H.*

THIS fine plant is quite distinct in appearance, and readily known by its much divided leaves; the leaflets are oval, toothed, and ciliated, of a fine shining green, and the whole plant not unlike a fern in aspect. From amidst these rich masses of shining leaves springs the sweet and abundant inflorescence; the flowers are small, but very freely produced in graceful panicles, of which the bracts, little flower-stems, and all the ramifications are, like the flowers, white. Although now so much grown in pots, it is perfectly hardy, and very suitable for borders and the margins of clumps, or for association with the larger classes of alpine plants, on or near rockwork. In the open air it flowers, according to climate and position, from May to July and August, and it is particularly fond of a sandy peat or very sandy loam, a sheltered position, and moist soil. It is, however, probable that a somewhat warmer climate than ours is necessary for its perfect development, for it certainly looks a handsomer plant when forced than when grown in the open ground. It is very easily forced, and it has latterly been so much admired that quantities of it are prepared and sold for forcing. Some years ago, a few fine plants of it were exhibited at our spring shows, if I mistake not, by Messrs. Veitch in the first instance, and since then it has been gradually making way. Previous to that time, however, it was popular in Paris, where many persons cultivate it in windows, and where it is often used with good effect in room decorations.

HOTTONIA PALUSTRIS.—*Water Violet.*

A BEAUTIFUL British water-plant, which I include here in consequence of having seen it thrive better on soft mud banks than when submerged. The deeply-cut leaves formed quite a deep green and dwarf turf over the mud, and from these arose stems bearing at intervals whorls of handsome pale-lilac or pink flowers. It might perhaps be more justly called the Water Primrose, as it is nearly allied to the Primulas. As water and bog may with the best taste be associated with rockwork, this plant might with advantage be grown either in the water or on a bank of soft thoroughly wet soil at its margin. It grows from nine inches to two feet high, flowers in early summer, and may be found in abundance near London on the banks of the Lea River, and probably in many other places, and is pretty freely distributed over England, though scarce in the Western Counties, and only found in the County of Down in Ireland.

HOUSTONIA CÆRULEA.—*Bluets.*

A DELICATE American herb, producing a profusion of pale sky-blue flowers, fading to white, and with yellowish eyes, crowding on thread-like stems to a height of one inch to two inches and a half, from close low cushions of leaves shorter than many mosses, less than half an inch high when fully exposed. It is usually considered somewhat difficult to grow, but this arises chiefly from its minuteness; in level exposed spots it does very well in moist peaty soil, the chief care required being to keep it quite clear of weeds or coarse-growing neighbours. It is said by some to be a biennial, but in cultivation does not seem to be so, increasing to any extent by careful division. It grows freely in pots or pans, in cold frames, pits, or houses kept near the glass, and in such positions is likely to ripen its seeds. Suitable for association with the smallest and choicest mountain-plants. I have grown this plant well in the open air in London: it withstood the evil influences of abundant showers of smut, and should therefore not be difficult to keep under more favourable circumstances.

HUTCHINSIA ALPINA.—*Alpine H.*

A VERY neat little plant, from moist and very elevated parts of nearly all the great mountain-chains of Central and Southern Europe, with shining leaves, deeply cut into narrow lobes, so as to resemble pinnate leaves, and pure white flowers, produced in clusters and abundantly, on stems about one inch high. Quite free in sandy soil, and easily increased by division or by seeds. Planted in an open spot, either on rockwork or in good free border soil, it becomes a compact dense mass of pure white flowers. Its proper home is on the select rockwork, though where borders of dwarf and choice hardy plants are established and carefully attended to, it may be grown in them with perfect success.

HYACINTHUS AMETHYSTINUS.—*Amethyst Hyacinth.*

BEAUTIFUL deep sky-blue bells, five to fifteen, rather loosely and gracefully disposed on stems from eight inches to a foot high, ascending in numbers when the plant is grown in very light rich soil, in a not too cold or exposed position. It is quite hardy, a native of the Pyrenees and Southern Europe, flowering in early summer, and is in stature and general appearance somewhat like a graceful Scilla, but at once distinguished by its pretty bells not being divided into segments as they are in Scilla. It is worthy of association with the choicest hardy bulbs and inhabitants of the rock-garden.

IBERIDELLA ROTUNDIFOLIA.—*Round-leaved I.*

A DISTINCT spreading plant, rarely more than a few inches high, and producing pretty, rosy-lilac, sweet-scented flowers in abundance in April, May, and June. The leaves are thick, smooth, leathery, and of a glaucous olive-green, and the flowers are produced in short racemes or corymbs, and usually attain a height of from three to six inches. Flowering with the vernal Gentian, the Bird's Eye and Scotch Primroses, the alpine Silene, and the little yellow Aretia, it is admirable for association with such plants. It grows naturally very high on the Alps, but thrives in loamy soil on rockwork, does not seem difficult to cultivate, and is easily raised from seed. A native of the Alps of Switzerland, Savoy, and Austria. It is occasionally found with white flowers in a wild state. Dr. Hooker has recently figured this plant in the 'Botanical Magazine' from specimens received from Zurich. This would seem to be a distinct variety from that which I have seen elsewhere, having numerous pale-lilac flowers, with yellow eyes, borne in stout crowded racemes, whereas those of the form introduced by Mr. Jas. Backhouse have the flowers somewhat larger, but in lax few-flowered heads. = *Thlaspi rotundifolia.*

IBERIS CORIFOLIA.—*Coris-leaved Candytuft.*

A VERY dwarf kind, about half the size of *Iberis sempervirens*, attaining a height of only three or four inches when in flower, and perfectly covered with small white blooms early in May. Few alpine plants are more worthy of general culture, either on rockwork or in the mixed border, for the front rank of which it is admirably suited. It is probably a small variety of the Ever-green Candytuft, but for garden purposes it is distinct enough. A native of Sicily, and probably of other parts of Southern Europe, easily propagated by seeds, cuttings, or division, and thriving in any soil.

IBERIS CORREÆFOLIA.—*Correa-leaved Candytuft.*

THIS plant is now becoming very popular in London gardens, and generally goes by the name of *I. gibraltarica*, from which it is quite distinct. It is readily known from any other cultivated species by its entire and rather large leaves, by its compact head of large very white flowers, and by flowering later than the other common white kinds. Both the individual flowers and the corymb are larger than in the other species, and the blooms stand forth more boldly and distinctly from the smooth dark-green leaves. It is an invaluable hardy plant, and particularly useful in consequence of coming into full beauty about the end of May or beginning of June, when the other kinds are fading away. It is indispensable for rockwork, the mixed border, the spring garden, and may also be naturalised with good effect in bare rocky places. It is particularly well suited for planting on the margin of choice shrubberies, bringing them neatly down to the grass line, and may also be used as an edging to beds. Of its native country we know nothing; but my friend Mr. Jennings, now of the Wellington Nurseries, informed me that it was raised in, and first sent out from, the botanic garden at Bury St. Edmunds, and it is probably a hybrid. Mr. J. G. Baker considers it to come nearest to *I. Pruiti*, of the Nebrode Mountains, in Sicily. Its native country, like its name, is not certainly known. Readily increased by cuttings, and also by seeds.

IBERIS GIBRALTARICA.—*Gibraltar Candytuft.*

THIS kind has hitherto been but very rarely cultivated; indeed, it was lost to our gardens till recently sent from Guernsey by Mr. Wolsey, who found it in some cottage garden there. It is larger in all its parts than the other cultivated kinds, has oblong spoon-shaped leaves, nearly two inches long by half an inch or more wide, and distinctly toothed; the large flowers, often reddish lilac, being arranged in low close heads, and appearing in spring and early summer. It is an ornamental species, but will never rival the well-known white border kinds. I am doubtful of its hardiness, and should advise its being wintered in pits or frames till sufficiently abundant to be tried in the open air, but am informed that it has stood without injury this last, severe, winter at York. It should be planted on sunny spots on rockwork or banks. A native of the South of Spain; increased by seeds and cuttings.

IBERIS TENOREANA.—*Tenore's Candytuft.*

A NEAT species, with toothed leaves, which, with the stems, are hairy, and a profusion of white flowers changing to purple. As the commonly cultivated kinds are pure white, this one will be the more valuable from its purplish tone, added to its neat habit. It, however, has not the perfect hardiness and fine constitution of the white kinds, and, so far as my experience goes, is very apt to perish on heavy soils in winter; but on light sandy soils it will prove a gem, and also for well-drained positions on rockwork. Where rockwork does not exist, it should be placed on raised beds or banks. A native of Naples, and easily raised from seed.

IBERIS SEMPERVIRENS.—*Evergreen Candytuft.*

THIS is the common rock or perennial Candytuft of our gardens, as popular as the yellow Alyssum and the white Arabis. Half shrubby, dwarf, spreading, evergreen, and perfectly hardy, it escaped destruction where many herbaceous plants were destroyed; and as in April and May its neat tufts of dark green are transformed into masses of snowy white, its presence has been tolerated longer than many other fine old plants. Occasionally, even in gardens entirely devoted to a few "bedding plants," it may be seen on the margin of a shrubbery border or in some neglected spot. No hardy flower is more worthy of being universally grown in gardens, from that of the cottage to the largest in the land. It is one of the very best plants in the country for growing on the margins of the mixed border or properly-finished-off shrubbery, the rockwork, rootwork, and also for naturalisation in bare rocky places. Where a very dwarf evergreen edging is desired for a shrubbery, or for beds of shrubs, it is one of the most suitable plants known, as on any kind of soil it quickly forms a spreading band almost as low as the lawn-grass, finishing off the plantation very neatly at all times, and changing into dense wreaths of snowy-white flowers around the borders in spring and early summer. When in tolerably good soil, and fully exposed, it forms spreading tufts often more than a foot high, and they last for many years. Like all its relatives, it should be exposed to the full sun rather than shaded, if the best result is sought. A native of Greece, Asia Minor, Italy, Southern France, and Dalmatia, and readily increased by seeds, cuttings, or division.

I. Garrexiana is a variety of the Evergreen Iberis, not sufficiently distinct to be worthy of cultivation; in fact, it and several other Iberises prove to be mere varieties, and very slight ones, of *I. sempervirens* when grown side by side.

IONOPSIDION ACAULE.—*Violet Cress.*

This, being an annual plant, is only introduced here in consequence of its peculiar beauty and suitableness for adorning bare spots on rockwork devoted to very minute and delicate alpine plants. As it sows itself, the cultivator will have no more trouble with it than with a hardy perennial. It frequently flowers at one inch high, and rarely exceeds two inches, the small flowers being of a pale violet tinge, and the leaves roundish and compactly arranged. From the neatness of its habit, it has been called the Carpet-plant. It is too minute for borders, though it will grow freely in them, but will be found most at home on the shady side of the rockwork, and in spots where a coarse vegetation will not prevent its growth. It does very well in pots, and those who are fond of window-gardening may grow it well outside their windows in pans or small pots. It will flower a couple of months after being sown; and when sown in spring in the open ground, the self-sown seeds of the summer flowers soon start into growth, and the second crop flowers in autumn, and far into winter. A native of Portugal and Morocco.

IRIS CRISTATA.—*Crested I.*

A very diminutive and charming species, usually running about with its creeping and rooting stems exposed on the surface, not rising above the ground more than a few inches, having flowers, however, as large as many of the tall and coarser species. Notwithstanding this, it has never become popular, and was indeed, till recently, scarcely to be found. It flowers in May; blue, with spots of a deeper hue on the outer petals, and a stripe of orange and yellow variegation down the centre of each. The plant is readily distinguished, at any season, from any other dwarf species by the creeping stems (rhizomes) growing well above the ground. This feature, indeed, is so marked as to suggest the desirability of frequently replanting it, and even young tufts push so boldly out of the ground that a top-dressing of an inch of fine soil placed around them cannot fail to help the roots to descend more freely. It loves, and flourishes luxuriantly on, rich but free and light soil, in a warm position. I have never seen it do so well as in the Glasnevin Botanic Gardens, which points to the fact that somewhat moist districts will suit it, but I have seen it thrive both to the north and south of London. Charming for association with the dwarf Crimean Iris, the alpine Catchfly, and any other dwarf gems among the later spring flowers of the choice mixed border. On rockwork it thrives best on level earthy spots, and where it does well and increases freely in rich light moist soil, it will form a pretty edging for beds of dwarf shrubs or American plants. A native of mountainous regions in North America, with all the gem-like loveliness of the choicest Swiss alpine flowers; was introduced by Mr. Peter Collinson, so long ago as 1756, and figured in Sir James Smith's 'Rare Plants' in 1792.

IRIS NUDICAULIS.—*Naked-stemmed I.*

THIS species, at present scarcely grown in this country, I first observed in the Jardin des Plantes at Paris. It is one of the most attractive of all Irises, about the same height as the Dwarf Iris, but with larger blooms and stouter habit; the leaves are lance-shaped and bent, those of the stem somewhat spoon-shaped; flowers of bluish violet, external divisions spoon-shaped, not, or but very slightly, wavy, the internal ones oval and longer, the central blades violet; appearing in May. It flourishes in ordinary garden soil, is valuable for the spring or rock garden or the mixed border, and comes from Southern Europe. A vigorous grower, and easily increased by division.

IRIS PUMILA.—*Dwarf Crimean I.*

OFTEN flowering at four inches, the dwarf Iris, even in favourable soils, rarely exceeds ten in height; the stems usually bear one or two deep-violet flowers, of which the external divisions are large and oblong, the internal ones dilated, broad at the top, narrow at the base, and wavy at the edge; the flowers very large and beautiful. Blooms in April and May, and is useful in a variety of ways, as for edgings of one or of several colours; for beds of distinct or alternated colours in the spring garden; for a place among the choicest and lowest plants of the mixed border, on lower and flatter parts of the rockwork in wide-spreading tufts, on old ruins, on walls, and even on the tops of thatched houses—a position which it may be seen sometimes embellishing on the Continent. It thrives in ordinary garden soil, the lighter and deeper the better; the finest specimens I have ever seen were in a deep sandy peat, and they were twice the ordinary size. There are several varieties: yellow, white, light blue, and deep dark violet, respectively known under the names of *I. pumila lutea*, *alba*, *cærulea*, and *atro-cærulea*. Each of the varieties is worthy of cultivation, and easily increased by division of the rhizomes.

IRIS RETICULATA.—*Early Bulbous I.*

DISTINCT from all other Irises, and perhaps the most valuable of all, considering its early bloom, delicious violet scent, and rich and brilliant colour. The root is a tuber; leaves four-angled and rather tall when fully developed; and the flowers, borne on stems three to six inches high, are of the deepest and most brilliant purple, each of the lower segments marked with a deep orange stain, contrasting richly with the other parts of the flower. It blooms in early spring, long before any other Iris shows itself, and loves a deep sandy soil, and a warm well-drained position. There is no more beautiful plant for a sunny bank on the lower slopes of the rock-garden, and it will be found desirable in other positions when sufficiently plentiful. A native of Southern Europe, Syria, Asia Minor, and adjacent countries. Increased by division of the tubers. Where seed is produced, it should be saved and sown, as the plant is, at present, comparatively rare.

ISOPYRUM THALICTROIDES.—*Meadow-rue I.*

A GRACEFUL little plant allied to the Meadow-rues, but with white flowers prettier than those of the Thalictrums. It is, however, chiefly valuable for its maidenhair-fern-like foliage, and is worthy of culture in the flower-garden for this alone. When it is grown for the sake of its dwarf elegant leaves only, the flower-stems should be pinched out. It will prove attractive as a fern-like edging for the little beds of very dwarf succulents that are now deservedly becoming popular. It is also well suited for rockwork, and particularly so for the front edge of the mixed border, is hardy, and easy to grow on any soil. Comes from the Pyrenees and mountainous parts of Greece, Italy, and Carniola, is easily propagated by division or by seed, and produces its flowers white, with a faint green tinge, in early summer. The leaves rarely rise more than a few inches high, the flower-stems from ten to fourteen inches.

JASIONE HUMILIS.—*Dwarf J.*

NOT a showy but an interesting and pretty little plant, suited for the select rockwork. The leaves are in rosettes, very slightly toothed, stem-leaves larger, waved, and with sharp teeth; all the leaves ciliate at the base. The heads of light-blue flowers borne on stems from one to two or three inches high, are from half an inch to three-fourths of an inch across. A native of the Pyrenees; increased by division. This plant is related to *Jasione perennis*, but is more ornamental.

JEFFERSONIA DIPHYLLA.—*Twinleaf.*

A PLANT very little known in England, and, where grown, usually regarded as merely a botanical curiosity; but when planted in very sandy peat associated with subjects like the Epimediums, *Rhexia virginica*, and *Spigelia marilandica*, it becomes a pretty spring flower, as well as interesting from its curiously paired leaves. The flowers are white, with yellow stamens, about an inch across, and freely produced when the plant is in vigorous health. A good plant for peaty and somewhat shady spots on rockwork, and in a minor degree for the margins of beds of dwarf American plants, planted in sandy peat, and flowers somewhat about the same time as the Bloodroot, in early spring. A native of rich shady woods in North America. If seeds are saved, they ought to be sown a few days after they are gathered; but generally careful division of the root, in winter, must be resorted to where it is desired to increase the plant.

LEIOPHYLLUM BUXIFOLIUM.—*Sand Myrtle.*

A NEAT and pretty tiny shrub, forming compact bushes from four to six inches high, and densely covered with pinkish-white flowers in May, the unopened buds being of a delicate pink hue. It is particularly suited for grouping with diminutive shrubs, such as the Partridge Berry, the sweet *Daphne Cneorum*, the tiny Andromedas, and Willows like *S. reticulata* and *serpyllifolia*, that rise little more than an inch or two above the ground. Requires peat, and, if planted on rockwork, should have a bed of that material beneath it. A native of sandy "pine barrens" in New Jersey, and easily to be had in our nurseries under the name of *Ledum thymifolium*. There is probably more than one variety in cultivation.

LEONTOPODIUM ALPINUM.—*Lionsfoot*.

A NATIVE of high sloping pastures on many parts of the great continental mountain ranges; a very popular plant with the peasantry in many parts, and often sold in little bouquets in Germany, where it is called "Edelweiss," which, translated, means *nobly white*. The flowers are small, yellowish, and not ornamental; the leaves covered with white down, like those of many mountain composite plants, but it is at once distinguished by that to which it owes its popularity—a beautiful whorl of oblong leaves, springing star-like from beneath the closely set and somewhat inconspicuous flowers, and almost covered with pure white, dense, short down. It is a perfectly hardy perennial, growing from four to eight inches high, and thriving in firm, sandy, or gritty and well-drained, soil in thoroughly exposed parts of the rockwork, and is one of the most interesting and desirable inhabitants of the rock-garden.

LEUCANTHEMUM ALPINUM.—*Alpine Feverfew*.

A VERY dwarf plant, with small fleshy leaves, deeply cut, and hoary, and not rising more than half an inch above the surface. It bears pure white flowers more than an inch across, and with yellow centres, produced in abundance, and supported on hoary little stems, from one to three inches long. It is a rather quaint and pretty plant, and well deserves cultivation on rockwork, in bare level places on poor, sandy or gravelly, soil. Sometimes known as *Chrysanthemum alpinum* and *Pyrethrum alpinum*. A native of the Alps of Europe. Readily increased by division or by seed.

LEUCOJUM ÆSTIVUM.—*Summer Snowflake*.

A RELATIVE of the Snowdrop, but not venturing above ground nearly so soon; though, as it flowers around London early in April, even during cold seasons, there seems little reason why it should be called the Summer Snowflake. From a dense crowd of daffodil-like leaves, more than a foot long, the flower-stems, from a foot to eighteen inches high, spring, bearing at the apex a cluster of snowdrop-like, slightly drooping, blooms. The flower is of pure sparkling white, springing from a very dark smooth green seed-vessel, and the tip of each petal being of a pleasing soft green, both inside and out, the effect is very pretty. It is a meadow plant, a native of some parts of England, and seems to be at home in every kind of garden soil. It is particularly well suited for naturalisation in spots where the Daffodil thrives, and by the sides of wood walks, where mowing is not resorted to, and is well worthy of a place in the mixed border, or on the margins of shrubberies. Though it occurs in the British Flora, it is found in only a few places in England. Readily propagated by division.

Mr. Syme says in 'English Botany': "*L. æstivum*, in its typical form, is less often met with in cultivation than its sub-species, *L. Hernandezii*, a native of Southern Europe, which often does duty for *L. æstivum* in botanic gardens, and is sold by seedsmen under the name of *L. pulchellum*. This form flowers from three weeks to a month earlier than *L. æstivum*, and has the flowers smaller, little more than half an inch long, the perianth segments more incurved, so that the perianth is somewhat ovoid, and after flowering urceolate."

LEUCOJUM VERNUM.—*Spring Snowflake.*

A DWARF, stout, broad-leaved plant, like a Galanthus, but with larger and handsomer flowers, and appearing about a month later than the Snowdrop; deliciously fragrant; the segments white, an inch long, and each distinctly marked with a green or yellowish spot near the point, drooping and usually produced singly on stems from four to six inches high. It is certainly more worthy of cultivation than the Snowdrop, and that is as high praise as we can give to any dwarf spring-flowering plant. It has long been known as a continental plant, and was valued and grown in our gardens when hardy flowers were more esteemed than they are at present; but, singularly enough, its existence, as a true native, was not known, with certainty, till a year or two ago, when it was found, in abundance, on the "Greenstone heights, in the neighbourhood of Britford." It is not by any means a common plant, and those who have it would do well to place it in positions where it is likely to thrive and increase—in light, rich, well-drained soil on rockwork, or in borders, and as, after the plant has flowered, the leaves attain the length of nearly a foot, and are nearly or quite three quarters of an inch across, a sheltered position, where they may not be torn by winds, will be desirable. There is no more fitting ornament for rockwork, and every lover of spring flowers or alpine plants should try to increase and popularise it.

LILIUM TENUIFOLIUM.—*Small Scarlet Lily.*

ALTHOUGH the Lilies generally grow too tall for association with the plants included in this book, this is such a tiny and exquisite kind that it must not be omitted from any choice collection of alpine plants. It will do for our puny artificial rockworks what the Orange Lily does for the huge rocks of the Piedmontese valleys. The stem is often scarcely six inches high, but sometimes twice that height, bearing one, two, or more flowers of a deep orange scarlet, the petals recurved so that they nearly touch their bases. The fiery, slightly pendulous, flowers seem very large for such weak stems, clothed with leaves not one-twelfth of an inch in diameter. The bulb is pear-shaped, about the size of a walnut, and with white scales, by means of which, like all Lilies, it may be propagated. A native of the Caucasus, and in this country requiring a sandy peat, or very free loamy soil, in a well-drained and warm position. It will associate with the choicest summer-flowering alpine plants on level spots on rockwork. When planted in borders, it should be in some spot not likely to be dug or disturbed by persons not aware of its value. I am inclined to think that there are several different plants going under this name; one dwarf but very sturdy kind, which I obtained from Holland under this name, is certainly distinct. Mr. Wm. Thompson, of Ipswich, states that the seeds of this brilliant plant vegetate with as much certainty as those of any ordinary annual, and this should induce us to raise it in that way.

LINARIA ALPINA.—*Alpine Toadflax.*

A TRUE alpine plant, from the Alps and Pyrenees, found on moraines and débris of the mountains; allied to the Snapdragon and the Ivy-leaved Linaria, but quite different in aspect, forming dense, dwarf, smooth and silvery tufts, covered with bluish-violet flowers, with two bosses of intense orange in the centre of the lower division of each. Its habit is spreading, but neat and very dwarf, rarely rising more than a few inches high. On the Alps I have seen it flowering profusely at one inch high, the leaves, which attain a length of three quarters of an inch in our gardens, being almost rudimentary and scarcely perceptible beneath the flowers, which quite obscure stem and leaves, being larger proportionately than on the cultivated plant. It is usually a biennial; but in favourable spots, both in a wild and cultivated state, becomes perennial. Its duration, however, is not of so much consequence, as it sows itself freely, and is one of the most charming subjects that we can allow to "go wild" in sandy, gritty, and rather moist earth, or in chinks of rockwork. In moist districts it will sometimes even establish itself in the gravel walks. It is readily increased from seed, which should be sown in cold frames, in early spring, or in the places it is destined to embellish out of doors.

No other cultivated Linaria approaches the preceding in beauty or character, but *L. origanifolia* is pretty and worthy of a place in large collections; and our common wild Ivy-leaved Linaria, *L. Cymbalaria*, that drapes over so many walls so gracefully, has a white and a pretty variegated variety. The old plant itself would be fully described here were it not that it usually takes possession of old walls and other places suitable for its growth. The singular and handsome Peloria variety of our common *L. vulgaris*, with five spurs and a regular five-lobed mouth, is also well worthy of a place, being remarkably curious as well as ornamental.

LINNÆA BOREALIS.—*Twinflower*.

An interesting trailing evergreen, with opposite, round-oval leaves, slightly toothed at the top, and bearing delicate, fragrant, and gracefully drooping pale pink flowers, which are produced in pairs. This plant is named after the great Linnæus, with whom it was an especial favourite, as it generally is with cultivators of alpine or rock plants. A native of moist mossy woods, in

Northern Europe, Asia, and America, and sometimes of cold bogs or rocky elevated situations in Britain, occurring in fir woods in a few places in Scotland and Northern England. It loves a sandy peat and moist soil, and may be grown on the rockwork as a trailer, the shoots being allowed to fall down over the faces of the rocks, or in mossy rocky ground among bushes, on the fringes of the artificial bog, or in some half-shady position in the hardy fernery. It usually enjoys a somewhat shady position, but, if in proper soil, will bear the sun. Where there is not a place for it in the rock-garden, it may be grown in large pots of moist peaty earth, allowed to stand in water in the open air during the summer months. On a northern window-sill it will do very well, placed in a cold frame in winter, and it may be tastefully used in the indoor, as well as in the hardy, fernery. Readily increased by division.

LINUM ALPINUM.—*Alpine Flax.*

A DWARF and quite smooth Flax, growing only from three to eight inches high, and bearing very large dark-blue flowers in summer. It is most readily distinguished by its external sepals being acuminately pointed, and the internal ones obtusely pointed. A charming rock-plant, native of the Alps, Pyrenees, and many hilly parts of Europe, thriving well in warm well-drained spots on rockwork, in a mixture of sandy loam and peat. There are several varieties—*alpicola, collinum,* and *crystallinum; L. austriacum* is intimately related to it, and scarcely sufficiently distinct from a horticultural point of view.

LINUM ARBOREUM.—*Evergreen Flax.*

THIS is the neat, glaucous-leaved, dwarf, spreading shrub, with a profusion of clear handsome large yellow flowers, an inch and a half across, sometimes seen in our gardens under the name of *L. flavum.* Although said to be tender in the colder and drier parts of the country, it thrives well in others in the open air, even as a border-plant, and in all is well worthy of a position on rockwork. A native of hilly parts of South-Eastern Europe, Asia Minor, and North Africa; usually propagated by cuttings. It is sometimes grown as a frame and greenhouse plant, but should be tried everywhere in warm spots on dry borders, banks, or rockworks. It begins to bloom in early summer, and I have seen it flowering for months at a time.

LINUM CAMPANULATUM.—*Yellow Herbaceous Flax.*

AN herbaceous plant, with golden-yellow flowers in corymbs on stems from twelve to eighteen inches high, distinct from anything else in cultivation, and well worthy of a place in collections of alpine and herbaceous plants. A native of the South of Europe, flowering in summer and flourishing freely in dry soil on the warm sides of banks or rockwork, and propagated by seeds. *Linum flavum* is said to be different from this by its shorter sepals, and several minor characters; but Messrs. Grenier and Godron found these very inconstant, and differing very much, in the French plant. = *L. flavum*.

LINUM NARBONNENSE.

A BEAUTIFUL and distinct sort, bearing during the summer months a profusion of large, light sky-blue flowers, with violet-blue veins. A fine ornament for borders, the flower-garden, or the lower flanks of rockwork, on rich light soils, forming lovely masses of blue from fifteen to twenty inches high. A native of Southern Europe, distinguished from its relatives by its sepals tapering to a long point, its anthers being three times as long as broad, its long thread-like stigmas, and its large flowers.

LINUM PERENNE.—*Perennial Flax.*

A PLANT found in some parts of Britain, particularly in the Eastern counties, but very rare. Usually grows in dense tufts from twelve to eighteen inches high, with bright cobalt-blue flowers more than an inch in diameter, the stamens in some being longer than the styles, in others shorter, the petals overlapping each other at the edges. Mr. Syme considers it probable that *L. alpinum* and *L. Leonii* are forms that may be included under *L. perenne*. *L. perenne album* is also an ornamental plant, and there is also a variety with blue flowers variegated with white, known in gardens as *L. Lewisii variegatum*, but this marking is not very conspicuous or constant. A pale-rose variety is also announced in some of the catalogues. *L. sibiricum* and *L. provinciale* are also included under *perenne*. Of very easy culture in common garden soil, it is a useful border-plant, and may also be used in rough rocky places.

LITHOSPERMUM PETRÆUM.—*Rock Gromwell.*

A NEAT and dressy dwarf shrub, somewhat like a Lilliputian Lavender-bush, with leaves from half to three quarters of an inch long, rarely more than one-eighth of an inch broad, and of a greyish tone, like those of Lavender. Late in May or early in June all the little grey shoots of the dwarf bush begin to exhibit a profusion of small, oblong, purplish heads, and early in July the plant is in full blossom, the full-blown flowers being of a beautiful violet blue, with protruded anthers of a deep orange red, the buds of a reddish lilac. The flowers are barely more than a quarter of an inch long, and tubular, not at all open, like those of the valuable *Lithospermum prostratum*, but as every shoot is crested by a densely-packed head of flowers, the effect is very pretty and distinct. The best position for this plant is on the rockwork, somewhere near or on a level with the eye, on a well-drained, deep, but rather dryish sandy soil on the sunny side. A native of dry rocky places in Dalmatia and Southern Europe; propagated by cuttings, or seeds if they can be obtained.

LITHOSPERMUM PROSTRATUM.—*Gentian-blue Gromwell.*

A PERFECTLY hardy little evergreen spreading plant, having rich and lovely blue flowers, with faint reddish-violet stripes, about half an inch across, produced in great profusion where it is well grown. A native of Spain and the South of France, easily propagated by cuttings, very hardy, and peculiarly valuable as a rock-plant from its prostrate habit, and the fine blue of its flowers—a blue scarcely surpassed by that of the Gentians. It may be planted so as to let its prostrate shoots fall down the sunny face of a rocky nook, or allowed to spread into flat tufts on level parts of the rockwork. On dry and sandy soils it forms an excellent border-plant, and where the soil is deep and good, as well as dry and sandy, it becomes a round spreading mass, a foot or more high. It is in such soils suited for the margins of beds of choice and dwarf shrubs, either as an edging or as a single plant, or in groups. In heavy or wet soil it should be elevated on rockwork or banks, and planted in sandy earth. It is sometimes grown as *L. fruticosum*, but the true *L. fruticosum* is a little bush, whereas our plant is prostrate. It flowers in early summer, often continuing a long time; the leaves are nearly oblong in outline, and covered with short bristle-like hairs.

Lithospermum purpureo-cœruleum, a British plant, *L. Gastoni*, and *L. canescens*, are also worthy of culture in large collections.

LOISELEURIA PROCUMBENS.

IN a wild state on the Alps, or on mountain moors, this is a wiry trailing shrub, growing quite close to the ground, the plants occasionally forming a rather dense tuft, bearing small reddish flowers in spring, when the snow melts. Being found in the Scotch Highlands, it is usually a highly esteemed variety with lovers of alpine plants, though it is not very attractive at any time, so far as the bloom goes. It is very rarely seen in a thriving state under cultivation, and most of the plants transferred from the mountains to gardens usually perish. This sometimes occurs from the strongest-rooted and finest specimens being selected for transplantation, instead of the younger ones. I never saw it in such perfect health in a garden as in that of the late Mr. Borrer, in Sussex, where it flourished in compact masses thrice its usual size, in deep sandy peat. Its true garden home is the rockwork, and it will seem well worthy of a place to most lovers of rare British plants. = *Azalea procumbens*.

LOTUS CORNICULATUS.—*Birdsfoot Trefoil.*

A WELL-KNOWN native plant, occurring in almost every lawn, meadow, or pasture, forming tufts of bright yellow flowers, the upper part often red on the outside. Although such a very common British plant—common also in many and distant parts of the world—I do not hesitate to introduce it here, believing that in many rock-gardens a place may be spared for it where its full beauty may be seen. It is best planted so that its shoots may fall in long and dense tufts over the face of rocks or stumps. It varies a good deal, but the common, low-tufted, spreading form, abundant in pastures and on sunny banks everywhere in Britain, is the most ornamental.

LYCHNIS ALPINA.—*Alpine L.*

A DIMINUTIVE form of *Lychnis viscaria*, but in a wild state, seldom rising more than a few inches high, and not viscid. In

Britain, says Mr. Bentham, "it is only known on the summit of Little Kilrannock, a mountain in Forfarshire," but in 1866, under the safe guidance of Mr. James Backhouse, I had the pleasure of seeing it abundantly in Cumberland in very lonely and high mountain gorges. We found it on the face of a dry crumbling crag quite 500 feet long and of great height, and generally in such positions that extermination is impossible, at least until such times as travelling botanists are provided with wings. In some places where the rocks overhung, it was in full health, where a drop of rain could scarcely ever fall upon it; but many plants which had sprung from seeds fallen from these cliffs were growing freely in moist shattered rock. The form seemed to be somewhat larger than that usually grown as *L. alpina*. In cultivation it is a pretty and interesting, if not a brilliant, plant, and may be grown without difficulty on rockwork or in rather moist sandy soil.

LYCHNIS LAGASCÆ.—*Rosy L.*

A LOVELY dwarf alpine plant, with a profusion of bright rose-coloured flowers, with white centres when young, each about three quarters of an inch across, and quite obscuring the small and slightly glaucous leaves. In consequence of its exceeding brilliancy of colour, and slightly spreading, though firm, habit, it is peculiarly well suited for adorning fissures on the exposed faces of rocks, the colour telling a long way off, while it is also a gem for association with the smallest alpine flowers, being beautiful when closely examined, as well as attractive at some distance. It is a native of the sub-alpine region of the North-Western Pyrenees, and was introduced two or three years ago by Mr. J. C. Niven, of the Hull Botanic Garden, in whose collection I first had the pleasure of seeing it grown in pots, in which it seems as easy of culture as on rockwork in any free sandy or gritty soil. A thoroughly exposed position in the open air, however, should always be preferred, as the plant is so free to grow, as well as neat and hardy. The figure in the 'Botanical Magazine' gives no idea of the brightness of colour of tufts of this plant grown in fully exposed positions. It is distinct from, and more beautiful than, any other alpine or dwarf Lychnis. The pale-flowered *L. pyrenaica*, which comes nearest to it, is not, since the introduction of the present subject, worthy of a place in any but a botanical or a very full collection. Flowers in early summer, and, when not drawn and weakened from shade, or by being placed in frames, is in perfect condition at about three inches high, and is most readily increased by seeds. = *Petrocoptis Lagascæ*.

LYCHNIS VISCARIA.—*German Catchfly*.

A BRITISH plant, found chiefly in Wales and about Edinburgh, but widely distributed in Europe and Asia in dryish places. It has long grass-like leaves, and very showy panicles of rosy-red flowers, on stems from ten to nearly eighteen inches high, and abundantly produced in June. The variety called *splendens* is the most worthy of garden cultivation, being of a brighter colour. *L. v. alba* is a charming white variety, also worthy of a place, and *L. v. flore pleno*, the double Catchfly, is a fine variety, with more rocket-like blooms. They are excellent plants for the rougher parts of rockwork, and as border-plants on dry soils. I have seen the double variety used with good effect as an edging-plant about Paris. Any of the kinds are worthy of being naturalised on dryish slopes or rather open banks, on which they seem to form the largest, healthiest, and most enduring tufts. Easily propagated by seed or division.

Lychnis Haageana, with shaggy stems and bracts, and flowers of a splendid scarlet, two inches across ; *L. flos-Jovis*, a downy plant, with rich purplish flowers ; *L. Coronaria*, the popular and handsome Rose Campion ; *L. fulgens*, with vermilion-coloured flowers, from Siberia ; and the double varieties of *L. diurna* and *vespertina*, although for the most part handsome plants, are too large for association with any but the coarsest rock-plants.

LYCOPODIUM DENDROIDEUM.—*Ground Pine*.

A CLUB-MOSS, in habit like a Lilliputian Pine-tree, and of all its family by far the most worthy of a place in the rock-garden. The little stems, ascending to a height of six to nine inches, from a creeping root, are much branched, and clothed with small, bright, shining green leaves ; fruit-cones yellow, long, cylindrical, and, like the stems, erect. A native of moist woods in North America and high mountains of the Southern United States.

I have never seen this plant perfectly grown except in Mr. Peek's garden, at Wimbledon, where it flourishes as freely as in its native woods, in a bed of deep sandy peat fully exposed to the sun. Few plants are more worthy of being established in a deep bed of moist peat in some part of the rock-garden, where its distinct habit will prove attractive at all seasons. It is apparently difficult to increase, and as yet exceedingly rare in this country. In attempting its culture the chief point is the selection of sound well-rooted plants to begin with; small specimens may retain their verdure after the root has perished, and thus often deceive. Should fresh spores be obtained, it would be a good plan to sow them in some moist spot in a half-shady position, the soil being made level and firm, and surfaced with a little silver sand.

LYSIMACHIA NUMMULARIA.—*Creeping Jenny.*

WERE I deterred from including plants here by reason of their commonness, this would not be mentioned, but as the Daisy has a place, Creeping Jenny, which gracefully suspends its drooping shoots with bright yellow flowers, as thickly strung as beads, from the little window gardens of thousands of poor city dwellers, must not be omitted. Were it a new plant, and not one found mantling over the ditch side in various parts of England, we should probably think it not too dear at half a guinea a root, as there is no hardy flower more suitable for any position in which long-drooping flower-laden shoots are desired, whether on points of the rockwork, or rootwork, or in rustic vases, or rapidly sloping banks. Creepers and trailers we have in abundance, but few which flower so profusely as this. Grows in any soil; in that which is moist and deep, the shoots will attain a length of nearly three feet, flowering the whole of their extent. Rarely or never seeds, but is as easily increased by division as the common Twitch, flowering in early summer and often throughout the season, especially in the case of young plants. There is a pretty variety with variegated leaves.

MAZUS PUMILIO.—*Dwarf M.*

A DISTINCT little New Zealander, vigorous in habit and creeping underground, so as rapidly to form wide and dense tufts, yet rarely reaching more than half an inch in height. The flowers, produced on very short stems, so as barely to show above the leaves, are pale violet, with white centres; the leaves nearly entire, or lobulate and obtuse, with a striking tendency to lie flat on or embrace the surface of the soil. It thrives in pots, cold frames, or in the open air, and is best placed in firm, open, bare spots on rockwork in free sandy soil in warm positions. It is not a showy but an interesting plant, easily increased by division. Flowers in early summer.

MECONOPSIS ACULEATA.—*Prickly Poppy.*

THIS is the only one of the splendid Indian species of this family that I have seen in flower, and that but once in the Royal Gardens at Kew. It is a singularly beautiful plant with purple petals like shot silk, which contrast charmingly with the numerous yellow stamens; the flowers two inches across, on stems about two feet high; the leaves heart-shaped in outline, somewhat five-lobed, but variable, and covered with rigid hair-like prickles. It is found at elevations of 11,000 to 14,000 feet on the Himalayas, and is probably hardy, though we have as yet but a slight knowledge of it. A warm well-drained position on the sunny side of rockwork is most likely to agree with it. Flowers in summer, and may be raised from seed.

MECONOPSIS CAMBRICA.—*Welsh Poppy.*

A HANDSOME poppy-like plant, forming large pale-green tufts of slightly hairy divided leaves, with handsome large pale-yellow flowers, known immediately from those of Poppies by the stigmas being supported on a short but distinct style. A native of rocky woods and shady places, in some of the Western Counties, and also found in Wales and Ireland. In the Lake Country it may be seen gathered in crowds round the gate-lodges, and running along by boundary walls, making quite an attractive show of bloom. It is well worthy of being introduced in semi-wild places, and is also suited for an extensive rock-garden. Flowers in early summer, and is increased from seed with facility.

MELITTIS MELISSOPHYLLUM.—*Balm M.*

A DISTINCT-LOOKING plant of the Salvia order, with slightly hairy ovate leaves, about two inches long, clothing the stem to its apex, and from one to three flowers arranged in the axils of the opposite leaves. The flowers are usually nearly or quite an inch and a half long, and opening at the mouth to a little more than an inch deep. The lower lip is the largest, and is usually stained with a deep purplish rose, except a narrow margin, which is a creamy white. The peculiarly handsome lip reminds one of the flowers of some of our handsome exotic orchids rather than those of a labiate plant. It varies a good deal in colour; sometimes the lip has not the handsome stain above alluded to, and sometimes the whole flower is of a reddish-purple hue. *M. grandiflora* of Smith is merely a slight variety differing in colour from the normal form. The plant is entirely distinct from any other in cultivation, and is well worthy of a position by shady wood and pleasure-ground walks. It naturally inhabits woods, and even when one finds it on the lower flanks of some great alp, it is seen nestling among the shrubs and low hazel-trees; woody spots near a fernery or rockwork would suit it to perfection, and it grows very readily among shrubs, and also in the mixed border. Found in a few localities in Southern England, and widely distributed over Europe and Asia. Readily increased by seed or division, and flowers in May about London.

MENZIESIA CÆRULEA.—*Yew-leaved M.*

A NATIVE of the northern and arctic parts of Europe, Asia, and America, and in Scotland found on the Sow of Atholl, in Perthshire. Grows from four to six inches high, with pinkish-lilac flowers in small umbellate clusters. The leaves are strap-shaped and obtuse, crowded, and very finely notched; the flower-stems and segments of the calyx hairy. It is not so beautiful or brilliant in colour as *M. empetriformis*, but merits a place in full collections and those in which rare and interesting British plants are esteemed. Flowers rather late in summer and in autumn.

MENZIESIA EMPETRIFORMIS.—*Empetrum-like M.*

A TINY shrub, neat in habit and of exquisite beauty, producing numbers of rosy-purple bells in clusters on a dwarf heath-like bush, seldom more than six inches high, the small closely placed leaves having toothed margins. This plant, very rarely seen in gardens, is one of the brightest of gems for the rock-garden, thriving best in a rather moist sandy peat soil in fully exposed positions. I have seen it cultivated with most success in nurseries in the neighbourhood of Edinburgh. It flowers in summer, and is sometimes known as *Phyllodoce empetriformis*. This, unlike the rather tall and spreading *M. polifolia*, may be associated with the dwarfest alpine plants. A native of North America.

MENZIESIA POLIFOLIA.—*St. Dabeoc's Heath.*

A SPREADING heath-like shrub, attaining a height of from twelve to twenty inches; the leaves narrow when young, and with their margins rolled back, but becoming broader when old, all green on the upper and white on the under sides; the numerous erect flowering-stems bearing beautiful crimson-purple blooms in graceful, one-sided, drooping racemes. This plant is found rather abundantly in Connemara, in Ireland, and is also a native of South-Western France and Spain, but not of England or Scotland. It is often grown in collections of American shrubs, and is admirably suited for the extensive rock-garden, loving a moist peat soil. There is a pure white variety much less common and no less beautiful than the typical form, also a charming plant for the rock-garden or collection of dwarf shrubs. They flower in summer, and may be obtained from most nurserymen. The white variety is sometimes sold under the name of *M. globosa*.

MERTENSIA VIRGINICA.—*Virginian Cowslip.*

THIS is readily distinguished from allied plants by the smoothness of all its parts, by its slightly glaucous hue, and large leaves, the lower ones being four to six inches long. The flowering stems grow to a height of from ten to eighteen inches, suspending blooms of a peculiarly beautiful purple blue, trumpet-shaped, and about an inch long. Flowers from the beginning of April to May or early June, and loves a soil cool and light, and a half-shady position. Suitable for the mixed border or rockwork, the margins of beds of dwarf shrubs, the fringes of plantations of American plants in peat soil, or even for naturalisation in half-shady spots in wood or copse. However, it is so uncommon at present that it can hardly be spared for the last purpose.

Known in old gardens as *Pulmonaria virginica*, which name is also often erroneously applied to one of the forms of the much dwarfer and very different *Pulmonaria officinalis* of our woods and gardens. In its native country it is occasionally found with white flowers, but I have not seen a white-flowered form in cultivation.

MESEMBRYANTHEMUM ACINACIFORME.—*Wall M.*

AN Ice-plant from the Cape of Good Hope, with purple flowers of extraordinary beauty, four to five inches across, and abundantly produced. Stem prostrate, young shoots compressed and angular; leaves two to three inches long, sometimes margined with a red line. The peculiar interest that this plant offers to lovers of alpine and rock plants is that it may be grown successfully on old walls and on warm parts of rockwork in dry soil, and in warm and sunny spots in old quarries, chalk pits, and like places. Doubtless many will be as proud of succeeding with this African plant as with those that bloom where the icicles drip.

MUSCARI BOTRYOIDES.—*Grape Hyacinth.*

AMONG spring bulbs few exceed this in loveliness; it is quite as valuable as the Siberian Squill, and should be in every garden. At one time it was much grown, but latterly it has been far from common. The flower-stems, from six inches to a foot long, spring from among the channelled leaves, bearing in March and April little racemes of blossoms tinted with one of the sweetest tones of deep skyblue that ever stained a spring flower. The tiny mouth of each almost globose bloom has six teeth or diminutive segments, which, being white, give each individual pip, as well as the whole spike, a peculiarly dressy character. There is a white variety also well worthy of cultivation. There is scarcely a position in a garden that may not be embellished by this plant, but its most appropriate place is here and there along the shrubbery borders, where, being hardy and free, it will thrive without attention, care being taken, however, to prevent its disturbance or destruction by digging in winter. A native of Southern and Middle Europe, readily increased by division, and will be the better for being raised and divided occasionally, say every third or fourth year.

MUSCARI MOSCHATUM.—*Musk Hyacinth.*

THIS is so deliciously sweet-scented and withal so decidedly ugly that it ought long ago to have been a favourite with authors of books on the "Language of Flowers," so suggestive is it of merit under the plainest of exteriors. When it comes in flower in March or April, according to warmth of season or position, the flowers, larger than those of the blue kind, are of such an indescribably unattractive tone of livid greenish yellow that many persons do not notice them lying among the leaves, and wasting their sweetness on the winter-beaten earth. For its fragrance it deserves general culture, and should be valued for imparting to bouquets the sweetness in which many brilliant flowers are deficient, and also for scenting the air in the open garden in spring. The most suitable position for it is on bare, open spots by wood walks where it would not be disturbed, or along the low margins of shrubberies, or in any position where its want of brilliancy may not lead to its destruction. There are several varieties of this plant—one, *M. luteum*, being of a clear waxy sulphur colour, and almost as handsome in tone as it is deliciously sweet. A native of South Europe.

MUSCARI RACEMOSUM.—*Common Grapeflower.*

THIS, the "dark blue grape-flower" of Parkinson, is a very old inhabitant of our gardens, and is also a native plant, or one that has escaped from cultivation. The dull green, channeled, weakly leaves attain a length of from twelve to eighteen inches, the flower-stem growing from nine inches to a foot high, the flowers being arranged on it in a dense terminal raceme. The flowers, very dark purple, and slightly covered with what grape-growers call bloom, almost remind one of berries. They smell "like unto starch when it is new made and hot." It is a vigorous grower, and will, as our interesting old friend, John Parkinson, remarks, "quickly choak a ground, if it be suffered long in it. For which cause most men do cast it into some by-corner if they mean to preserve it, or cast it out of the garden quite."

M. pallens, comosum monstrosum, commutatum, neglectum, and *Heldreichii,* are also in cultivation, the last, which is as yet very rare, being the best.

MYOSOTIS ALPESTRIS.—*Alpine Forget-me-not.*

A BRITISH alpine plant, found in one or two places in Scotland and Northern England, and inferior to no gem of the high Alps in vivid colour and beauty. It forms close tufts of dark-green hairy leaves, rather narrow and pointed, healthy plants rising to a height of only about two inches. About the end of April a few flowers of a beautiful blue, with a very small yellowish eye, begin to appear among the leaves, and as the weather gets warmer, the little flower-stems gradually rise, and soon the plants become compact masses of blue, remaining in perfection all through the early summer. Fortunately, it is very easily raised, and comes quite true from seed. It loves to be pinched in between lumps of millstone grit, whether in pots or on rockwork, and sometimes grows freely in more ordinary circumstances, but is apt to perish in winter if allowed to grow too grossly. It is quite distinct from, and much finer than, the dwarf mountain form of the Wood Forget-me-not, often met with on the Alps, the leaves always being in very dense tufts close to the earth, while the smallest specimens of *M. sylvatica* seen on the mountains do not branch below the surface, but are rather slender and erect in habit. It is also a true perennial, while the Wood Forget-me-not usually perishes after blooming. The garden home of the Alpine Forget-me-not is on the most select spots in the rock-garden—where it grows finest, perhaps, on ledges with a northern aspect, though it thrives perfectly in open sunny spots; the soil to be moist throughout the warm season. = *M. rupicola.*

MYOSOTIS AZORICA.—*Azorean Forget-me-not.*

THIS is at once recognised among other Forget-me-nots by the flowers being of a rich indigo-blue throughout, and rich purple when they first open. It was first brought home by Mr. H. C. Watson, author of the 'Cybele Britannica,' who found it near cascades and on wet rocks with a north-eastern aspect, in the Westerly Azores. It is a little tender, but so beautiful and distinct from our European blue and yellow-eyed Forget-me-nots that it is worthy of being annually raised, in case old plants should perish during winter. Easily increased by seed. It is best raised in autumn, and kept through the winter in dry frames, pits, or a greenhouse, or in very early spring in a gentle heat, and planted out about the beginning of May in a somewhat shaded or sheltered position, in light but deep and moist soil, in which it will form round spreading tufts. Peculiarly well adapted for half-shady quiet spots in the rock-garden, or among low shrubs near it, and also for the mixed border.

MYOSOTIS DISSITIFLORA.—*Early Forget-me-not.*

THIS beautiful plant has now been some years in cultivation, and there can be no doubt that, when it is better known to cultivators, it will be more grown than any other species of Forget-me-not. It bears more resemblance to the Wood Forget-me-not than to any other; but is much earlier in flower, blooming in January and February, and lasting till early summer. It is less hirsute, has the ribs on the stem much less strongly marked, and the leaves more gradually pointed. The pedicels are much more distant from each other, and, after the flower falls, lengthen considerably, and become twisted to one side, with a tendency to approach and embrace the stem. Early in the season, and in poor ground, it sometimes opens with pink flowers; but where the plants are healthy and the ground suitable, it soon expands into tufts of the loveliest deep skyblue. In dry ground it is apt to go off with the droughts of spring or early summer; but when placed in some moist cranny in a rockwork, it continues in flower for a long time, and accompanies the Wood Forget-me-not in its beauty, though it begins to show much earlier. It is a biennial plant, and in districts where the air is pure and somewhat moist, it flourishes to great perfection without any trouble, sowing itself on rockwork or on borders. I have grown it to great perfection in a moist bed of peat in a well exposed position, the ground being regularly watered in summer; there it continued a mass of beautiful blue till the end of June, growing continuously all the time, and sending up successive crops of flowers throughout the spring and early summer. This or a like plan is the one to adopt where the soil or climate are too dry for it. I can imagine no more valuable subject for naturalisation in rocky spots or in copses. For this treasure to our gardens we are indebted to Mr. J. Atkins, of Painswick, who found it on the Alps near the Vogelberg, and grew it for several years in his garden before it was in cultivation elsewhere. From him I obtained it, and soon afterwards it passed into general cultivation under the name of *M. montana.*

MYOSOTIS PALUSTRIS.—*Water Forget-me-not.*

To remind even those who are least observant of our wild flowers of the beauty of the Water Forget-me-not by our streams, canals, &c. would be wasting words; but I may advocate its cultivation as a garden-plant. It may be grown easily anywhere by the side of a stream, or pond, or moist place, where it does not already exist in a wild state, by merely pricking in bits of the shoots and afterwards leaving them to nature; and perhaps this is the best way in most places, particularly where the ordinary soil is warm and dry. But in many districts the climate and soil are moist and congenial enough to grow it well away from the water, and in such it is often desirable to have a little bed or two of a plant so great a favourite with all, and so useful for mingling with more showy flowers indoors. I have never seen the flowers so large as among Rhododendrons growing in beds of moist peat, the moist peat and shade just suiting it. It thrives, however, in ordinary soil in many gardens, as at Trentham, where it is used as an edging round the children's gardens. Grows as far north as the Arctic Circle, and is a native of North America as well as of Europe and Asia.

MYOSOTIS SYLVATICA.—*Wood Forget-me-not.*

A NATIVE of woods, mountain pastures, and shady situations in the North of Europe and Asia, and in the great central chain from the Pyrenees to the Caucasus, and also a British plant, though rare, limited to Scotland and the North of England. Met with in the woods in early summer, a mass of blue, it is one of the most charming sights afforded by any plant. It has been extensively grown in spring flower gardens during the past few years, and is the best blue flower used therein. In a wild state it is said to be perennial, but in gardens usually proves a biennial, and should be sown every year in early summer, putting the plants into the beds in autumn as soon as the summer flowers are gone. A few years ago it was generally grown and known as *M. arvensis*, a worthless weed, and may yet be known by that name in gardens. A good plan with this plant is to scatter some of the seeds in the woods and half wild places, or even to put out young plants in the autumn. There is a white variety, which is not so pretty as the blue, but which is yet worth growing where much variety is sought.

NARCISSUS BULBOCODIUM.—*Hoop-petticoat Daffodil.*

A DISTINCTLY beautiful kind, with its rich golden-yellow cup usually erect, gradually and regularly widening from the base to the margin and longer than the divisions, ascending, one on each stem, to a height of four to ten inches, from tufts of half-rounded, dark green, and somewhat rushy-looking, erect leaves. A native of Southern France and Spain, requiring in our gardens a little more attention than is necessary with Daffodils generally, but so handsome when well grown that it fully repays for it. It attains the greatest perfection in very light and deep sandy loam, in sunny sheltered positions, where its leaves may not suffer from cutting winds. In many sheltered gardens it may thrive on level ground in light earth, but in all it would be improved by being placed on low warm sheltered banks in the rock-garden, where few things are more attractive in early spring.

NARCISSUS JUNCIFOLIUS.—*Rush-leaved Daffodil.*

A SWEET-SCENTED, very dwarf, and pretty species, from the Pyrenees; bulbs about half an inch through; leaves three or four in a bundle, round and rush-like, four to six inches long, scarcely so long as the flower-stems. Flowers solitary, or in gardens occasionally two or three on a stem, both crown and cup of golden yellow, which is a distinct feature in the plant; appearing in early spring with the Primulas and Gentians in our gardens, in April and May in its native habitats. One of the smallest and most beautiful Daffodils known, and one of the most valuable for rockwork, thriving freely in gritty or sandy earth, and perfectly hardy; but as it flowers at a cold season, it is better planted on warm and slightly sheltered slopes on banks. Increased by division.

NARCISSUS MINOR.—*Least Daffodil.*

THIS is Parkinson's "least Spanish yellow bastard" Daffodil. "The leaves seldom exceeding the length of three inches, and very narrow withal, of a grayish green colour: every flower standing upon a small and short footstalk, scarce rising above the ground; so that his nose, for the most part, doth lie on or touch the ground." The bulb is small, of a darkish brown, often flowering at three inches high, and rarely attaining six inches under favourable conditions; the cup or crown of the flower much developed, much cut or fringed at the summit, of an orange yellow, and longer than the pale sulphur-coloured divisions. Flowers in March and April, is perfectly hardy, and thrives freely in fine sandy soil, but, being small and delicate in its parts, repays for a little care. It should be associated with very dwarf spring-flowerring plants, and in sunny but fully exposed positions, where it may enjoy the full sun without having its leaves and flowers destroyed by violent winds. If the patch of earth in which it is planted be surfaced with some very minute verdant plant like the Spergula, the flowers will not be so likely to be disfigured by having their "noses" so very near the ground, and the effect will be much improved. A native of Southern Europe; increased by careful division of established tufts.

NARCISSUS PULCHELLUS.—*Pretty Daffodil.*

A DRESSY and very elegant dwarf Daffodil, with slender rounded leaves and flowers borne upon stems six to nine inches high, each bearing three or four flowers, the cup scarcely half an inch in diameter, of a pale lemon tint, the divisions of the flower of a clear yellow and boldly recurved, or held back like those of a Cyclamen. It is deliciously scented, flowers in April and May, and deserves a place in every rock-garden, associated with other choice dwarf and spring-flowering bulbs and alpine plants, and thriving best in rich sandy loam. It is no doubt a variety of *Narcissus triandrus*. I have only seen it in cultivation in the Botanic Garden at Chelsea.

I have given only a few of the best of the dwarfest Daffodils; but there are many lovely kinds, somewhat larger, which, if too large for the select rock-garden, are indispensable in collections of spring flowers, and may be planted with the best effect in tufts among herbaceous plants and low shrubs in the fringes of the rock-garden, or by the sides of woods near at hand; among the finest kinds may be named *N. odorus, bicolor, major, Jonquilla,* the Poet's Narcissus (*N. poeticus*) and its varieties, also sundry fine hybrids and seedling varieties.

NERTERA DEPRESSA.—*Fruiting Duckweed.*

THE flowers of this diminutive plant are very inconspicuous, but when in fruit, it is best compared to a small Duckweed growing on firm earth, and bearing numbers of little oranges! They not only occur on the surface of the tufts, but by pushing the fingers between the small dense leaves the bright berries are found in profusion hidden among them. It is quite distinct in aspect from any other plant in cultivation, and, though no more ornamental in flower than a Duckweed, deserves culture for the pretty fruit. Should be associated with the dwarfest alpines on open parts of the rockwork, in firm, free, and moist soil. A native of New Zealand and the Andes of South America. Easily increased by division.

NIEREMBERGIA RIVULARIS.—*Water N.*

OF quite a different type to the other members of its family seen in our gardens, the stems and foliage of this trail along the ground as dwarfly as those of the New Holland Violet, while from amongst them spring erect, open, cup-like flowers of a creamy-white tint, just barely pushed above the foliage. Sometimes the blossoms are faintly tinged with rose, are usually nearly two inches across, with yellow centres, and continue blooming during the summer and autumn months. Their distant effect, suggesting Snowdrops at first, is very pleasing, and they are no less pretty when you come near and stand over them. It is said to abound by the side of the Plate River, but only within high-tide mark, its flowers rising so high among the very dwarf grass that the plant is discerned from a great distance. No collection of rock or herbaceous plants can be complete without it, while the tasteful flower-gardener may well use it in his smaller designs. Why not add it to the charms of our best form of spring garden—those parts of a country place planted with beautiful spring bulbs, flowers hardy enough to take care of themselves, and lovely as things in a really wild state, peeping through the young grass in spring or early summer, without the usual gardenesque accompaniments of naked muddy earth and prim surroundings? Rooting much at the base, it is easily increased by division.

ŒNOTHERA MARGINATA.—*Large Evening Primrose.*

ALTHOUGH a humble plant—even when vigorous not more than six to twelve inches high—this blooms as nobly as any luxuriant native of the tropics, the individual flowers being four to five inches across, of the purest white, changing, as the flower becomes older, to a very delicate rose, the blooms coming well above the toothed or jagged leaves as the evening approaches, and remaining in all their glory during the night, emitting a delicious magnolia-like odour. This plant, tastefully arranged in the rock-garden or in some quiet border, should prove one of the greatest charms of the garden in every northern and temperate clime—as welcome among the night bloomers as the nightingale among night singers. It is perennial, quite hardy, increased by suckers from the root, which are freely produced. Cuttings also root readily. "Begins to flower in May, continuing till the weather gets hot about July, when it seems to like a rest, and again blooms in September and October." Young vigorous specimens in rich ground would probably flower continuously throughout this period. Mr. Robert Stark, of Edinburgh, a well-known lover and cultivator of rare hardy plants, obtained roots of this, when in Canada, from a botanist in the Western States of America, and it is to him we are indebted for its introduction to our gardens.

ŒNOTHERA MISSOURIENSIS.—*Missouri Evening Primrose.*

A NOBLE, yellow, herbaceous plant from North America, with prostrate, rather downy stems, entire leaves, their margins and nerves covered with silky down, and with rich clear golden-yellow flowers, from four to nearly five inches in diameter, so freely produced that the plant may be said to cover the ground with tufts of gold. There is no more valuable border-flower, and well placed on rockwork it is a glorious ornament, especially when the luxuriant shoots are allowed to hang down. As the seed is but rarely perfected, it is better increased by careful division, or by cuttings made in April. When used as a border-plant, it does nor make such a free growth in cold clayey soils as it does in warm light ones. The blooms open best in the evenings. = *Œ. macrocarpa.*

ŒNOTHERA SPECIOSA.—*Showy Evening Primrose.*

A VERY handsome plant, with an abundance of large white flowers, which afterwards change to a delicate rose, in these respects somewhat resembling Œ. *taraxacifolia*, but the plant is erect, with almost shrubby stems. It forms neat tufts, usually from fourteen to eighteen inches high, is a true perennial, and exceedingly valuable for borders, or the lower and rougher parts of rockwork. A native of North America; increased by division, cuttings, or seeds, but not seeding freely in this country, and flourishing vigorously in well-drained rich loam. Plants of this type, somewhat above the size we require for the choice rock-garden, may be very tastefully used near it on banks rocky or otherwise. About Paris it is used a good deal for bedding, replanted every year, and is very effective for this purpose, as the bloom continues from midsummer till October.

ŒNOTHERA TARAXACIFOLIA.—*Dandelion-leaved Œ.*

ONE of the most popular and beautiful of all our dwarf hardy plants, with rather stout stems, that freely trail over the ground like any humble weed in this cold climate of ours, but bearing a profusion of flowers, large and delicately tinted as those of a tropical Dipladenia, which can only be grown in this country at considerable and constant expense of time and money. The leaves are deeply cut, somewhat like those of the Dandelion, but of a greyish tone; the flowers several inches across, white, changing to pale delicate rose as they become older. The plant is quite hardy and perennial, but on some very cold soils perishes in winter. Where it does so, and where the plant is much admired, it should be raised annually from seed. It will thrive in almost any garden soil, but best in one rich and deep, and may be used with the best result as a drooping plant in the rock-garden border. Plants raised in early spring and pricked over bare surfaces of rose-beds, &c. embellish them finely, flowering profusely the first year. A native of Chili, flowering all the summer and autumn, and seldom rising more than six inches above the ground.

Œnothera acaulis differs from this fine plant chiefly in having much smaller flowers; it is, therefore, barely worthy of a place except in botanical collections.

OMPHALODES LUCILLÆ.—*Lucilia's O.*

A seldom seen and very charming sister of the sweet little *Omphalodes verna*, but with a dwarf crop of very glaucous smooth leaves, in tone resembling those of the Oyster-plant, and with flowers of a beautiful light skyblue, with a faint stain of something akin to the palest lilac. A native of Mount Taurus. It will probably one day become one of the most admired plants on our rockeries, as it is the very type of sweet modest beauty. I have not observed enough of the plant to speak more fully of its wants or habits, having only seen it on the rockwork in the botanic garden at Geneva, but have little doubt that it will be found to thrive on sunny flanks of rockwork in perfectly well drained fissures of light soil. It deserves careful treatment till sufficiently abundant to be tried in various positions.

OMPHALODES VERNA.—*Creeping Forget-me-not.*

A NEAR relative of our Forget-me-nots, quite as beautiful as and on the whole more useful than any of them in consequence of the facility with which it creeps about in shady places. Its handsome deep and clear blue flowers with white throats, produced in early spring, immediately remind one of the finest Forget-me-nots, but are larger and of a more intense blue, and immediately distinguished by the stems sending out runners from their base, apart from other characteristics. A native of mountain woods on several of the great continental chains, and indispensable for every rock and spring garden. I know of no plant more worthy of naturalisation in half-wild places under trees or shrubs, or even beneath Rhododendrons and the like. In elevated woods I have seen it run about as vigorously as any native plant, but it will thrive well by almost every woodwalk, and prove one of the brightest ornaments of spring in any position in which it grows. Easily increased by division. Tufts of it taken up and forced in midwinter form beautiful objects in pots or baskets in the conservatory.

ONONIS ARVENSIS.—*Wild Liquorice.*

ONE of the prettiest of our wild plants, and well worthy of cultivation on banks and rough rockwork. It is a somewhat variable plant, usually forming dense spreading tufts, clammy to the touch from being thinly overspread with glutinous hairs, and becoming covered with racemes of pink flowers in summer. There is a white variety even more valuable for cultivation, and worthy of a better position and soil than the common form, which grows in any soil. No plants can be more readily increased from seed or by division. This plant is distinct from the spiny *Ononis campestris*, which forms stems nearly two feet high, sometimes even more.

ONONIS ROTUNDIFOLIA.—*Round-leaved Rest-harrow.*

THIS species is easily known by its roundish trifoliate leaves, margined with triangular teeth, and thickly furnished with gland-tipped, slightly viscid hairs, and large and handsome rose-coloured flowers, with the upper petal or standard veined with crimson, usually in pairs on one petiole springing from the axil of almost every leaf of the upper portion of the stem. A distinct and pretty plant, hardy, and easily cultivated, flowering in May and June and through the summer. It attains a height of from twelve to twenty inches, according to soil and position, increasing in height as the season advances. Suitable for the mixed border, or rougher parts of rockwork; comes from the Pyrenees and Alps of Europe, and is easily propagated by seeds or division.

OPHRYS APIFERA.—*Bee Orchis.*

ONE of the most singularly beautiful of our native plants, very rarely seen in gardens; it varies from six inches to more than a foot in height, with a few glaucous leaves near the ground; the lip of the flower being convex, of a rich velvety brown with yellow markings, so that it bears a fanciful resemblance to a bee. Found abundantly in various parts of England and Ireland, though far from common or generally distributed, and met with generally on chalky hills or banks, or on a limestone, and occasionally on a clayey or sandy, soil. It is usually considered very difficult to grow, but this is by no means the case, and it may be grown easily in rather warm and dry banks in the rock-garden, planting it in a deep little bed of calcareous soil, if that be convenient; if not, using loam mixed with broken limestone. It will be found to thrive best if the surface of the soil in which it grows be carpeted with the Lawn Pearlwort or some other very dwarf plant, and failing these, with an inch or so of cocoa-fibre and sand, to keep the soil somewhat moist and compact about the plants. Flowers in early summer. I am not aware that it has yet been increased in gardens.

ORCHIS MACULATA.—*Spotted O.*

This is the handsomest British Orchis. Usually prettily tinted in the poorest and driest soils, but a very different object in a rich Buckinghamshire meadow (where I have seen it sometimes so thick that the spikes were as plentiful as those of the grass flowers) to what it appears on a starved Surrey heath. If well grown in moist and rather stiff garden loam, it will surprise and please even those who know it well in a wild state. Obtain it at any season, and plant twelve or twenty tubers in a patch, taking them up as tenderly as possible, and plant them carefully in a half-shady and sheltered position in moist deep soil. When the plants form dense patches two or three years afterwards, they will prove more beautiful than many orchids from warmer climes. Flowers in summer, and may be associated with the Cypripediums, or planted in tufts in borders, or on the margins of shrubberies.

Orchis foliosa, latifolia, and *laxiflora*, are also worthy of a place; the two last thrive best in moist boggy soil, and should be placed in the artificial bog.

OROBUS CYANEUS.—*Blue Bitter Vetch.*

A dwarf vetch-like plant, with large, handsome, bluish flowers among masses of light green leaves, with two or three pairs of leaflets, flowering in spring, the plant growing little more than six inches high. I have only observed this plant growing on very cold stiff ground scarcely acceptable to coarse weeds, and there it was quite hardy and flowered regularly, so that it is probable it would do much better on light good soils. Propagated by seeds or division; comes from the Caucasian mountains, and is best suited for warm, sheltered, sunny spots on rockwork. It is sometimes met with under the name of *Platystylis cyaneus*, under which name it was figured by Sweet.

OROBUS VARIEGATUS.—*Variegated Bitter Vetch.*

A COMPACT plant, with two firm and opposite keels on its wiry stems, which ascend in a zigzag manner to about a foot in height, bearing leaves with two or three pairs of leaflets, and rather closely arranged racemes of flowers supported on a footstalk a couple of inches long. The flowers, though small, are beautifully variegated, the upper petal being a fine rose-colour with a network of full purplish-crimson veins, the points of the wings being blue. It is a very hardy, easily grown plant, suited for the front margin of the mixed border or the rockwork, and may be increased by seeds or division. A native of Southern Italy and Corsica; not unlike *Orobus vernus* in habit, but at once distinguished by its much smaller flowers in shorter, densely packed, racemes.

OROBUS VERNUS.—*Spring Bitter Vetch.*

THIS is one of the most charming border-flowers that begin to open in that sweet season, the end of April and beginning of May. From black roots spring rich healthy tufts of leaves with two or three pairs of shining leaflets, the flower buds showing soon after the leaves, and eventually almost covering the plants with beautiful blooms, purple and blue, with red veins, the keel of the flower tinted with green, and the whole changing to blue. It is no fastidious alpine beauty, that, when carried to our gardens in the cultivated plains, sickens and dies for want of the pure cool mountain air and moisture, but a vigorous native of Southern and Central Europe, well able to make the most of our warm deep sandy loams, growing in almost any soil, and perfectly hardy everywhere. It is one of the best ornaments of the mixed border in cultivation, useful for those parts of the rockwork where a bold and vigorous vegetation is desired, and also a noble subject for those who wish to naturalise fine hardy plants in open spots along their wood or shrubbery walks. It may be seen in great abundance in the Royal Botanic Gardens at Edinburgh, where Mr. M'Nab arranges it in lines and various other ways, in which we are not accustomed to see it. It varies a good deal—all the better, of course—the most marked of the known varieties or sub-species being *ruscifolius* and *flaccidus*. Flowers in April and May; grows from ten inches to eighteen inches high; is propagated very readily from seed, and also by division of the root, and attains its greatest development in deep warm soils in sunny, sheltered positions.

Some other species of Orobus useful for borders and rockwork are in cultivation : *O. pubescens*, *O. canescens*, *O. varius*, and *O. Fischeri* being among the best. All are of easy culture in ordinary garden soil, and are increased by seeds or division of the root.

OTHONNA CHEIRIFOLIA.—*Barbary Ragwort*.

A PLANT of very distinct character; the leaves and shoots quite smooth and glaucous, and the habit neat and spreading. Forms whitish-green and rather ornamental tufts from eight inches to a foot high, or perhaps more on very rich soils, and flowers sparsely on heavy and cold soil, but on light soils it blooms often somewhat freely in May, the blooms of a rich yellow, about an inch and a half across, but not ornamental. Chiefly useful for its distinct type of leaf and aspect on the rough rockwork or mixed border. A native of Barbary; propagated by cuttings.

OXALIS BOWIEI.—*Bowie's Wood Sorrel*.

A BEAUTIFUL Wood Sorrel from the Cape of Good Hope, often grown in pots in our greenhouses and frames, but quite hardy, and never seen in its true beauty except when growing on very sandy or warm soil in the open air. It grows all round the wall of one of the stoves in the Botanic Garden at Chelsea on a very sandy soil, and flowers most profusely, as it does in various other places in Southern England, and would do in most places on dry soil if commonly planted. Few dwarf plants are more attractive when its large and handsome rose-red flowers, six to ten on stems usually reaching from eight inches to a foot high, are open. It should be on every rockwork, and is best planted on sunny slopes in well-drained sandy loam, and, if convenient, against rocks or stones. Where walls rise directly from gravel walks, a pleasing line of the plant may be established, and, if the bottom be dry, it will flower charmingly every year. In cold soils I have observed it grow freely enough, but not flower well. It is very easily increased by division, and flowers in autumn.

OXALIS FLORIBUNDA.—*Many-flowered Wood Sorrel.*

A FREE-FLOWERING kind of Wood Sorrel, apparently quite hardy in all soils, and producing numbers of rose-coloured flowers with dark veins, for months in succession. The leaflets, three in number, are roundish-egg-shaped, concave at the apex, and hairy. There is a white-flowered variety as free to flower and in every way as valuable as the rose-coloured form. Both are very useful for rockwork, for the margins of borders, and are easily increased by division. This appears to be the commonest kind of Oxalis in cultivation. A native of South America, and hardy enough to encourage one to attempt to naturalise it on any rocky place or about ruins.

There are other species of Wood Sorrel obtainable, and worthy of a place, especially on very dry sandy soils, and among them, *O. lobata*, a yellow, and *O. speciosa*, a rose-coloured kind, are perhaps the best. The elegant *O. rosea* is an easily grown annual kind, suitable for grouping with choice rock-plants.

OXYTROPIS PYRENAICA.—*Pyrenean Oxytrope.*

A VERY dwarf species, with pinnate leaves, composed of from fifteen to twenty pairs of leaflets, each about a quarter of an inch long, slightly concave, and clothed with a short silky down. These barely rise above the ground, as the short stems are nearly prostrate, and seldom exceed a few inches in height ; the flowers, borne in heads of from four to fifteen, are of a purplish lilac—the upper petal or standard barred with white in the centre. Not a showy but withal a desirable little plant for those parts of rockwork devoted to very dwarf subjects. A native of the Pyrenees, rare in gardens, increased by seed or division, and should be planted on well-exposed and bare parts of rockworks, in firm, sandy, or gravelly soil. Flowers in early summer.

PACHYPHYTUM BRACTEOSUM.—*Silverbracts.*

I SHOULD not have mentioned this plant here had it not been used with a singularly good effect of late associated in several instances with dwarf hardy succulents. It is a fleshy succulent plant, a native of Mexico, belonging to the Crassula order, with leaves about half an inch thick, somewhat spoon-shaped, but narrowing very gradually from the wide apex to the thick base, with a large, flat, and thick point, rounded on the under side, slightly concave above; the whole plant has a whitened tone, being densely coated with minute powdery matter, except on the old and weather-beaten leaves, the young leaves and those towards the centre of the rosette being faintly but sweetly suffused with pink. The stout little stems, bearing their columns of fat leaves, usually attain a height of from three to eight inches, this depending a good deal on the size they are when placed in the open air; the strongest of them sending up stems from twelve to sixteen inches long, weighed down at the top with a quaint but singularly pretty inflorescence. The line of flowers springs from the upper part of the stem, but immediately from beneath the footstalk of each bloom a neatly carved leaf (bract), a little more than half an inch, is given off; and, as these all grow in a downward direction, there is a line of small silvery leaves on the under side of the arching inflorescence. The flowers themselves are well enclosed in a calyx with sepals of a bluish-silvery tone, fleshy like those of the bracts. The freshly opened flowers are usually in the drooping portion of the raceme, so that they are not seen till slightly raised up, or the plant, if in a pot, placed on nearly a level with the eye; but when they are seen, they afford a pleasing surprise, being of a deep crimson with yellow stamens. The unopened part of the inflorescence is composed of the overlapped small leaves of the calyx and bracts, somewhat resembling a hop fruit in outline, but, like all the rest, of a peculiarly attractive hue.

This plant, first seen about London in the tiny collections of succulents grown for the market in small pots, is now found to be a most valuable addition to the flower-garden in summer, and also merits a place in every greenhouse, and in every collection of window-plants. It has survived out of doors one or two winters about London, but for flower-garden purposes it is best kept over the winter in dry frames, pits, or on a shelf in the greenhouse. On the rockwork, planted out in sunny nooks in light soil, it may be ventured out all the winter. It is seen to great advantage on glistening carpets of the dwarf mossy Saxifrages, or diminutive plants of like habit. Readily propagated by merely pulling off the leaves, inserting them in pans or boxes, and placing them in a dry warm pit or frame, or any like place that may be to spare. A little plant soon appears at the base of each leaf. If it be desired to advance the young plants rapidly, it may be done by keeping them in a brisk dry heat near the glass. In the open air it is usually grown for the effect of its leaves, and does not often flower except in the greenhouse.

PAPAVER ALPINUM.—*Alpine Poppy.*

THIS has large and beautiful white flowers, with yellow centres, and with smooth, or hairy, dissected leaves, cut into fine acute lobes. A native of the higher Alps of Europe, this plant may sometimes be seen in good condition in our gardens, but it is liable to perish as if not a true perennial. It varies a good deal as to colour, there being white, scarlet, and yellow forms in cultivation. The variety *albiflorum* of botanists has white flowers, spotted at the base, while the variety *flaviflorum* has showy orange flowers, grows three or four inches high, and is hairy. This last variety is also known as *P. pyrenaicum.*

PAPAVER NUDICAULE.—*Iceland Poppy.*

A FINE dwarf kind, with deeply lobed and cut leaves, and large rich yellow flowers on naked stems, reaching from twelve to fifteen inches high. A native of Siberia and the northern parts of America, and a handsome plant for borders or rockwork, easily raised from seed, and forming rich masses of cup-like flowers, but, like other dwarf Poppies, does not seem to be permanent, and should be raised annually. There are several varieties.

PARADISIA LILIASTRUM.—*St. Bruno's Lily.*

WHEN the traveller first crawls down from the cold and snowy heights of an alp into the grateful warmth and English-meadow-like freshness of a Piedmontese valley, most likely the first flower he notices in the long and pleasant grass of the lower flanks of the valley is a lily-like blossom, standing about level with the tops of the blades of Grass and Orchises. The blooms, about two inches long, so clearly and delicately white that they might well pass for emblems of purity, have each division faintly tipped with pale green, and from two to five flowers occur on each stem. It does not grow in close tufts as in our borders, but one or perhaps two stems spring up here and there all over the meadows, and the effect of the half-pendent " Lilies " is then fascinating. If it were an English flower, it would be called the Lady of the Meadows. It is easy of culture on light deep warm soil, and should be in every collection. Slight shelter would prove beneficial, and that may readily be afforded by planting it among dwarf shrubs and the rather tall subjects on the flanks and lower parts of rockwork. It is also a beautiful ornament for the mixed border, and would prove worthy of naturalisation in open spots in semi-wild places and in unmown parts of the large pleasure-ground. It will be found to flourish in British as well as in alpine grass, and is easily propagated by division or by seeds. = *Czackia Liliastrum*.

PARNASSIA CAROLINIANA.—*Carolina P.*

A NATIVE of North America, chiefly in mountainous places, on wet banks, and in damp soil. Much larger than the British Parnassia, the stem reaching from one to nearly two feet high, the flowers from an inch to an inch and a half across, the leaves thick, leathery, and roundish-heart-shaped, one usually occurring on the stalk low down, and clasping the stem; the sterile stamens three in each set, instead of nine to fifteen, as in *P. palustris*. It is a very desirable plant for artificial bogs, succeeding in deep moist soil, and flowering in autumn. Increased by seed or division: at present rare in gardens.

P. asarifolia, a native of high mountains in Virginia and North Carolina, does not differ much from the preceding, but has the leaves rounded and kidney-shaped, with larger flowers, and requires much the same treatment.

PARNASSIA PALUSTRIS.—*Grass of Parnassus.*

A WELL-KNOWN and admired native mountain plant, with handsome white flowers an inch or more in diameter, and with very distinctly marked veins, smooth, rather pale-green leaves, heart-shaped at the base, from less than an inch to an inch and a half long, the flower-stems wiry, angular, and from four inches to about a foot high. An interesting native as well as ornamental plant, growing naturally in bogs, moist heaths, and high wet pastures. A very desirable addition to the artificial bog, or to very moist spots in or near the rock-garden, and may also be grown in pots placed halfway in any fountain or other basin devoted to aquatic plants. Plants or seeds may be easily obtained in its wild habitats, and the seed should be sown, in moist spots, as soon as gathered.

PELARGONIUM ENDLICHERIANUM.—*Endlicher's P.*

THIS is interesting as the only species of Pelargonium that comes so far north as Asia Minor, and from being a hardy plant. But it is also a remarkably showy and handsome one, with deep rose-coloured flowers, boldly upheld on stems about eighteen inches high, the two upper petals being very large. The leaves are roundish, and form rather close tufts on the surface. I first saw it in the Jardin des Plantes, at Paris, where it had remained several very severe winters in the open air, thus proving perfectly hardy. A sunny nook on a rockwork would suit it well, or a position on a warm dry bank, sheltered from the north, but it will probably be found to thrive without these attentions. Readily increased by seed or division.

PENTSTEMON GLABER.—*Smooth P.*

OF dwarf habit, mostly quite decumbent, but occasionally more or less erect, with lance-shaped, entire, smooth leaves, and terminal racemes of bright-blue flowers, the throat being usually of a pinkish lilac, but the flowers vary in hue. Perfectly hardy, and succeeds in any friable soil. A native of the plains adjacent to the Rocky Mountains. The better known *P. speciosus* is now regarded as a variety only of this plant, growing taller and erect, but the forms run into each other; and *P. cyananthus* is only another form. Flowers in early summer, and may be easily raised from seed, the seedlings blooming the second year. Being of a prostrate habit, it is well suited for warm and well-drained slopes and ledges in the rock-garden.

PENTSTEMON PROCERUS.—*Tufted P.*

PERHAPS the most suitable of any easily-obtained Pentstemon for the margins of borders or for planting on rockwork, spreading into wide evergreen tufts not more than a couple of inches high in any kind of soil, and sending up in early summer numerous spikes of flowers, not large or showy individually, but freely produced, and of a veined purple, in little racemes, but standing so close to the stem that they seem in whorls. The stems, at least those of the variety cultivated about London, rarely rise more than a foot high; leaves lance-shaped and quite entire, the lower ones with a stalk, the upper stalkless. Few plants are more easily increased by division of the root. Flowers in early summer. A native of North-West America and the Rocky Mountains.

PENTSTEMON SCOULERI.—*Scouler's P.*

A SHRUBBY species, with a somewhat spreading compact habit, narrow sharply notched leaves, the upper ones entire, and with large, pale lilac-purple flowers. It is very hardy, grows freely in ordinary soil, and flowers in summer. It is said to attain a height of between two and three feet, but I have not seen it nearly so tall. Propagated by seeds or cuttings, and well suited for extensive rockworks associated with small shrubs and the finest types of herbaceous vegetation.

In addition to the preceding, there are various large Pentstemons more fitted for borders than rockwork, but also useful in this way where there are very large rock-gardens, especially the fine and somewhat tender *Pentstemon Jeffreyanus*, with gentian-blue flowers, the free-flowering lilac-rose *Pentstemon diffusus*, the white and the red varieties of *P. Hartwegi* (= *P. gentianoides*), and its other innumerable and beautiful forms.

PETROCALLIS PYRENAICA.—*Beauty of the Rocks.*

TRULY a "rock beauty!" as everybody must confess who sees its fresh light-green tufts, not more than an inch high, and cushioned snugly amidst the broken rocks. From these stains of light green spring in April peculiarly innocent-looking flowers, reminding one of Lilliputian "Ladies' Smocks," and supported on stems that rise little more than half an inch over the thrice-divided leaves, except where the plant is in a shady position, when they push up a little taller, and are more attenuated. When well grown, its faintly-veined pale-lilac flowers seem to form a little cushion, so cold and delicate-looking that I have known people to grow and admire it for years without ever suspecting it to be capable of emitting an odour of any kind; but it breathes a delicious, if faint and delicate, sweetness. Only suited for careful culture on the well-made rockwork, being of a fragile nature, though perfectly hardy. It should be planted in sandy fibry loam, in rather level warm spots on rockwork, where it could root freely into the moist soil, and yet be near the congenial influences of the broken rocks and stones, down the buried sides of which it can send its roots. It should always have a sunny position. I have seen it grown as a border-plant in a moist part of Ireland, but in the hands of a very careful cultivator, who grew it in very fine soil on a select border, and took up, divided, and carefully replanted the tufts every autumn. It may also be grown in pots plunged in sand in the open air, and in frames in winter; but I have always noticed that it becomes drawn and delicate under glass protection of any kind. Easily increased by careful division, and may also be raised from seed. A native of Northern Italy, the Tyrol, and various other elevated parts of Southern Europe, as well as the Pyrenees.

PHLOX DIVARICATA.—*Spreading P.*

LARGER than the Creeping Phlox or Moss Pink, attaining a height of about one foot, and bearing large lilac-purple blossoms: the leaves are rounded at the base, oblong-egg-shaped or oblong-lance-shaped in outline. The plant thrives very well on rockwork in good garden soil, and flowers in summer. A native of North America, and increased by division.

PHLOX REPTANS.—*Creeping Pink.*

WITH the large flowers and richness of colour of the taller Phloxes, this mantles over borders and rockworks with a healthy soft green about an inch or two high, and sends up numbers of stems from four to six inches high at the end of April or beginning of May, each producing from five to eight deep purplish-rose flowers. It is by no means fastidious as to soil or situation, but will be found to thrive best in peat or light rich soils. As it creeps along the ground, and gives off numbers of little rootlets from the joints, it is propagated with the greatest ease and facility. A person with the slightest experience in propagation may convert a tuft of it into a thousand plants in a very short time. It is almost indispensable for the rock-garden, makes very pretty edgings round the margins of beds, and also capital tufts on the front edge of the mixed border. It may also be used in the spring garden and for vase decoration, and is a native of

North America, inhabiting damp woods. It is, perhaps, better known in gardens as *P. stolonifera* and *P. verna* than by its proper name.

PHLOX SUBULATA.—*Moss Pink*.

A MOSS-LIKE little evergreen with stems from four inches to a foot long, but always prostrate, so that the dense matted tufts are seldom more than six inches high except in very favourable rich and moist, but sandy and well-drained, soil, where, when the plant is fully exposed, the tufts attain a diameter of several feet, and a height of one foot or more. The leaves are awl-shaped or pointed, and very numerous; the flowers of pinkish purple or rose colour, with a dark centre, so densely produced that the plants are completely hidden by them during the blooming season. It occurs in a wild state on rocky hills and sandy banks in North America, and there are few more valuable plants for the decoration of the spring garden borders or rock-work, being at once hardy, dwarf, neat in habit, profuse in bloom, forming gay cushions on the level ground, or pendent sheets from the tops of crags or from chinks on rockwork. It is easily increased by division, forming roots freely at the base of the little stems, and usually thrives in ordinary garden soil, particularly in deep sandy loam. Excessive drought seems to injure it, but it is less likely to suffer when rooted beneath the stones on rockwork. There is a white variety (*P. subulata alba*), known in many gardens as *P. Nelsoni*, which is also a beautiful plant.

PINGUICULA GRANDIFLORA.—*Irish Butterwort*.

LEAVES in rosettes, light green, fleshy, and glistening, oval-oblong and obtuse, broadest in the middle; flowers handsome, two-lipped, spurred like the Horned Violet, more than an inch long, nearly or quite an inch across in well-grown specimens, of a fine violet blue. Mr. Bentham unites this with the much less beautiful *P. vulgaris*, but Mr. Syme says: "I cannot conceive how anyone who has seen the plants alive can consider them as the same species;" and as *P. grandiflora* has flowers twice as large as *vulgaris*, and is a much handsomer plant, it is certainly distinct, and the kind by far best worthy of cultivation. It inhabits bogs and wet heaths in the South-west of Ireland, and delights in moist mossy spots on the northern and shady slopes of the rock-garden, or in more open places in very moist peat soil. In the former situations it may be associated with the Wintergreens, the Twinflower, or the Starflower, and in the artificial bog with *Rhexia virginica* and the Bavarian Gentian. Increased by small green bulbils, which are given off at the base of the rosettes.

PLUMBAGO LARPENTÆ.—*Hardy P.*

A DWARF herbaceous plant, originally cultivated in stoves and greenhouses, but now found to be perfectly hardy, and a first-rate ornament for rockwork, banks, or sunny borders. Its numerous wiry stems, covered regularly from top to bottom with light green leaves nearly two inches long, and margined with hairs, are half prostrate, but, being very profuse, form neat and full tufts from six to ten inches high, according to soil and position. In September these become nearly covered with flowers arranged in close trusses at the end of the shoots, and of a fine cobalt blue, afterwards changing to violet—the calyces being of a reddish violet. The bloom usually lasts till the frosts. I have seen this plant flourish in very cold soils, but it is in all cases desirable to give it a warm sandy loam or other light soil and a sunny warm position, as under these conditions the show of bloom is much finer. In consequence of the semi-prostrate habit, it is well suited for planting above the upper edges of vertical stones or slopes on rockwork, and it may also be used with good effect as a border-plant, or as an edging-plant in the flower-garden, particularly in the case of slightly raised beds. A native of China; very easily increased by division of the root during winter or early spring.

POLEMONIUM CÆRULEUM.—*Jacobsladder*.

THIS old garden favourite, with its tender green leaves and rather showy blue flowers, so widely diffused over the northern regions of the world, is chiefly mentioned here in consequence of the striking beauty of its variegated variety (*P. cæruleum variegatum*). It is feathery, silvery, and so graceful that sometimes people mistake it for a "variegated fern," and most valuable for mixed borders, rockwork, the flower-garden, or almost any position. It is so desirable as an edging-plant in the flower-garden that its propagation and culture are of importance. On good garden soils generally it will be found almost as easy of culture as the common form. It is, however, apt to go off on a very wet clayey soil, but flourishes finely in deep rich but well-drained loam. As regards the propagation, it is effected by simply digging up well-established old plants, pulling them in pieces, and then planting them immediately in a nursery-bed of good soil. This is best done in early autumn, so that it may be nicely established in the nursery-beds before midwinter. Where plants are merely required for borders and rockwork, the old stools simply require to be taken up, divided, and replanted, where desired, in the old-fashioned way of dealing with herbaceous plants. As the variegated variety is grown for its leaf beauty alone, it is scarcely necessary to add that the flower-stems should be removed when they appear. There are several other species in cultivation, but they are scarcely of a sufficiently perennial or ornamental character.

POLYGALA CALCAREA.—*Chalk Milkwort.*

A NATIVE plant found in Kent, Surrey, Gloucester, Berks, and a few other places in the South of England, generally on chalky débris, and very pretty, usually with blue but sometimes with pink or whitish flowers, about a quarter of an inch long, in compact racemes; the leaves deep shining green and smooth, and the shoots six inches long in well-grown specimens. Mr. Syme says this has no connecting links with the common Milkwort (*P. vulgaris*). It is known by the flowering shoots rising from rosettes of leaves, and by the leaves on those shoots becoming abruptly smaller and narrower than those below them. It is the handsomest and the easiest to grow of the British species, and does very well on rockwork in sunny chinks, planted in calcareous soil, forming neat dressy tufts of violet-blue and white flowers, and blooming profusely in early summer. It should be allowed to sow itself if possible, or the seed may be gathered from wild plants and carefully sown in sandy soil. Plants carefully taken up from their native positions have also been established in gardens.

POLYGALA CHAMÆBUXUS.—*Box-leaved Milkwort.*

A VALUABLE little creeping shrub, a native of the Alps of Austria and Switzerland, where it often forms but very small plants; in our gardens, however, on peaty soil, and in some fine sandy loams, it spreads out into compact tufts covered with cream-coloured and yellow flowers, afterwards changing to a bay colour in the lower division. A variety with purple wings and calyx is not in cultivation, I believe. This plant was cultivated 200 years ago, at Oxford, but is now comparatively rare in our gardens. It succeeds best in a peat soil, loves moisture, and is admirably fitted for association with dwarf alpine shrubs on rockwork or in peat beds.

POLYGONUM VACCINIFOLIUM.—*Rock Knotweed.*

ALTHOUGH it comes of rather a poor and weedy race, this is a neat and ornamental trailing plant, scrambling freely over stones, and producing many bright-rose spikes of flowers in summer and autumn. It comes from elevations of from 11,000 to 13,000 feet on the Himalayas, which may perhaps have had much to do in refining its character and making it so unlike the Knotweeds that garnish the slime of our ditches and canals. Easily increased by division or cuttings, and thrives in common garden soil. Suited for banks, the fringes of shrubberies, rockwork, and the less important parts of the alpine garden.

POTENTILLA ALBA.—*White Cinquefoil.*

A PRETTY species, with the leaves in five stalkless leaflets, green and smooth above, and quite silvery, with dense silky down, on the lower sides. It is a very dwarf kind, neat and not rampant in habit, with white strawberry-like flowers, nearly an inch across, with a dark orange ring at the base. A native of the Alps and Pyrenees, of the easiest culture in ordinary soil, and a fitting ornament for borders or rockwork, flowering in early summer, and easily increased by division.

POTENTILLA ALPESTRIS.—*Alpine Cinquefoil.*

A RARE native plant, closely allied to the spring Potentilla (*P. verna*), but with flower-stems more erect, forming tufts nearly a foot high when well grown, the stalks of the leaves being nearly six inches long, so that the whole plant is much larger. Found on ledges of rock and elevated slopes in Scotland and Northern England, the root-leaves in five wedge-shaped divisions notched at the top, and of a bright shining green colour, the flowers of a bright yellow, about an inch across. Well worthy of a place on rockwork, it matters little how cold the position, and will enjoy a moist deep soil. *P. verna* is also worthy of a place in the garden, and is of the easiest culture. It is not a very common plant, but is found in a good many parts of the country on rocks and dry banks.

POTENTILLA CALABRA.—*Calabrian Cinquefoil.*

A VERY silvery species, particularly on the under sides of the leaves; the shoots prostrate, with lemon-yellow flowers about three-fourths of an inch across. This species is chiefly valuable from the hue of its leaves; it flowers in May and June, and flourishes freely in sandy soil. It is worthy of a place in the rock-garden and wherever dwarf Potentillas are grown. A native of Italy and Southern Europe.

POTENTILLA NITIDA.—*Shining Cinquefoil.*

A BEAUTIFUL little plant, only a couple of inches high, with silky-silvery leaves of three leaflets each, rarely more; the flowers of a pretty and delicate rose, the green sepals showing between the petals. This native of the Alps is well worthy of a place in the choice rock-garden, and is of the easiest culture and propagation.

POTENTILLA PYRENAICA.—*Pyrenean Cinquefoil.*

A DWARF but vigorous and showy species, with fine, large, deep golden-yellow flowers, the petals very round, full and over-lapping. A native of high valleys in the Central and Southern Pyrenees, easily increased by division or seeds. It will grow without any particular attention on rockwork or in the mixed border. It is known from other species by the stipules being adherent to the leaf-stem for nearly their whole length.

PRIMULA AMŒNA.—*Pleasing Primrose.*

AN uncommon and beautiful species, allied to our own wild Primrose, but quite distinct. The flowers are large, of a pleasing purple hue, and have a habit of coming out before the snow has left the ground—thus dethroning, in a way, the common Primrose. In leaf it is not unlike *P. denticulata*, and the fact that it possesses the vigour of that plant, and also has much larger flowers, makes it very welcome. It is so much earlier than the common Primrose that, while that species is in full flower, *amœna* has quite finished blooming, and sent up almost the same kind of strong tuft of leaves which the common Primrose does after its flowers are faded. I have seen it flourishing quite freely in common borders, on a chalky soil, and doubt not that it will prove one of the most valuable additions to the early spring garden and mixed border that has been made for many years. As the leaves are rather large, a sheltered and slightly shaded position will most tend to the perfect health and development of the plant. It is charming for rockwork or well-arranged borders, and, when plentiful enough, will, no doubt, be used in other ways. It is readily propagated by division of the root, and is a native of the Caucasus. The corolla is purplish lilac in bud, or when recently expanded, turning bluer after a few days. The umbel is many-flowered, the blooms larger than those of *P. denticulata*, borne about six or seven inches high ; the leaves woolly beneath and toothed. There is a stemless variety, which would probably prove a great addition to our gardens.

PRIMULA AURICULA.—*Common Auricula.*

THIS flower was quite a favourite in England in old times, and Parkinson, writing more than two hundred years ago, enumerates twenty-one varieties, and says there were many more ; and in 1792 the catalogue of Maddock, the florist, named nearly five hundred sorts. In our own time they have come to be almost forgotten, and are rarely seen except in the garden of an occasional enthusiastic florist. Of course I speak of the plants as florists' flowers ; the common kind has always been, and I trust will always remain, a popular cottage-garden and border flower. *Primula Auricula* lives in a wild state on the high mountain ranges of Switzerland, France, Austria, and the Caucasian chain, and has probably a much wider distribution. It is one of the many charming Primulas which rival the Gentians, Pinks, and Forget-me-nots, in making the Flora of alpine fields so exquisitely beautiful and interesting. Possessing a vigorous constitution, and sporting into a goodly number of varieties when raised from seed, it attracted early attention from lovers of flowers; its more striking variations were perpetuated and classified, and thus it became a "florists' flower." I do not desire to approach the subject from the florists' point of view, believing that to be a narrow and to some extent a base one; so much so, indeed, that I cannot regret that their practices and laws about the flower have taken but weakly root. There are many things with which man interests himself that require study and knowledge before they can be thoroughly enjoyed; but among them flowers should not and cannot be included. To lay down mechanical rules to guide our appreciation of flowers must for ever be the shallowest of vanities. To the eye direct they appeal with "most persuasive reasons," and rarely in vain; so that the regulations which lay down in what way their innumerable and inimitable dyes should arrange themselves are as needless as they are absurd. But, without seeking to conform or select them according to mechanical rules, we may preserve and enjoy all their most attractive deviations from the "original," or, more correctly, the best known wild forms of the species.

The varieties of cultivated Auriculas may be roughly thrown into two classes: 1st, self-coloured varieties, with the outer and larger portion of the flower of one colour or shaded, the centre or eye being white or yellow, and the flowers and other parts usually smooth and not powdery; 2nd, those with flowers and stems thickly covered with a white powdery matter or "paste." The handsomest of the not-powdery kinds, known by the name of "alpines," to distinguish them from the florists' varieties, are the hardiest of all. The florists' favourites are always readily distinguished by the dense mealy matter with which the parts of the flower are covered. They are divided by florists into four sections: green-edged, grey-edged, white-edged, and selfs. In the green-edged varieties, the gorge or throat of the flower is usually yellow or yellowish; then comes a ring, varying in width, of white powdery matter, surrounded by another of some dark colour, and beyond this a green edge, which is sometimes half an inch in width. The outer portion of the flower is really and palpably a monstrous development of the petal into a leaf-like substance, identical in texture with that of the leaves. The "grey-edged" have also the margin of a green leafy texture, but so thickly covered with powder that this is not distinctly seen. This, too, is the case with the "white-edged," the differences being in the thickness and hue of the "paste" or powdery matter. In fact, the terms green-edged, grey-edged, and white-edged, are simply used to express slight differences between flowers all having an abnormal development of the petals into leafy texture. It is a curious fact that between the white and the grey the line of demarcation is imaginary, and both these classes occasionally produce green-edged flowers. The "selfs" are really distinct, in having the outer and larger portion of the corolla of the ordinary texture, a ring of powdery matter surrounding the eye.

The enumeration and classification of such slight differences merely tend to throw obstacles in the way of the flower being generally grown and enjoyed in gardens. By all means let the florists maintain them, but those who merely want to embellish their gardens with some of the prettier varieties need not trouble themselves with named sorts at all. One fact concerning the florists' kinds should, however, be borne in mind—they are the most delicate and difficult to cultivate. The curious developments of powdery matter, green margins, &c. have a tendency to enfeeble the constitution of the plant. They are, in fact, variations that, occurring in nature, would have little or no chance of surviving in the struggle for life. The general grower will do well to select the free sorts—alpines, as they are called—or even good varieties of the common border kinds. An especial merit of these is that they may be grown in the open air on rockwork and borders, while the florists' kinds must be grown in frames.

At the risk of earning the contempt of the florist, I must first allude to the culture of the free-growing kinds, those most likely to be enjoyed in all classes of gardens. It is very simple : light vegetable soil and plenty of moisture during the growing season being the essentials. In many districts the moisture of our climate suits the Auricula to perfection, and in such may be seen great tufts of it grown in gardens without any attention. In others it must be protected against excessive drought by putting stones round the plants, and cocoa-fibre and leaf-mould are also useful as a surfacing. However, none but good varieties of the "alpine" section would justify even this trouble ; and, wherever practicable, we should prefer to place these on rockwork, on spots where they could root freely into rich light soil, and have some shelter. They would cause little or no trouble, save taking up, dividing, and replanting every second or third year, or as often as they become too crowded or lanky. The very common kinds may be planted as edgings, or in beds in the spring garden. Some have used them as edging plants for the sake of their whitish leaves alone. In all places where the plant naturally does freely, improved varieties should be substituted for the common old border kind. The Auricula does not do so well as could be desired around London—in the northern suburbs, at all events; and this is to be attributed more to the soil than the air perhaps, as I have seen most vigorous examples in deep areas in London, the surface of the pots being piled up with tea-leaves.

Auriculas are easily propagated by division in spring or autumn—best in early autumn. They are also easily raised from seed, which ripens in July, the common practice being to sow it in the following January in a gentle heat. It should be sown in pans thinly. The plants need not be disturbed till they are big enough to prick into a bed of fine rich and light soil, on a half-shady border in the open air. It is a most desirable practice to raise seedlings, as in this way we may obtain many beautiful varieties. When a desirable variety is noticed among the seedlings, it should be marked and placed under conditions best calculated to ensure its health and rapid increase, and propagated by division as fast as possible.

As to the florists' varieties, endless precise descriptions of the modes of culture considered necessary to success with these have been given by good cultivators; but the essential points may be summed up in a few words. They require protection in frames or pits during the winter and spring months; and they may be placed in the open air in summer and early autumn. Their suitable winter quarters are shallow pits, in which they are placed as near to the light as may be convenient, the lights being left off in mild weather, and air being given at all times, except in severe frosts. Air at night as well as by day is decidedly beneficial. The aspect of the pit or frame may be the usual one for the winter months; but as soon as the plants begin to show flower, they ought to be removed to one with a northern exposure, so that the bloom may be prolonged. Here, with abundance of air, they form objects of much interest and beauty through the month of April and the first weeks of May. After flowering they should be potted in May, and kept shaded till they have recovered. The potting is usually a process of carefully shaking away all the soil, and putting the plant in fresh compost, and the practice is well-founded, for it is the habit of this plant and its wild allies to put forth young roots higher up on the stem every year, and the encouragement of these young roots is sure to lead to a good result. Four-inch pots are generally used, and are quite large enough where the annual disrooting system is practised, one sucker of a kind being placed in the centre of each pot. I, however, doubt the wisdom of applying this system of potting to every plant, and should select those that had sound roots, and were set firmly and low in the earth, and, disturbing the ball but little, give them a careful shift into a five-inch pot. In case of growing the alpine kinds in pots—and they are quite as well worthy of it as the others—instead of confining ourselves to one plant in a small pot, we should put five or six of a kind in a six-inch pot, one in the centre and four or five round the side, thus forming a handsome specimen. Or the same principle might be carried out in pans, and with the free-growing florists' varieties as well as the alpines. In summer all the plants should be placed in the open air, and on boards, slates, a bed of coal-ashes, or some substance that can prevent the entrance of worms into the pots. Some careful growers guard the plants from heavy rains; this is unnecessary if the pots are perfectly drained and everything else as it ought to be. The florists rarely plunge them; but, if plunged in a bed of clean sharp sand, or any like material placed on a well-drained bottom, and free from the earthworm, they would be in a safer and certainly less troublesome condition, because free from the vicissitudes that must attend all plants exposed in a fragile porous shell containing but a few inches of soil.

Prescriptions for the ailing human creature were never measured out with more care than the mixtures of soil recommended by some authors for Auriculas; and prescriptions never did less good than these, or were less founded on philosophical principles. The perfect development of the choicest florists' kinds is secured by a simple mixture of one part good turfy loam, one part leaf-mould, and another composed of well-decayed cow manure and silver or sharp river sand. It should be observed that some pot their plants in August, but just after the flowering is the best time, as, if disrooted in the autumn, the plants have not that accumulated strength for flowering which they possess when the blooming time is preceded by a long period of undisturbed growth.

PRIMULA CORTUSOIDES.—*Cortusa-like Primrose.*

THIS is entirely distinct in appearance from any of the species commonly grown, the leaves being comparatively large and soft, not nestling firmly on the ground like many of the European species, but elevated on stalks two to four inches in length ; the deep rosy clusters of flowers being produced on stalks from six to ten inches high. In consequence of its taller and freer habit the plant is liable to be much injured and disfigured if placed in an exposed spot or open border, therefore the first consideration should be to give it a sheltered position, in a sunny nook on rockwork, surrounded by low shrubs, &c., or in any position where it will not be exposed to cutting winds, and at the same time not be shaded to its injury. The soil should be light and rich, and a surfacing of cocoa-fibre or leaf-mould would be beneficial in dry positions. It is one of the most beautiful and easily raised of the Primulas, being readily increased from seed, and quite hardy, at least where any pains are taken to give it a well-drained and suitable position. It forms a charming ornament for the lower and less exposed parts of rockwork, for a sunny sheltered border near a wall or house, or for the margin of the choice shrubbery. A native of Siberia.

Primula cortusoides amœna is considered a variety of the preceding, but, if so, it is a very distinct one, being much larger in all its parts than its relative, and bearing large trusses of beautiful, deep rose-coloured flowers. The cultural directions given for the preceding will apply to it. It is, however, so conspicuously beautiful that many people grow it in cold frames and pits, and show it among their choice greenhouse flowers. Handsome specimens may be grown in good loam and leaf-mould in six-inch pots. It came to us from Japan, but is suspected to be a native of Siberia. Mr. Wm. Thompson thinks that, "apart from the size of the flowers and the breadth of the foliage, the creeping root, the exclusively vernal habit of the plant, the pseudo-lobed or grooved seed-vessel, and the roundish flattened form of the seed, especially the two last features, warrant the belief in its distinctness from *P. cortusoides.*"

PRIMULA DENTICULATA.—*Denticulated Primrose.*

A HIMALAYAN species, with neat dense umbels of many small bright lilac flowers, on stalks from eight inches to a foot high, springing from oblong-lanceolate toothed leaves, hairy on both sides, but densely so beneath, by which it is distinguished from its relative, *P. erosa.* It is often grown in pots, but is perfectly hardy on rockwork in deep rich light loam with a dry bottom, selecting a position sheltered on the coldest sides by stones or other plants. It succeeds as a border-plant in some districts on well-drained light loams. When grown in pots, it may be brought into flower about six weeks earlier than it usually flowers (April and May), and this is the chief reason for growing it in pots, except where it cannot be grown in the open air. Propagated by division or by seeds. Although hardy, the leaves are injured by the first sharp frosts, so that it is well to keep it in well-drained warm positions. There is a handsome dwarf variety known as *P. denticulata nana.*

PRIMULA EROSA.—*Himalayan Primrose.*

AN interesting kind, occasionally grown about London in pots, and sometimes under the name of *P. Fortunei*, with shining leaves, quite smooth, and sometimes quite powdery, which, with its smoothness, distinguishes it at a glance from *P. denticulata.* The purplish blossoms with yellow eyes in flattish heads expand in early spring, and are borne on stems usually very mealy. Drs. Hooker and Thompson noticed it blooming at great elevations among the snow on the Himalayas, and, as might be expected from this, it is quite hardy in this country, and the true way to enjoy its beauty is to place it in a sunny but sheltered nook on rockwork, in deep sandy loam, lightened with peat and leaf-mould, and with the drainage perfect. It should never be allowed to suffer from drought in summer.

PRIMULA FARINOSA.—*Bird's-eye Primrose.*

SLENDER powdery stems, from three to twelve inches high, springing from rosettes of musk-scented leaves, with their undersides clothed with a silvery-looking meal, bear the graceful lilac-purple flowers of the Bird's-eye Primula. No sweeter flower holds its head up to kiss the breeze that rustles over the elevated bogs and mountain pastures of Northern England, and it should be doubly interesting to the British cultivator because it is a true native, found in Lancashire, North Yorkshire, Durham, Northumberland, Westmoreland, and Cumberland, and in Scotland, at Bridgehouse near West Linton, on the south side of the Pentland Hills. It is, however, a local plant. To find it profusely inlaid over moist parts of the great hill-sides on an early summer morning as one ascends the Helvellyn range for the first time, is to a lover of our wild flowers a great pleasure, and one that will be long remembered. I have mostly seen it in very moist spots where running water spreads out all over the surface, still, however, continuing to flow; but it is also found under somewhat different conditions. I have seen it attain a height of nearly eighteen inches in a cultivated state, and it is also occasionally seen very tall and strong in a wild one, though generally it is little over half a foot. In our gardens, as in Nature's gardens, it loves a moist vegetable soil. On the west coast, and in moist and elevated parts of the country, I have seen it flourishing on rockwork and slightly elevated beds without any attention; but in most parts of the country a little care will be necessary to ensure its perfect health. On rockwork a moist, deep, and well-drained crevice, filled with peaty soil or fibry sandy loam, will suit it to perfection. It is easy to cultivate in pots, the chief want, whether in pots or in the open, being abundance of water in summer, and where this does not fall naturally, it ought to be supplied artificially. When planted on rockwork in the drier districts, it would be well to cover the soil with cocoa-fibre or leaf-mould, which would protect the surface from being baked and from excessive evaporation; broken bits of sandstone would also do. It varies a little in the colour of the flower, there being pink, rose, and deep crimson shades.

P. farinosa acaulis is a very diminutive variety of the preceding. The flowers are not freely upheld on stems like those of the common wild form, but nestle down in the very hearts of the leaves, and both flowers and leaves being very small, when a number of plants are grown together on one sod, or in one pan, they form a singularly charming little cushion of leaves and flowers not more than half an inch high. The same positions will suit it as have been recommended for the Bird's-eye Primula, but being so very diminutive, it ought to have more care, whether grown on rockwork or in pots. If any weeds or coarse plants were allowed to vegetate over or near it, it would of course suffer.

PRIMULA GLUTINOSA.—*Glutinous Primrose.*

A DISTINCT little Primrose, rare in gardens, and growing abundantly in peaty soil at elevations of 7000 or 8000 feet on mountains near Gastein and Salzburg, in the Tyrol, and in Lower Austria. The leaves are nearly strap-shaped, but widening towards the top, where they are somewhat pointed and neatly and regularly toothed. The stem is as long again as the leaves, growing from three to five inches high, bearing from one to five blossoms, which have been described to me by those who have seen it wild in quantity as of a peculiar purplish mauve, with the divisions rather deeply cleft. Suitable for rockwork or pots in moist peaty or very sandy soil.

PRIMULA INTEGRIFOLIA.—*Entire-leaved Primrose.*

A MOST diminutive Primrose, easily recognised by its smooth, shining, entire, ciliated leaves, lying quite close to the ground, and in spring, when in bloom, by its handsome rose flowers, with the lobes deeply divided, one to three flowers being borne on a dwarf stem, which elevates them but little above the leaves, and these flowers are often large enough to obscure the plant that bears them. It is very common on the higher parts of the Pyrenees, and I met with it in abundance in elevated pastures in North Italy. Scores of plants sometimes grew together in a sod, like Daisies, wherever there was a little bank or slope not covered by grass; and it was also plentiful in the grass, growing in a sandy loam. There should be no difficulty in growing this plant on flat exposed parts of rockworks, the soil moist and free, but firm. The best way would be to try and form a wide tuft of it, by dotting from six to a dozen plants over one spot, and, if in a dry district, scattering a little cocoa-fibre mixed with sand between them, to prevent evaporation. This, or stones, will help till the plants become established. It flowers in early summer, and is increased by division and by seeds. *P. Candolleana* is another name for this, and *P. glaucescens* is at best but a variety of it.

PRIMULA LATIFOLIA.—*Broad-leaved Primrose.*

A VERY handsome and fragrant Primrose, with from two to twenty violet flowers in a head, borne on a stem about twice as long as the leaves, or sometimes more; the leaves slightly and distantly toothed, both the surfaces and the margins being covered with glandular hairs. This is less viscid, larger, and more robust than the better known *P. viscosa* of the Alps, the leaves sometimes attaining a length of four inches and a breadth of nearly two. It grows to a height of from four to eight inches, flowers in early summer, comes from the Pyrenees, the Alps of Dauphiny, and various mountain chains in Southern Europe, and in a pure air will thrive on sunny slopes of rockwork in sandy peat, with plenty of moisture during the dry season, and perfect drainage in winter. It, like *P. viscosa*, will bear frequent division; and it may also be well and easily grown in cold frames or pits.

PRIMULA LONGIFLORA.—*Long-flowered Primrose.*

RELATED to our Bird's-eye Primrose, but distinct from it, and considerably larger than those of the best varieties of that species; the lilac tube of the flower being more than an inch long, which immediately distinguishes it. It is not at all difficult to cultivate on rockwork or in pots, and the treatment and position recommended for *Primula farinosa* will suit it perfectly. In colour it is deeper than the Bird's-eye Primrose, and is a native of Austria.

PRIMULA MARGINATA.—*Margined Primrose.*

ONE of the most attractive of the family, and readily distinguished by the silvery margin on its greyish, toothed, smooth leaves, caused by a dense bed of white dust which lies exactly on the edge of the leaf; and by its sweet, soft, violet-rose flowers, appearing in April and May. I have grown this plant successfully in the open air in London, and in parts of the country favourable to alpine plants it will prove almost as free as the common Auricula. Even when not in flower, the plant is pleasing, from the tone of the margin and surfaces of the leaves. Our wet and green winters are doubtless the cause of this and other kinds becoming rather lanky in the stems after being more than a year or so in one spot. When the stems become long, and emit roots above the surface, it is a good plan to divide the plants, and insert each portion firmly down to the leaves. This will be all the more beneficial in dry districts, where the little roots that issue from the stems would be more likely to perish. It is a charming ornament for rockwork, where it thrives freely, and a first-class subject for the front margin of the select mixed border. In the open ground a few bits of broken rock, placed around each plant, or amongst the plants if they are planted in groups or tufts, will do good by preventing evaporation, and also acting as a protection to the plant, which rarely exceeds a few inches in height. It may also be grown easily in pots; and, where plentiful, it is much better to put one plant in the centre, and half a dozen round the side, of a six-inch pot than depend on one tuft in the centre. A native of the Alps of Tauria and Dauphiny, and various ranges in the South of Europe, but not of the Pyrenees. Readily increased by division.

PRIMULA MINIMA.—*Fairy Primrose.*

FOR its size a singularly ornamental kind, with very small leaves, prostrate, and rather deeply notched at the ends; but the flowers make up for the very diminutive leaves, being not unfrequently nearly an inch across, and quite covering the minute rosette from which they spring. It is a native of the Alps of Austria, and flowers in early summer, the stem rarely bearing more than one, but occasionally two flowers, rose-coloured, or sometimes white. Bare spots in firm open parts of the rockwork are the best places for it, the soil to be very sandy peat and loam; it is peculiarly suited for association with the very dwarfest and choicest alpine plants. It may be propagated by division or by seed, and comes from the mountains of Southern Europe.

P. Floerkiana is very like the Fairy Primrose, probably only a variety of it, and in the flowers only differing by bearing two, three, or more, instead of a single bloom. There is also a difference in the leaves, which in *P. minima* are nearly square at the ends, but in *P. Floerkiana* are roundish there, and notched for a short distance down the sides. It is a native of Austria, and will be found to enjoy the same conditions and positions on rockwork as the preceding. Of both it is desirable to establish wide-spreading patches on firm bare spots, scattering half an inch of silver-sand between the plants to keep the ground cool.

PRIMULA MUNROI.—*Munro's Primrose.*

This has not the brilliancy or dwarfness of the Primulas of the high Alps, nor the vigour of our own wild kinds, but it has the merit of distinctness, and is of the easiest culture in any moist boggy soil. It grows at very high elevations on the mountains of Northern India, in the vicinity of water, and bears creamy-white flowers, with a yellowish eye, more than an inch across on stems five to seven inches high, springing from smooth green leaves a couple of inches long, nearly as much across, and with a heart-shaped base. The flowers are very sweet, and altogether it highly merits culture in a moist spot on the select rockwork, or in a bog, and flowers from March to May. *P. involucrata* is a very closely allied kind, from the same regions, somewhat smaller, the leaves not heart-shaped at the base, and the plant not quite so ornamental. It thrives under the same conditions as its relative.

PRIMULA NIVEA.—*Snowy Primrose.*

A dwarf and neat species, freely producing trusses of lovely white flowers, quite distinct in aspect from any other in cultivation, happily very easy of culture, and may be grown in pots or in the open ground. If in pots, it should be frequently divided; for it has a tendency, in common with other choice Primulas, to get somewhat naked about the base of the shoots, and, as these protrude rootlets, the whole plant is likely to go off if not taken up and divided into as many pieces as possible. Every shoot will form a plant, inasmuch as each is usually furnished with little rootlets, which take hold of fresh soil immediately. Many people keep plants of Primulas like this for years in the centre of the same pot, whereas by dividing them and placing them down to the leaves in fresh soil much finer specimens may be obtained. In a wild state the natural moisture, and perhaps the accumulating débris of the mountains, enable them to use those exposed rootlets and thrive; but in cultivation I have found it an excellent plan to divide such fine Primulas as this, and plant them down to the leaves when their stems have grown at all above the soil; and I have no doubt that careful annual division would suit them well. On dry ground, it would have little or no chance unless surrounded by a few stones, which would prevent evaporation, and at the same time save it from being lost among coarser things. The ground would also be the better of being covered with an inch or so of cocoa-fibre. In moist and elevated regions there would be less trouble, but in all care should be taken to give the snowy Primrose what it deserves—a select position on rockwork or in the border, a light free soil, and plenty of water during the warm season. It flowers in April and May, is a native of the Alps, and is by some supposed to be a variety of *P. viscosa*.

PRIMULA OFFICINALIS.—*Cowslip*.

THE Cowslip, that familiar inhabitant of our meadows and pastures, would be well worthy of introduction to our gardens were it not a common native; but the many handsome kinds that have sprung from it are far more valuable from a gardening point of view than the original form. Many of the plants passing under the name of Polyanthuses may often prove hybrids of parentage difficult to identify, and some botanists consider the Primrose, Cowslip, and Oxlip as varieties of one species, *P. veris* of Linnæus, who also entertained this opinion. Mr. Darwin, however, has recently proved that they have as good right to be called distinct species as any other subjects held to be distinct. He concludes that, "although we may feel confident that *Primula veris, vulgaris,* and *elatior,* as well as the other species of the genus, are all descended from some primordial form, yet, from the facts which have been given, we may conclude that they are now as fixed in character as are very many other forms which are universally ranked as species. Consequently, they have as good a right to receive distinct specific names as have, for instance, the ass, quagga, and zebra." He has proved the Polyanthus to be a variety of the Cowslip, having raised a great number of crosses between it and the Cowslip, and found the mongrels perfectly fertile.

Polyanthuses present a wonderful array of beauty, and are not at all sufficiently appreciated. For rich and charmingly inlaid colouring they surpass all other flowers of our gardens in spring. It would require pages to describe even the good varieties. At one time the Polyanthus was highly esteemed as a florists' flower, and none in existence better deserved the attention and esteem of amateurs; but nearly all the choice old kinds are now lost, and florists who really pay the flower any attention are very scarce indeed. In consequence, however, of the great facility with which varieties are raised from seed, nobody need be without handsome kinds, and raising them will prove interesting amusement for the amateur. The laws of the florists are in this case of little more value than usual, but I quote Maddock, because in the following passage he is describing a very beautiful type of the numerous variations of this flower : " The ground colour is most to be admired when shaded with dark rich crimson resembling velvet, with one mark or stripe in the centre of each division of the limb, bold and distinct from the edging down to the eye, where it should terminate in a fine point." He further says : " The pips should be large, quite flat, and as round as may be consistent with their peculiarly beautiful figure, which is circular, excepting those small indentures between each division of the limb which divide it into five or six heart-like segments. The edging should resemble a bright gold lace, bold, clear, and distinct, and so nearly of the same colour as the eye and stripes as scarcely to be distinguished. In short, the Polyanthus should possess a graceful elegance of form, a richness of colouring, and symmetry of parts not to be found united in any other flower." Here, however, as in most similar cases, the general cultivator will do well to select the most diverse of the varieties that he may raise from seed or otherwise become possessed of, and not be tied by any conventional rules.

As to the capabilities of the various kinds of Polyanthus, it would be difficult to name a hardy flower so generally useful. The finer varieties are worthy of a place on rockwork amidst the choicest alpine plants ; the showier ones do for bedding in the spring garden. Numbers of the vigorous varieties so easily raised from seed will form the most appropriate ornaments that can be massed alongside of shady walks in pleasure-grounds. Some may be employed as edgings. Many varieties are worthy of being abundantly naturalised in pleasure-grounds and along wood walks ; and, as we all know, the enthusiastic florist grows the finer ones in pots. They are perhaps scarcely so much to be recommended for using in masses in the spring garden as are the finer varieties of the Primrose—requiring, in fact, to be seen rather closely to be admired ; but wherever flowers are placed for their individual beauty rather than for their mere effect as colouring agents, they are invaluable, and should be seen in strong tufts round every shrubbery border.

The cultivation of Polyanthuses is, happily, almost as simple as that of meadow grass. They grow vigorously in almost any kind of garden soil, best in that which is somewhat rich and moist. They thrive in the full sun, but enjoy best a partially shaded and sheltered position, and are somewhat impatient of heat and drought. When grown for bedding purposes, they are, like the Primroses, &c., removed from the flower-garden to the kitchen-garden or nursery in early summer, and again conveyed there when the summer bedding plants have passed away. Some varieties, a good deal larger in their parts than the common type, have been raised of late, and these are also very easy of culture and very vigorous; and there are none, or very few, double varieties, as in the case of the Primrose. There are, however, some which are curious and interesting from the duplication of the calyx or corolla, and these are popularly known as "hose-in-hose" Polyanthuses; they grow with the same facility as the others. Where soil is prepared for the choicer varieties, any good loam with a free addition of decomposed leaf-mould and decomposed cow-manure and sand will form an admirable compost.

The Polyanthus may be raised with great facility from seed, which should be sown immediately after it is gathered from the plants, say, about the end of June. It will grow with vigour if kept till the following spring, but by sowing it immediately nearly a year is gained. The amateur wishing to raise choice kinds had better sow the seed in rough wooden boxes or in pans, but for all ordinary purposes a bed of finely pulverised soil in the open air will answer to perfection. Sowings in early spring are better made in rough shallow boxes or pans placed in cold frames, as time will be gained thereby; but the best plan is not to lose the time by allowing the seed to remain idle in the drawer all the autumn and winter, but to sow it directly it is ripe, and by doing so have strong plants the following spring.

The common Oxlip is an hybrid more or less intermediate between the Cowslip and the Primrose. It differs from the true or Bardfield Oxlip by having much larger and brighter-coloured flowers on longer footstalks, and by showing in the throat of the flower the five bosses characteristic of the Primrose and Cowslip. Some varieties approach the Cowslip and others the

Primrose in character. There are various forms, not usually so ornamental as the varieties of the Polyanthus, but more worthy of culture than the true Oxlip of the Eastern Counties. In a wild state the common Oxlips are rather rare, and, according to Mr. Syme, "generally occurring whenever the Primrose and the Cowslip grow together, but never found in districts inhabited by only one of the parents." When cultivated in gardens, the positions and treatment that suit the Polyanthuses and Primroses will also suit the Oxlip.

P. suaveolens of Bertolini is a variety of the Cowslip found in many parts of the Continent, and not sufficiently distinct or ornamental to merit cultivation.

P. elatior is the true as distinguished from the common Oxlip. It is not a very ornamental species, the flowers being of a pale buff yellow, and it is readily distinguished by its funnel- and not saucer-shaped corolla, which is also quite destitute of the bosses which are present in the Primrose and Cowslip. It is found in woods and meadows on clayey soils in the Eastern Counties of England, particularly in Essex, Suffolk, and Cambridgeshire. It is of easy culture, and most suitable for botanic gardens, or collections of interesting plants, being neither distinct nor ornamental enough for very limited collections of ornamental kinds only. This plant is also known by the name of the Bardfield Oxlip.

PRIMULA PALINURI.—*Large-leaved Primrose.*

THIS is quite removed from other cultivated Primroses, inasmuch as it seems to grow all to leaf and stem, whereas many of the other kinds often hide their leaves with flowers. It is at first sight more like a specimen of *Sempervivum arboreum* than a Primrose, the leaves, almost as strong as those of young cabbages, being toothed and springing from a thick stout stem, which generally rises sufficiently above the ground to be visible. In April the bright yellow flowers appear in a bunch at the top of a powdery stem, emit a cowslip-like perfume, and are sufficiently ornamental, though they rarely seem to fulfil the promise of the vigorous-looking plant. I have seen it flourish healthfully in rich light soil as a border-plant in various parts of these islands, so that nothing more need be said of its culture, and established plants are easily increased by division. It is well suited for some isolated nook on rockwork, where there is an unusually deep bed of soil. A native of Southern Italy.

PRIMULA PURPUREA.—*Purple Primrose.*

A HANDSOME Primrose, from elevations of 12,000 feet or more on the Himalayas, and allied to *P. denticulata*, though far finer: the flowers, of an exquisite purple, are larger, in heads about three inches across, and the leaves entire, by which it may be distinguished from its near relations. Sheltered and warm positions, but not very shady, on rockwork, or in the open parts of the hardy fernery, will best suit it, the soil being a light deep sandy loam, well enriched with decomposed leaf-mould. I have never seen it thrive so well as when planted in nooks at the base of rocks which sheltered it, where it enjoyed more heat than if exposed; but probably it will be found to thrive under ordinary circumstances when plentiful enough for trial in various positions. At present it is scarce.

PRIMULA SCOTICA.—*Scotch Bird's-eye Primrose.*

THIS, one of the most lovely of its family and of the choicest little gems in the British Flora, is a near ally of the Bird's-eye Primrose of the moist and boggy mountain sides of the North of England. Its rich purple flowers, with large yellowish eye, open in the end of April, supported on stems from half an inch to an inch high, growing an inch or two taller as the season advances. It is said by some botanists to be simply a variety of the Bird's-eye Primrose, but the seedlings show no tendency to approach the larger and looser *P. farinosa*, and Mr. Syme, who has carefully observed the living plant both in a wild state and cultivated in his own garden, declares it to be "perfectly distinct." The leaves are very powdery on the under side, broadest near the middle, shorter, and less indented than those of *P. farinosa*, which are broadest near the end; and the whole plant is about large enough to associate with a dwarf moss or lichen. It is rather difficult to obtain, unless one has an opportunity of getting it from its native localities in Scotland; but it can be had from several English and Scotch nurserymen who cultivate such subjects. A native of the counties of Sutherland and Caithness, and of the Orkney Isles, growing in damp pastures. The best place to select for its cultivation is on a properly made rockwork in some spot where it would have perfect drainage, and not be injured by strong-growing subjects shading it. The soil should be a friable loam, mixed with sandy peat or a little cocoa-fibre, and made perfectly firm. If placed on the level ground or on a raised border, a few pieces of broken porous rock should be placed firmly in the ground around it, so as to show half their size above the surface, prevent evaporation, and also act as a guard to the very diminutive plant; and the same plan might be followed to some extent on a rockwork. If a coating of dwarf moss is spread over the earth after a time, I should not remove it, believing the tiny plant to enjoy such a carpet, whether grown in pots or the open air. Although so small, it is, when in health, a vigorous Lilliputian, and seeds very freely, the self-sown seedlings having often formed with me good plants on the mossy surface of the ground or pots. I have grown it in the open air in the suburbs of London; but as a rule it is best for all who do not try it in a pure atmosphere to grow it in well-drained pots or pans, using the same kind of soil, and protecting the plants in a cool shallow frame in winter, placing the pots out of doors in summer plunged in coal-ashes or sand. In all cases the plant should be abundantly watered in dry weather, whether in spring, summer, or autumn. Easily propagated by seeds, which should be sown soon after they are ripe in shallow pans of sandy peat or fibrous loam mixed with cocoa-fibre, and placed in an open pit or shallow cold frame.

PRIMULA SIKKIMENSIS.—*Sikkim Cowslip.*

This, one of the most remarkable of Primroses, has a leaf-development not unlike that of *P. denticulata*, but when well grown, it throws up strong flower-stems from fifteen inches to two feet high, bearing numerous bell-shaped, pale-yellow flowers, without a spot of any other colour, the pedicel mealy, the blooms of an agreeable and peculiar perfume. Some of the stems bear a head of more than five dozen buds and flowers, and each individual flower is nearly an inch long and more than half an inch across. It is perfectly hardy, and loves deep well-drained and moist ground; spots in the lower parts of rockwork near water, or in deep boggy places, suit it best; begins to flower in May, and remains in flower for many weeks. It is a plant which, when plentiful, ought to be popular everywhere. It is said to be the pride of all the Primroses of the mountains of India, inhabiting wet boggy localities, at elevations of from 12,000 to 17,000 feet, and covering acres of ground with its yellow flowers. Must be propagated by division, as it rarely or never matures its seeds in this country.

PRIMULA STUARTII.—*Stuart's Primrose.*

A NOBLE and vigorous yellow Primrose, a native of the mountains of Northern India, to some parts of which, according to Royle, it gives a rich yellow glow. It grows about sixteen inches high, has leaves nearly a foot long, lanceolate, mealy below, smooth above, and sharply serrated; the umbels being many-flowered. Like *P. denticulata* and the purple Primrose, the place most suitable for this is some perfectly drained and sheltered spot on slightly elevated rockwork; if convenient, plant it against the base of rocks, which will shelter it from cutting winds, though, when sufficiently plentiful, this precaution may be dispensed with. A light deep soil, never allowed to get dry or arid in summer, will suit it well.

PRIMULA VISCOSA.—*Viscid Primrose.*

THIS is the lovely little Primrose that travellers who visit the Alps in early summer see opening its clear rosy-purple flowers with white eyes at various altitudes: sometimes, in crossing a high pass, it comes into view, plant, flower and all, not bigger than a shilling, but still bravely flowering—indeed, nearly all flower; while on sunny slopes and in the valleys it may be seen nearly as large as the Auricula. It is known by its dark-green obovate or suborbicular leaves with close-set teeth, covered with glandular hairs, and viscid on both sides; the flower-stems, which elevate the sweet blooms barely above the foliage, being also viscid. It is well adapted for rockwork, on which it may be grown in any position in light peaty or spongy loam, with about one-half its bulk of fine sand, provided its roots are kept moist during the dry season. A native of the Alps and Pyrenees; easily increased by division, and may also be raised from seed. Varieties are sometimes found with white flowers, but rarely. It is sometimes grown under the name of *P. villosa*. The handsome purple Primroses known in gardens under the name of *P. ciliata* and *P. ciliata purpurea* are varieties of this, the last said to be a hybrid between it and an Auricula.

PRIMULA VULGARIS.—*Common Primrose.*

The Gentians and dwarf Primulas do not do more for the Alps than this charming wilding for the hedge-banks, groves, open woods, and borders of fields and streams of the British Isles. No need to say anything of its appearance, which must be familiar to every person who has seen a wild flower in a country lane. In some places it varies a good deal in colour, and some of the prettiest of the wild varieties are worthy of being introduced into shrubberies and semi-wild places, and also into gardens. Although it does not vary so much as the Cowslip, yet it does so in a degree that ought to make it much more valuable in the hands of flower-gardeners than it is at present. All the varieties would perhaps find their most appropriate home in our woods and shrubberies; but so long as rich and lovely colour and fragrance are esteemed in the flower-garden in spring, some of the more distinctly toned varieties should be sought after. Varied hues of yellow, red, rose, lilac, bluish-violet, lilac-rose, and white, have already been raised, and no doubt many others will be raised in future, particularly if the good single varieties should become popular plants for the spring flower garden. Striking and desirable variations from the commoner types will then be much more likely to be preserved. For bedding purposes these single varieties will always prove more useful and effective than the old double kinds, because more vigorous and easily increased. All the varieties are readily increased by division of the offsets, or by seeds, which are produced in abundance. Planted in woods and shrubberies, the plants will take care of themselves—a quality that adds to their charms.

When raised for flower-gardening purposes, some system of culture must be pursued; a very simple one will secure the best results both as to the production of vigorous free-blooming plants and an abundant stock. After the summer occupants of the flower-beds are faded and removed in autumn, the Primroses and other spring flowers are planted in the beds as the taste of the grower may direct. About the middle or the end of May it will be time to think of preparing the beds for their summer ornaments, and by that time the Primroses will have begun to fade after furnishing a long and abundant bloom. Then take them up, divide the offsets singly, doing this in a shed or shady position if the day be sunny. Tufts of new or scarce varieties, or those of which a large stock is required, may be divided into the smallest offsets; where much increase is not required, the plants should simply be parted sufficiently to allow of their healthy development. As soon as they are parted, they should be planted in the kitchen-garden or some by-place. The richer and moister the soil is the better they will grow, and if the position be a half-shady one, it will be an improvement. The alleys between the Asparagus-beds would do admirably for them, in case room was scarce, and more convenient positions could not be found. It would be desirable to shade them for a few days after planting, if the weather be very bright, and simply by spreading boughs or old garden mats over them; and it need hardly be added they should be thoroughly watered soon after planting. They should be planted in lines at ten or twelve inches apart each way if the plants be strong and regular in their development, and, if small offsets, closer in the lines in proportion to their size. By autumn they will make fine plants, and may then be taken up, preserving as much of the root as will come up with ordinary care, but not of necessity any soil or ball, and transferring them to the beds in the flower-garden or pleasure-ground.

However, the forms of the plant most precious for the garden are the beautiful old double kinds. No sweeter or prettier flowers ever warmed into beauty under a northern sun than their richly and delicately tinted little rosettes. Once they were grown in every garden; then the day came when they, in common with many hardy flowers, were cast aside to make way for gaudier things; but now people are beginning to grow them again, and are enquiring where they can get some old and half-lost kinds they used to know long ago. The best known and most distinctly marked kinds are the double lilac, double purple, double sulphur, double white, double crimson, and double red. These and several other closely allied forms are occasionally honoured with Latin names descriptive of their shades of colour. In catalogues of the present day I find the following:—*Primula vulgaris alba plena, lilacina plena, purpurea plena, rosea plena, rubra plena, sulpharea plena;* but we had better speak of them in plain English and confine the Latin term to the species.

The double kinds, more delicate and slower-growing than the single ones, require more care, and in their case the development of healthy foliage after the flowering season should be the object of those who wish to succeed with them. Shelter and partial shade are the two conditions chiefly necessary to secure this. Open woods, copses, and half-shady places are the favourite haunts of the Primrose in a wild state. In them, in addition to the shade, it enjoys shelter not merely from tall objects around, but also from the long grass and other herbaceous plants growing in close proximity; and we should also take into account the moisture consequent upon such companionship, and let these facts guide us in the culture of the double kinds. As will be readily seen, a plant exposed to the full sun on a naked border would be under a very different condition to one in a thin wood; the excessive evaporation and searing away of the leaves by the wind would be quite sufficient to account for the failure of the exposed plant. It is therefore desirable, in the case of the beautiful double Primroses, to plant them in borders in slightly shaded and sheltered positions, using light rich vegetable soil, and, if convenient, keeping the earth from being too rapidly dried up by spreading cocoa-fibre or leaf-mould on it in summer. It would be better to plant them permanently in some favourite spot and leave them alone than to change them repeatedly from place to place. They may, however, be employed as bedding plants, and successfully treated as recommended for the single varieties, but for this purpose are not so useful or pretty as when seen in good single tufts.

They are increased by division of the roots, and to take them up in order to divide these is the only disturbance they should suffer. The double Primroses well grown and the same kinds barely existing are such very different objects that nobody will begrudge giving them the trifling attention necessary to their perfect development. Occasionally they may be seen flourishing by chance in some cottage or old country garden, where they find a home more congenial than the prim and bare fashionable flower-garden of our own day.

In addition to the Primroses above described, *P. spectabilis*, a handsome kind with purplish flowers and smooth leaves; *P. carniolica; P. intermedia*, allied to *farinosa* and *longiflora; P. sibirica*, allied to *P. cortusoides; P. pedemontana* and *P. Dinyana*, are amongst the best of the sorts in cultivation, but most of them are very rare. The singular *P. verticillata*, of Arabia, covered with silvery farina and with yellow flowers in whorls, is not hardy. *P. Thomasini, macrocalyx, intricata, suaveolens, stricta, auriculata, Allioni, Ciusiana*, and *mistassinica*, are either not sufficiently distinct from other and superior sorts or are not ornamental enough to be recommended for general cultivation.

PRUNELLA GRANDIFLORA.—*Large Self-heal.*

A HANDSOME and vigorous plant, with rather soft, sparsely-toothed, ovate leaves, the lower ones about two inches and a half long, and with channeled footstalks longer than the leaves, the upper leaves smaller, and with the footstalks shorter than the leaves. It is, however, not likely to be mistaken for any other plant, and is readily distinguished by its large flowers from the common British Self-heal, which is unworthy of cultivation. There is a white as well as a purple variety, both handsome plants, that thrive in almost any ground, but prefer a moist and free soil and a position somewhat shaded. They are apt to go off in winter on the London clay, at least on the level ground. Well suited for the mixed border, for banks, or for naturalisation in copses, &c. A native of Continental Europe; flowering in summer.

PULMONARIA OFFICINALIS.—*Lungwort.*

A SOMEWHAT rare British plant, but often grown in cottage gardens, and, if not sufficiently ornamental for the rock-garden, very desirable for borders and spring gardens. It has coarsely hairy root-leaves, nearly oval in outline and acute at the apex, abundantly marked on the upper surface with white blotches; the flowers at first rose and then blue, on stems from six inches to a foot high, abundantly produced in spring. The blue-flowered *P. angustifolia* is by some botanists united with this species, and there is a white and a spotless variety in cultivation. The Lungwort and its varieties are very vigorous and hardy, thriving on any soil, and forming attractive beds in the spring garden. They are also worthy of being planted in semi-wild places, but do not deserve a place in the select rock-garden except where more valuable plants are very scarce.

PUSCHKINIA SCILLOIDES.—*Scilla-like P.*

A FASCINATING fairy plant, and the most delicately beautiful thing in the way of a spring bulb that I am familiar with; the flowers white, striped and tinged with a delicate blue, the small prostrate leaves concave; easily grown too—more so than the Squills; it does not last long in flower, but few spring flowers do. The best position for this is on low banks, in the rock-garden, or in positions where its delicate flowers may be seen somewhat beneath the eye, associated with dwarf Primulas, the Rush Daffodil, and other diminutive spring flowers. A native of the Caucasus, flowering in spring, easily increased by division of the root, and flourishing best in very sandy light soil.

PYROLA ROTUNDIFOLIA.—*Larger Wintergreen.*

A RARE native plant, inhabiting woods, shady, bushy, and reedy places; with leathery leaves, roundish or broadly oval, very slightly toothed, and erect stems, bearing long and handsome slightly-drooping racemes of pure white fragrant flowers, half an inch across, ten to twenty flowers being borne on a stem from six inches to a foot high. *Pyrola rotundifolia,* var. *arenaria,* is another very graceful plant, found wild on sandy sea-shores, and differing from the preceding in being dwarfer, deep green, and smooth, and generally with several empty bracts below the inflorescence. Both are beautiful plants for shady mossy flanks of rockwork, in free sandy and vegetable soil, and flourish more readily in cultivation than any species of their family. In America there are varieties of this plant with flesh-coloured and reddish flowers, none of which are in cultivation with us.

Pyrola uniflora, media, minor, and *secunda,* are also interesting British plants, of which the first, a very rare one in our Flora, is the most ornamental. *P. elliptica,* a native of North America, is also in our gardens, though rare. Any of these plants that can be obtained are worthy of a place in thin mossy copses on light sandy vegetable soil, or in moist and half-shady parts of the rockwork or fernery.

PYXIDANTHERA BARBULATA.—*Bearded P.*

A CURIOUS and minute American plant, plentiful in sandy dry "pine barrens" from New Jersey to North Carolina. It is an evergreen shrub, yet smaller than many mosses; the leaves narrow, awl-pointed, and densely crowded, bearded at the base; the flowers are placed singly, and are stalkless, but very numerous, rose-coloured in bud, white when open. The effect of the rosy buds and five-cleft white flowers on the dense dwarf cushions is singularly pretty. Generally found in low, but not wet, places, and usually on little mounds, it is a gem for the rockwork, on which it should be planted in pure sand and vegetable mould, and fully exposed to the sun. Flowers in early summer; increased by division, and is as yet very scarce.

RAMONDIA PYRENAICA.—*Pyrenean R.*

AN interesting, distinct, and rather ornamental Pyrenean plant, found on steep and somewhat shady rocks, and, according to Ramond, exclusively in valleys leading from north to south; having leaves in rosettes spreading very close to the ground, blistered, deeply wrinkled, and densely covered with short hairs —quite shaggy beneath and on the leaf-stalk. The flowers are borne on stems two to six inches long, from an inch to an inch and a half across, purple-violet, with orange-yellow centre. It does very well on rockwork in mossy fissures filled with well-drained peaty earth, and is easily grown in cold frames in well-drained pots, well watered during the warm months. Flowers in spring and early summer. Increased from seed with the greatest facility, and well worthy of general culture.

RANUNCULUS ACONITIFOLIUS.—*Fair Maids of France.*

THIS white-flowered Crowfoot, which grows from eight inches to a yard high in moist parts of valleys and woods in the Alps and Pyrenees, is too large and not sufficiently ornamental for cultivation in the rock-garden; but its double variety (*R. aconitifolius fl. pl.*) is a beautiful old border-flower, now rarely seen. The flowers are not large, but are so white and neat and pretty and double that they resemble a miniature double white Camellia, and are useful for cutting, apart from their appearance in the border. A rich light soil and partial shade will be found to suit it best.

Well worthy of a place on the shady side of rockwork, in the open parts of a rocky fernery, or on half-shady borders.

RANUNCULUS ALPESTRIS.—*Alp Crowfoot.*

A VERY pretty and diminutive species, growing from one to three or four inches high, and, when well grown, forming neat tufts, each stem bearing from one to three pure white flowers, which open in April. The leaves are of a dark glossy green, roundish-heart-shaped in outline, and deeply divided; the roots fibrous, numerous, and white. A native of most of the great mountain ranges of Europe, in moist rocky places on the higher pastures, and one of the most desirable alpine plants for the rock-garden. It is not difficult to grow in moist, sandy, or gritty soil, in positions thoroughly exposed to the sun, and abundantly supplied with moisture in summer.

R. Traunfellneri seems to be a diminutive of the preceding; the whole plant, even as I have observed it in cultivation, being not more than one inch high. The same treatment will suit it as the preceding; but, being smaller, it will require a little more care in selecting some firm spot fully exposed to the sun and air, but kept moist and firm with a surfacing of moist grit, sand, or small stones, till the plant is established into a little spreading tuft.

RANUNCULUS AMPLEXICAULIS.—*Stem-clasping Crowfoot.*

A BEAUTIFUL species, with large white flowers having yellow centres, one to five flowers being borne on each of the stems, which are clasped by smooth, gradually pointed, and sea-green leaves, while oval-pointed leaves spring from the root, and set off, so to speak, the snowy bouquet of flowers which adorns the plant when well grown. I know no more graceful plant for the rock-garden or the border—an alpine Buttercup, with the purity of flower of a Snowdrop, its leaves not cut into raggedness like those of many of its fellows, but as tenderly veined and carved as if its natural home was some warm eastern isle. A native of the Alps, Pyrenees, and other mountain ranges, usually growing about six inches high, flowering in gardens in April or May, and increased by seed or division. I have seen the plant thriving in the London clay in Regent's Park, so there should be little difficulty about its cultivation. It is worthy of the best positions on rockwork or in borders, and is also an elegant subject for growing in pots in cold frames.

RANUNCULUS GLACIALIS.—*Glacier Buttercup*.

A WELL-NAMED plant, as it is an inhabitant of very high places on the Alps, and may be often seen in flower near the snow; indeed, I have scraped away the snow in quantities to get at it. The flowers are rather large, and white-tinted, of a dull purplish rose on the outside; the calyx soft, with shaggy brownish hairs; the leaves smooth, deeply cut, and of a dark brownish green. From one to five flowers are borne on the stems, which in a wild state are rather prostrate, and seldom more than six inches long. This very interesting, as well as ornamental kind, will thrive best in a cool position in deep, gritty, peaty soil, with abundance of water during the warm months. I have seen it growing abundantly from under flat stones, and perhaps it would be well to try a few under like conditions on the rockwork. Possibly it will prove more easy of cultivation than is at present supposed, and there certainly should be no difficulty in succeeding with it in northern and elevated districts. On the Alps it blooms in early summer, in our gardens somewhat earlier. It is easily raised from seed, and in its native habitat spreads about freely. This is the plant which Mr. Ruskin met high up among the icy rocks, struggling successfully for life near the margins of the everlasting snowy solitudes of the Alps, and which pleased him so much there.

RANUNCULUS LYALLII.—*Rockwood Lily*.

" DR. HOOKER calls this plant, as well he may, the 'most noble species of the genus'—'the Water Lily of the shepherds.' Indeed, even in the dried specimens, of which there are many in the Kew herbaria, the resemblance to our common white Water Lily is striking. The plant is stated to grow in moist places in the Southern Alps, the Wurumui Mountains, in the glacier regions of the Forbes River, near Otago, and elsewhere in the Middle Island of New Zealand, at heights of from 1000 to 5000 feet above the sea. In habit it seems almost identical with our common marsh Marigold, but is twice or thrice larger. The leaves are circular, twelve to fifteen inches in diameter, very like those of the plant last mentioned, but peltate, as in the Nelumbium. The flowers are borne in panicles; each flower is of the purest waxy-white colour, three to four inches across, and in shape and aspect is like a Brobdingnagian Buttercup. We imagine there would be no great difficulty in growing the plant if we once got it here. To this end we should be disposed, in addition to more ordinary methods of transport, to try several means, such as sowing the seeds in a Wardian case, or placing them in a closed bottle in damp moss or moistened earth. At any rate, 'moist shady gullies' in New Zealand mountains must no longer be suffered to have the monopoly of so grand a plant as this."—'Gardener's Chronicle.' When it is readily obtainable in this country, no doubt it will form a grand object for moist, depressed, sheltered places in the neighbourhood of the rockwork or hardy fernery, in deep moist soil. It grows from two to four feet high.

RANUNCULUS MONTANUS.—*Mountain Buttercup.*

A PLANT with dwarf compact tufts of deep green, glossy leaves, deeply divided, and with obtuse lobes, covered in spring with a dense mass of brilliant yellow flowers, somewhat larger than those of our common Buttercup (*R. acris*). Although so like a Buttercup in tone, it is unlike it in its dwarf compact habit, usually flowering at three inches high, and, though growing freely enough, not spreading about with the coarse vigour of many of its fellows, each little stem bearing one flower. A native of alpine pastures on the principal great mountain-chains of Europe, growing freely in moist sandy soil on rockwork, on which it should be planted so as to form spreading tufts, or in lines in chinks or ledges. Readily increased by seed or division.

RANUNCULUS PARNASSIFOLIUS.—*Parnassia-like R.*

A HANDSOME and distinct kind, with beautiful pure white flowers like those of *R. amplexicaulis*, from one to a dozen or more being borne on each stem, which, according to soil and position, grows from two to eight inches high, and is somewhat velvety, and of a purplish tone. The leaves are entire-margined, but of a dark brownish-green, oval-heart-shaped, or occasionally kidney-shaped. in outline, sometimes slightly woolly along the margins and nerves, and not so graceful as those of *R. amplexi-caulis*. The roots are long, fibrous, and numerous, united in a stout stock, and the whole plant of a leathery, firm texture. It is as yet comparatively rare in our gardens, though abundant in many parts of the Pyrenees and Alps. No plant is more worthy of culture in the rock-garden, in deep, sandy, well-drained loam. There is a variety with narrow leaves, probably not at present in cultivation.

RANUNCULUS RUTÆFOLIUS.—*Rue-leaved Crowfoot.*

AN interesting species, with much and deeply divided leaves (the radical ones twice divided), reminding one somewhat of those of a very dwarf Aquilegia more than of a Crowfoot, and pretty white flowers, with orange centres, about an inch across. The stems vary from three to six inches high, bearing from one to three flowers, usually but one; the flowers are sometimes rose-tinted on the outside. Not difficult to cultivate in the same soil and position as are recommended for the Alp Crowfoot, and deserving a place in every collection. A native of elevated parts of the great continental ranges, increased by seed or division. It is sometimes made a separate genus of, under the name *Callianthemum rutæfolium.*

The preceding kinds I believe to be the most worthy of culture of all obtainable dwarf Ranunculi; there are, however, others worthy of a place in large collections, and of these, *R. pyrenæus, gramineus, Thora, Gouani, spicatus,* and *uniflorus,* are among the best. The double varieties of our common *R. acris, R. bulbosus,* and also a fine large double kind known as *R. bullatus fl. pl.,* are very ornamental; and though they may, with the exception of the last, not be fit for the choice rock-garden, they should be planted freely in borders, rocky, or semi-wild places, as the golden button-like flowers last a long time in bloom.

RHEXIA VIRGINICA.—*Meadow Beauty.*

A RARE American plant of the Melastoma order, but perfectly hardy, forming neat little bushes, from six to twelve inches high; the stems square, with wing-like angles; the leaves oval-lance-shaped, on very short stalks, and with bristly teeth; the flowers a beautiful rosy purple, appearing in summer and early autumn. A native of North America, from a considerable distance north of New York to Virginia, and westward to Illinois and the Mississippi, usually in sandy swamps. It is very rare, indeed, to see it well-grown in this country, though no plant is more desirable for the rock-garden or artificial bog. The only place in which I have noticed this plant invariably doing well is in Messrs. Osborn's nursery, at Fulham, planted in beds of deep and moist sandy peat, and I believe it would enjoy even a greater degree of moisture than it obtains there. Deep, sandy, boggy soil, with abundance of moisture at all times, will suit it well. Propagated by careful division.

RHODODENDRON CHAMÆCISTUS.—*Thyme-leaved R.*

A LILLIPUTIAN Rhododendron, rising scarcely a span high, and thickly clothed with small fleshy leaves, ciliated at the edge, and covered with exquisite flowers, of a lively purple colour, three or four together, divided into five segments at the mouth, about the size of those of *Kalmia latifolia*. This plant is very rarely seen in good health in gardens, and for its successful cultivation requires to be planted in limestone fissures, in peat, loam, and sand in about equal proportions. A native of calcareous rocks in the Tyrol, and one of the most precious of dwarf rock-shrubs, very suitable for association with such subjects as the small Andromedas and Menziesias. Flowers in early summer.

The well-known *Rhododendron ferrugineum* and *hirsutum*, each bearing the name of "alpine Rose," and which often terminate the woody vegetation on the great mountain chains of Europe, are easily had in our nurseries, and well suited for large rockwork, in deep peat soil, in which they attain a height of about eighteen inches. *R. Wilsonianum, myrtifolium, amœnum, hybridum, dauricum-atrovirens, Gowenianum, odoratum,* and *Torlonianum*, are also comparatively dwarf kinds, which may be tastefully employed in the rock-garden—the last two very sweetly scented. It is perhaps needless to add that they should not be too intimately associated with minute alpine plants.

SAGINA GLABRA.—*Lawn Pearlwort.*

A PLANT very generally known in consequence of being much talked of a few years since as a substitute for lawn-grass, and though it has not answered the expectations formed of it in that way, it is none the less a very beautiful and minute alpine plant exceedingly welcome on rockwork, and forming carpets almost as compact and smooth as velvet on level soil, dotted with numerous small but pretty and white flowers, the light, fresh green, moss-like carpet being starred with them in early summer. It is unsurpassed for forming carpets of the freshest and dwarfest verdure beneath taller, but comparatively small, beautiful and rare bulbs or other plants which it may be desired to place to the best advantage. It is most readily multiplied by pulling the tufts into small pieces, and replanting them at a few inches apart; they soon meet and form a carpet. It is also readily increased by seeds, but this mode is rarely worth resorting to, unless it is desired to propagate the plant largely for lawn-making. Although it does not generally form a permanent and satisfactory turf yet it is quite possible by selecting a rather deep sandy soil, and by keeping it perfectly clean and well rolled, to make a beautiful turf of it; but this is rarely worth attempting except on a small scale, and when it begins to perish in flakes here and there, it should be taken up and replanted. It is very commonly grown in gardens under the name of *Spergula pilifera*, but it really is a variety of *Sagina glabra* of the Alps, found on high mountains in Corsica. *Arenaria cæspitosa*, *Spergula subulata*, and *Spergula saginoides* (a British plant) are also names by which it is known in gardens and described in gardening periodicals.

SANGUINARIA CANADENSIS.—*Bloodroot.*

A CURIOUS, distinct, and pretty plant, with thick underground stems, from which spring kidney-shaped leaves cut into large wavy or toothed lobes, sea-green in tone, and full of an orange-red and acrid juice. The stems grow from four to six inches high, each bearing a solitary and handsome white flower when the plant is established in healthy tufts, which are very effective in early spring. I have always observed this plant grow best in somewhat shaded and moist positions and in rich but well-drained soil; it is well worthy of a place on the rockwork, but decidedly the best thing to do with it is to try and naturalise it on bare spots in open rich woods—positions like those in which it is found wild in North America. Propagated by division, and flowers in March.

SANTOLINA INCANA.—*Hoary S.*

A SMALL silvery shrub, with numerous branches and narrow leaves, the whole plant covered with dense white down, forming neat prostrate tufts of edgings in the flower-garden. The flowers are rather small, pale greenish-yellow, and in no way ornamental, but the plant is likely to be popular from its neat habit and silvery hue, and it grows readily in ordinary soil on the level border, and may be tastefully used on slopes of rockwork. It is considered a variety of the better known *S. Chamæcyparissus*, the Lavender Cotton. This and its other variety *squarrosa* are suitable for very large rockworks, banks, &c., but, forming spreading silvery bushes, two feet high, in suitable soil, are not suited for intimate association with very dwarf alpine plants.

Other species of Santolina are suited for like purposes, *S. pectinata* and *S. viridis*, for example, forming bushes somewhat like the Lavender Cotton. *Santolina alpina* is of more alpine habit, forming dense mats quite close to the ground, from which spring yellow button-like flowers on long slender stems. It grows in any soil, and may be used on the less important parts of the rock-garden. Cuttings of the shrubby species strike readily, and *S. alpina* is easily increased by division.

SAPONARIA OCYMOIDES.—*Rock S.*

A BEAUTIFUL trailing rock-plant, with prostrate stems and an abundance of rosy flowers, so densely produced as to completely cover the cushions of leaves and branches. It is easily raised from seed or from cuttings, thrives in almost any soil, and is one of the most valuable plants we have for clothing the most arid parts of rockwork, particularly in positions where a drooping plant is desired, the shoots falling profusely over the face of the rocks, and becoming masses of rosy bloom in early summer, and also excellent for planting on ruins and old walls, on which the seed should be sown in mossy chinks in spots where a little soil has gathered. It is also a valuable border-plant, forming roundish spreading cushions, with masses of flowers, and is well worthy of being naturalised in bare and rocky places. A native of Southern and Central Europe in stony and rocky places, and although it grows freely in poor soil when it is planted with the view of allowing it to fall freely over the face of a rock, a greater development will be secured by putting it in deep rich loam.

One or two other dwarf Saponarias have recently been introduced to our gardens, *S. lutea* and *S. cæspitosa*, for example. The former of these is not worthy of cultivation except in a botanic garden, and the last insufficiently tried.

SAXIFRAGA AIZOIDES.—*Yellow Mountain Saxifrage.*

A NATIVE plant, very abundant in Scotland, the North of England, and some parts of Ireland, in wet places, by the sides of mountain rills or streams, and often descending along their course, into the low country, producing at the end of summer or autumn an abundance of bright yellow flowers, half an inch across, and dotted with red towards the base. It forms dense, dwarf, bright green masses of leaves, and has leafy branched flower-stems, by which it is distinguished from the other yellow native Saxifrages. Although a moisture-loving mountain plant, it is quite easy to grow in lowland gardens, naturally doing best in moist ground. Wherever a small stream or rill is introduced to the rock-garden or its neighbourhood, it may be most appropriately used, and planted so as to form wide-spreading masses, as it does on its native mountains. Easily propagated by division or by seed. When the leaves are sparsely ciliated, it is, according to Mr. Syme, the *S. autumnalis* of Linnæus.

SAXIFRAGA AIZOON.—*Aizoon Saxifrage.*

NOT a pretty-flowering kind, having a greenish-white bloom, but it spangles over many a low mountain crest and high alp-flank in Europe and America with its silvery rosettes, and in our gardens these form such firm, compact, and roundish silvery tufts in any common soil that it deserves to be universally cultivated as a rock, border, and edging plant. Plants of it established two or three years form grey-silvery tufts a foot or more in diameter, and about six inches high, sometimes a few inches more —these great tufts not flowering so freely as the wild plants, which need not be regretted, as it is the silvery mass, and not the flowers, that is sought. The plant is easily distinguished by its rather oblong and obtuse leaves, bordered with fine teeth, the borders densely margined with encrusted pores, and stiffly ciliated at the base ; and by the flower-stems, which grow from six to fifteen inches high, being furnished with glandular hairs on the upper part, and usually smooth on the lower. As to its culture, nothing can be easier ; it is very often grown in pots, but grows as freely as any native plant, and best perhaps when exposed to the full sun. Easily increased by division. There are several varieties.

SAXIFRAGA ANDREWSII.—*Andrews's Saxifrage.*

THIS interesting British plant is considered by some botanists to be a garden hybrid, and with pretty good reason, judging by the leaves and flowers ; but nothing more has been ascertained about its history. Mr. Andrews found it first in Ireland, but it has not since been discovered. Among the green-leaved kinds there is no better. Its flowers are large and freely produced, but I never could see any good seed on it. The leaves are long, firm in texture, and with a membranous margin ; the prettily spotted flowers being larger than those of *S. umbrosa*, and the petals conspicuously dotted with red, which, with other slight characters, points to the probability of its being a hybrid between a London Pride and one of the continental group of encrusted Saxifrages. It is more worthy of a position on a rockwork than the London Pride or the Kidney-leaved S., but does quite freely on any border soil, merely requiring to be replanted occasionally when it spreads into very large tufts, or to have a dressing of fine light compost sprinkled over it annually.

SAXIFRAGA ARETIOIDES.—*Aretia-like Saxifrage.*

A VERY gem of the encrusted section, forming cushions of little silvery rosettes, almost as small and dense as those of *Androsace helvetica*, and about half an inch high. It has rich golden-yellow flowers, appearing in April, on stems a little more than an inch high, and reminding one of the flowers of *Aretia Vitalliana*. The stems and stem-leaves are densely clothed with short glandular hairs like those of a Drosera. Like most of its brethren, not difficult to grow, but requires a moist and well-drained soil, and, being so dwarf and tiny, must be guarded from being overrun by coarser neighbours. A native of the Pyrenees; increased by seed and careful division.

SAXIFRAGA ASPERA.—*Rough Saxifrage.*

A SMALL, grey, tufted, prostrate plant, with lance-shaped and ciliated leaves, the lower ones closely imbricated, the upper ones somewhat scattered, producing few flowers, rather large, but of a dull white colour, on stems about three inches high. *S. bryoides* is considered a variety of this, and forms a densely tufted diminutive plant, with pale yellow flowers, the rosettes of leaves being almost globular, and the plant not forming stolons or runners like the preceding. I have never seen either of these plants displaying any beauty of bloom, but both are worthy of growing for their moss-like character. Both are natives of the Pyrenees; *S. bryoides* in the most elevated regions. Both are quite easy of cultivation, growing freely in the open air in London, but rarely flowering there.

SAXIFRAGA BIFLORA.—*Two-flowered Saxifrage.*

A BEAUTIFUL dwarf species, allied to the British species *S. oppositifolia*, but larger in all its parts, and immediately distinguished by producing two or three flowers together, and by having its leaves thinly scattered, and not packed on the stems like those of that species. It is also a much larger plant, and has larger flowers, rose-coloured at first, changing to violet. I found it in abundance on fields of grit and shattered rock, in the neighbourhood of glaciers on very elevated parts of the Alps, in company with *Campanula cenisia*; and just without the margins of the vast fields of snow, under which, even in June, lay numberless plants waiting for an opportunity to open when the snow had thawed. It grew entirely in loose grit, so that with a little care, masses of the branched imbedded stems and long fine roots could be taken up entire. I believe it usually inhabits like positions in what may be termed moving débris, but like many other plants the conditions in which it occurs in a wild state are by no means necessary to succeed with it in gardens. It grows very freely on rockwork in gritty or sandy soil, in well-drained positions in rich light loam, may be increased by division, cuttings, or seed, and should be seen in every select rock-garden.

SAXIFRAGA CÆSIA.—*Silver Moss.*

THIS resembles an Androsace in the dwarfness and neatness of its tufts. I have met with it on the Alps, in minute tufts, staining the rocks and stones like a silvery moss, and on level ground, where it had some depth of soil, spreading into beautiful little cushions from two to six inches across. It is easily known, either in cultivation or in a wild state, by its exceeding dwarfness and neatness, and by its three-sided, keeled leaves, regularly margined with white crustaceous dots. It bears pretty white flowers, about the third of an inch in diameter, on thread-like smooth stems, one to three inches high. A native of the high Alps and Pyrenees, it thrives perfectly well in our gardens in very firm sandy soil, fully exposed, and abundantly supplied with water in summer. It may be also grown well in pots or pans in cold frames near the glass; but, being very minute, no matter where it is placed, the first consideration should be to keep it distinct from all coarse neighbours, and even the smallest weeds will injure or obscure it if allowed to grow. Flowers in summer, and is increased by seeds or careful division.

SAXIFRAGA CERATOPHYLLA.—*Horn-leaved Saxifrage.*

A FINE and ornamental species of the mossy section, with very dark highly-divided leaves, stiff and smooth, with horny points; the flowers pure white, and abundantly produced in loose panicles in early summer, the calyces and stamens covered with clammy juice. It quickly forms strong tufts in any good garden soil, and is admirably adapted for covering rockwork of any description, either as wide level tufts on the flat portions or pendent sheets from the brows of rocks. May also be used with good effect in borders. A native of Spain; increased by seed or division.

SAXIFRAGA CORDIFOLIA.—*Heart-leaved Saxifrage.*

ENTIRELY different in aspect to the ordinary dwarf section of Saxifrages, with very ample leaves, roundish-heart-shaped, on long and thick stalks, toothed; flowers a clear rose, arranged in dense masses, which are half concealed among the great leaves in early spring, apparently hiding under them from the cutting breath of March. *S. crassifolia* is allied to this, and is useful for similar purposes. They grow and flower in any soil or position, and are thoroughly hardy; but it is desirable to encourage their early-flowering habit by placing them in warm sunny positions, where the fine flowers may be induced to open well. They are perhaps more worthy of association with the larger spring flowers and with herbaceous plants than with dwarf alpines, and are well worthy of being naturalised on bare sunny banks, in sunny wild parts of the pleasure-ground, or by wood walks. They may also be used with fine effect on rough rock or root work, near cascades, or on rocky margins to streams or artificial water, their fine, evergreen, glossy foliage being quite distinct. They may, in fact, be called the fine-foliaged plants of the rocks. A native of Siberian mountains. *S. ligulata* (*Megasea ciliata*) is a somewhat tender species, and only succeeds out of doors in mild and warm parts of this country.

SAXIFRAGA COTYLEDON.—*Pyramidal Saxifrage.*

A NOBLE Saxifrage, which embellishes with its great silvery rosettes and elegant pyramids of white flowers many parts of the great mountain ranges of Europe, from the Pyrenees to Lapland, and easily known by its rather broad leaves, margined with encrusted pores, and its fine handsome bloom. It is the largest of the cultivated Saxifrages, and also the finest except *S. longifolia*, of which it has not the linear leaves. The rosettes of the pyramidal Saxifrage differ a good deal in size. When grown in tufts, they are for the most part much smaller, from being crowded, than isolated specimens. The flower-stem varies from six to thirty inches high, and about London, in common soil, will often attain a height of twenty inches, and in cultivation usually attains a greater size than on its native rocks; though in rich soil, at the base of rocky slopes in a Piedmontese valley, I have seen single rosettes as large as I have ever seen them in gardens. The plant is perfectly hardy, and second to none as an ornament of the rock-garden. It also thrives perfectly in common soil, and is in some places much grown in pots for the decoration of the greenhouse. When the pyramids of beautiful flowers are strong, or in an exposed position, it is well to put a neat and inconspicuous but firm stake to each. Nothing can be easier to propagate by division, or cultivate without any particular attention. It is sometimes known as *S. pyramidalis*, though some consider this at least a variety, having amore erect habit, narrower leaves, and somewhat larger flowers.

SAXIFRAGA CYMBALARIA.—*Golden Saxifrage.*

QUITE distinct in aspect from any of the family, and one of the most useful of all, being a bright and continuous bloomer. I have had little tufts of it, which, in early spring, formed masses of bright yellow flowers set on light green, glossy, small ivy-like leaves, the whole not more than three inches high. These, instead of falling into the sere and yellow leaf, and fading away into seediness, kept still growing taller, still elevating, and still preserving the same little rounded pyramid of golden flowers, until autumn, when they formed specimens about twelve inches high. It is an annual or biennial plant, which sows itself abundantly, coming up in the same spot, is peculiarly suitable for moist spots on or near rockwork, grows freely on the level ground, and might be readily naturalised on the margins of a rocky stream, and in various other places in large pleasure-grounds.

SAXIFRAGA DIAPENSIOIDES.—*Diapensia S.*

ONE of the very best of all the dwarf Saxifrages, and also one of the smallest, admirably suited for choice rockwork or culture in pans. I have grown it very well in an open bed in London, and it would flourish equally well everywhere if kept free from weeds, and in a well-exposed spot. The soil should be very firm and well-drained, though kept moist in summer. The flowers are of a good white, three to five on a stem, rarely exceeding two inches high, and often not more than an inch; the leaves grey, linear, and obtuse, packed into such dense cylindrical rosettes that established specimens feel quite hard to the hand. It comes near

S. aretioides, but is at once distinguished from that by its white, entire, and oblong petals. A native of the Alps of Switzerland, Dauphiny, and the Pyrenees.

SAXIFRAGA GEUM.—*Kidney-leaved London Pride.*

VERY like the London Pride in habit and flowers, but with the leaves roundish, heart-shaped at the base, on long stalks, and with scattered hairs on the surfaces; flowers about a quarter of an inch across, and usually with reddish spots. A native of various parts of Europe, useful for the same purposes and cultivated with the same facility as the London Pride, will grow freely in woods or borders, particularly in moist districts, and is worthy of naturalisation in the former. Like its neighbours, it is, of course, suitable for rockwork, but does not deserve that position so much as numbers of plants much more difficult to grow. *Saxifraga hirsuta* comes near this, and is probably a variety, the chief difference being that the leaves are longer than broad, less heart-shaped, and more hairy; it is suitable for like positions.

SAXIFRAGA GRANULATA.—*Meadow Saxifrage.*

A LOWLAND plant, with several small scaly bulbs in a crown at the root, and common in meadows and banks in England, with crenate leaves, thickly clothed with shaggy glandular hairs, and numerous white flowers, three quarters of an inch across. I should not name it here were it not for its handsome double form *S. granulata fl. pl.*, which I have often seen flowering profusely and prettily in little cottage gardens in Surrey. It is very useful in the spring garden as a border-plant, or on rougher parts of rockwork. Mr. Bentham considers that the small bulb-bearing *S. cernua* of Ben Lawers may be a variety of the Meadow S. As a garden-plant, *S. cernua*, however, is a mere curiosity, though it may be acceptable in botanical collections.

SAXIFRAGA HIRCULUS.—*Yellow Marsh Saxifrage.*

A REMARKABLE species, with a single bright yellow flower on each stem, or sometimes two or three, three quarters of an inch across, and tufts of obovate leaves, gradually attenuated into the stalk. Quite different in aspect from any other cultivated Saxifrage. A native of wet moors in various parts of England, not difficult to cultivate in moist soil, and thriving best under conditions as near as possible to those of the places where it is found wild. It is best suited for a moist spot near a streamlet of the rock-garden, or for the artificial bog.

SAXIFRAGA HYPNOIDES.—*Mossy Saxifrage.*

A VERY variable plant in its stems, leaves, and flowers, but usually forming mossy tufts of the deepest and freshest green, abundant on the mountains of Great Britain and Ireland, and common in gardens. In cultivation it attains greater vigour than in a wild state, and no plant is more useful for forming carpets of the most refreshing green in winter and almost in any soil. For this reason it is peculiarly well suited for planting in the low rocky borders often made in town and villa gardens, and should be largely used by those who desire to rest their eyes on glistening verdure during the winter months. It thrives either on rockwork or the level ground, in half-shady positions or fully exposed to the sun, forming the fullest and healthiest tufts in the latter case, and flowering profusely in early summer. Nothing can be easier to grow or increase by division. It is also suitable for forming dwarf verdant carpets in the flower-garden or on the rockwork with a view of placing one or more plants on the surface. Chiefly distinguished by its narrow and pointed leaves, sometimes entire, but often three- to seven-cleft, and stems, sepals, leaves, and shoots more or less covered with glandular hairs. Under this species may be grouped *S. hirta, S. affinis, S. incurvifolia, S. platypetala,* and *S. decipiens,* all exhibiting differences which some think sufficient to mark them as species. They present considerable differences in appearance when grown together in a garden, and many amateurs will, no doubt, think them worthy of a place; they all thrive with the same freedom as the Mossy Saxifrage, appearing to suffer only from drought or very drying winds. If, when first planted, a few largish stones are buried in the earth round each, the plants will soon lap over them, the stones will serve to preserve the moisture in each tuft, and the plants will be much less likely to suffer from drying winds. *S. cæspitosa,* a British plant, comes near to this, but is known at once by its obtuse lobes and more tufted habit, does not emit slender spreading shoots like *S. hypnoides,* and is scarcely so ornamental.

SAXIFRAGA JUNIPERINA.—*Juniper Saxifrage.*

THIS is one of the most distinct and desirable kinds in cultivation, having spine-pointed leaves, densely set in cushioned masses, looking, if one may so speak, like Juniper-bushes compressed into the size of small round pin-cushions, and with little seen but the prickly points of the leaves. The flowers are yellow, arranged in spikes on a leafy stem, and appear in summer. It thrives very well in moist, sandy, firm soil, and is well worthy of a place in the rock-garden, and also in every collection of alpine plants grown in pots. A native of the Caucasus; propagated by seed and careful division.

SAXIFRAGA LONGIFOLIA.—*Long-leaved Saxifrage.*

THE single rosettes of this are often six, seven, and eight inches in diameter, and, while retaining all the charms of its congeners, it boldly spreads forth into an object as striking as some of the prettier succulent plants of the New World or the Cape. I have, indeed, measured one specimen more than a foot in diameter. It may well be termed the Queen of the silvery section of Saxifrages, and by that section are meant those which have their greyish leathery leaves margined with dots of white, so as to give the whole a silvery character. This is so beautifully marked in that way that it is attractive at all seasons, while in early summer it pushes up massive foxbrush-like columns of flowers from a foot to two feet long, the stem covered with short, stiff, gland-tipped hairs, and bearing a multitude of pure white flowers.

A native of the higher parts of the Pyrenees; perfectly hardy in this country; not difficult of culture, and may be grown in various ways. On some perpendicular chink in the face of a rockwork into which it can root deeply, it is very striking when the long outer leaves of the rosette spread away from the densely-packed centre. It may also be grown on the face of an old wall, beginning with a very small plant, which should be carefully packed into a chink with a little soil. Here the stiff leaves will, when they roll out, adhere firmly to the wall, eventually forming a large silver star on its surface. It will thrive on a raised bed or border, surrounded by a few stones to prevent evaporation and to guard it from injury. It also thrives in a greenhouse or frame, and perhaps the readiest way of getting a weakly young plant from the nursery to develop into a sturdy rosette is to put it in a six-inch pot, well drained and filled with a mixture of sandy loam and stable manure, and placed in a sunny pit or frame, giving it plenty of water in spring, summer, and autumn. It is propagated by seeds, which it produces freely. In gathering them it should be observed that they ripen gradually from the bottom of the stem upwards, so that the seed-vessels there should be cut off first, leaving the unripe capsules to mature, and visiting the plant every day or two to collect them as they ripen successively.

S. lingulata is by some authors united with the preceding, from which it chiefly differs by having smaller flowers, by the leaves and stems being smooth and not glandular, by its shorter stems, and by the leaves in the rosette being shorter and very much fewer in number than in the Long-leaved S. It is also a charming rock-plant, and will succeed with the same treatment and in the same positions as the preceding. *S. crustata* is considered a very small variety of the Long-leaved S. with the encrusted pores thickly set along the margins; being several times smaller, it will require more care in planting, and to be associated with dwarfer plants.

SAXIFRAGA OPPOSITIFOLIA.—*Purple Saxifrage.*

IT is impossible to speak too highly of the beauties of this bright little mountaineer, so distinct in colour and in habit from the familiar members of its family. The moment the snow melts, its tiny herbage glows into solid sheets of purplish rose-colour; the flowers solitary, on short erect little stems, and often so thickly produced as to quite hide the leaves, which are small, opposite, and densely crowded. In a wild state on the higher mountains of Britain and the Continent, in which it has to submit to the struggle for life, it usually forms rather straggling little tufts; but on fully exposed parts of rockwork, rooted in deep, light, and moist loam, it forms rounded cushions, neat at all seasons, and peculiarly appropriate on slopes of rockwork, or fringing over the sides of rocks. Propagated by division, and flowering in early spring. There are the following varieties in cultivation : *S. opp. major*, rosy pink, large ; *S. opp. pallida*, pale pink, large ; *S. opp. alba*, white.

SAXIFRAGA RETUSA.—*Retuse-leaved Saxifrage.*

A PURPLISH species, closely allied to our own *S. oppositifolia*, but, in addition to the different character of the leaves, distinguished by the flowers having distinct stalks, and being borne two or three together on their little branches. The small, opposite, leathery leaves are closely packed in four ranks on the stems, which form dense prostrate tufts. A native of the Alps and Pyrenees, flowers in early summer, may be cultivated with great success in the same way as *S. oppositifolia*, and well merits a place in the rock-garden. It is also easily grown in a pan in a cold frame.

SAXIFRAGA ROCHELIANA.—*Rochel's Saxifrage.*

A VERY compact and dwarf kind, forming dense silvery rosettes of tongue-shaped white-margined leaves, with distinctly impressed dots. It is distinguished among the dwarf silvery Saxifrages by producing large white flowers on sturdy little stems in spring. I know no more exquisite plant for rockwork, for culture in pans, or for small rocky or elevated borders. Any free, good, moist, loamy soil will suit it, and I have seen it thriving very well on borders in London. It should always be exposed to the full sun, and deserves to be associated with the choicest spring flowers and alpine plants. A native of Austria; increased by seeds or careful division.

SAXIFRAGA SARMENTOSA.—*Creeping Saxifrage.*

A WELL-KNOWN old plant, with roundish leaves, mottled above, red beneath, with numbers of creeping, long, and slender runners, producing young plants strawberry fashion. Striking and singular in leafage, it is also ornamental in bloom, and growing freely in the dry air of a sitting-room, may be seen gracefully suspended in numerous cottage windows. It perhaps is most at home running wild on banks or rocks, in the cool greenhouse or conservatory; however, it lives in the open air in mild parts of England, and, where this is the case, may be used in graceful association with ferns and other creeping plants. A native of China, flowering in summer. Closely allied to *S. sarmentosa* is the delicate dodder-like Saxifrage, *S. cuscutaformis*, so called from having thread-like runners like the stems of a dodder, and distinguished by having much smaller leaves, and the petals more equal in size than those of *sarmentosa*, in which the two outer ones are much larger than the others. It will serve for the same purposes as the Creeping Saxifrage, but, being much more delicate and fragile in habit, will require a little more care. The plants grown in gardens as *S. japonica* and *S. tricolor* are considered varieties of the Creeping Saxifrage.

SAXIFRAGA UMBROSA.—*London Pride.*

THIS almost universally cultivated plant grows abundantly on the mountains round Killarney, though it was much grown in our gardens before it was recognised as a native of Ireland. It is needless to describe the appearance of such a very familiar plant; it is, however, distinguished from *S. Geum* by having oval-oblong leaves, narrowed and not heart-shaped at the base; its flowers, too, are a little larger and more freely dotted with red. It is much used as an edging plant in old gardens, and, being such a pleasing evergreen, should be freely used for embellishing the rough parts of rockwork, the fringes of cascades, &c. It is naturalised in several parts of England, and grows freely among dwarf herbage, or in rocky ground in woods. There are several varieties, as, for example, *S. punctata* and *S. serratifolia*, which are distinct enough when grown side by side, and submit to the same culture.

It is believed that the preceding are among those best worth growing. The following is a list of the other species or reputed species believed to be in cultivation now in this country. Those most worthy of culture are marked by an asterisk.

Saxifraga adscendens	Saxifraga cochleata	Saxifraga globifera
ajugæfolia	*crustata	Gmelini
ambigua	cuneifolia	*Guthrieana
androsacea	*daurica	hieraciifolia
aquatica	elatior	*Hostii
atropurpurea	elongella	*icelandica
*Bucklandii	erosa	infundibulum
bulbifera	exarata	*intacta
Burseriana	flavescens	*intermedia
calcarata	geranioides	japonica
*capillaris	germanica	lætevirens
condensata	*gibraltarica	lævigata
*contraversa	glacialis	*lævis

Saxifraga leptophylla	*Saxifraga pennsyl-	Saxifraga spathulata
*lutea viridis	vanica	sponhemica
*marginata	pentadactylis	*Stansfieldii
*media	petræa	stellaris
Mollyi	planifolia	stenophylla
multicaulis	pulchella	*Sternbergii
*muscoides	*purpurascens	Tazetta
*nervosa	pygmæa	*tenella
nivalis	*recta	thysanodes
ohioensis	recurva	tricuspidata
orientalis	reniformis	trifida
*palmata	Rhei	trifurcata
*paniculata	*rosularis	trilobata
parnassica	rotundifolia	villosa
*pectinata	rupestris	virginiensis
pedata	*Schraderi	Webbiana
pedatifida	sibirica	

SCHISTOSTEGA PENNATA.—*Iridescent Moss.*

MOSSES are small, but this is so very small that it would hardly be noticed by the naked eye were it not for the iridescent gleams of beautiful colour which it displays when viewed in positions where it flourishes. It is no exaggeration to state that some of the stones and sods on which it grows look as if sown with a mixture of gold and the material that goes to form the wings of green humming birds. It is almost startling to see this little gem for the first time, no "plant" being visible to the casual visitor, who wonders what produces the exquisite colour. It was supposed to require a particular kind of rock on which to grow; but I have been lately much pleased to see its wonderful coruscations spread over sods of turf and masses of peat, quite as well as on the chips of rock brought from its native place. Messrs. Stansfield have it in beautiful condition among the rocks in their cool fernery at Todmorden, and Messrs. Backhouse have it in perfection both in the open air, in a quiet, deep gorge of rocks, where it obtains sufficient moisture without being washed by rains, and also in their underground fernery, constructed for the rarer filmy ferns, which proves that there is no insurmountable difficulty in establishing it in like positions; and certainly the most graceful dwarf or biggest tree fern is not capable of adding to them a more decided charm than this most diminutive moss.

SCILLA AMŒNA.—*Pleasing Squill.*

A DISTINCT, early-spring-flowering kind, opening soon after *S. sibirica*, and readily known from any of its relatives by the large yellowish ovary showing conspicuously in the centre of the dark indigo-blue flowers. It is, though sufficiently attractive to merit a place in borders and collections of hardy bulbs, less ornamental than any other kind here mentioned, the flowers being arranged in a somewhat sparse and rigid manner, and having none of the grace characteristic of *S. campanulata*, or of the varieties of *S. nutans*, or the dwarfness and brilliancy of *S. sibirica*. The leaves, usually about half an inch across, attain a height of about one foot, and are very easily injured by cold or wind, so that a sheltered position is that best suited to its wants. It is not exactly suited for choice rockwork, though well worthy of a place in borders, and of being naturalised on sunny banks in semi-wild spots. A native of the Tyrol; increased from seeds or by separation of the bulbs.

SCILLA BIFOLIA.—*Early Squill*

ALTHOUGH not nearly so well known or popular as *S. sibirica*, this is quite as worthy of cultivation, producing in the very dawn of spring, indeed often in winter, rich masses of dark blue flowers, four to six on a spike, and forming very handsome tufts of vegetation from six to ten inches high, according to the richness and lightness of the soil, and the warmth and shelter of the aspect. It thrives well in almost any position, in ordinary garden soil, the lighter the better. Although it blooms earlier than *S. sibirica*, it does not withstand cold wintry and spring rains and storms nearly so well as that species, and therefore it would be well to place some tufts of it in warm sunny spots, either on rockwork or sheltered borders. A native of Southern and Central Europe. As shown by Dr. Masters, in the 'Gardener's Chronicle,' this species varies very much, and in consequence has gone under many names. The varieties are of more importance than is usually the case, and the following are those enumerated by Dr. Masters as most distinct, and therefore most worthy of cultivation :—*S. bifolia alba*, *S. bifolia candida*, *S. bifolia carnea*, *S. bifolia compacta*, *S. bifolia maxima*, *S. bifolia metallica*, *S. bifolia rosea*, *S. bifolia pallida*, and *S. bifolia præ-*

cox. The name *S. præcox*, which occurs so often in gardens and in nurserymen's catalogues, does not really belong to a distinct species, and, when most properly applied, refers to the last-mentioned variety of *S. bifolia*, which usually flowers somewhat earlier than the common form.

SCILLA CAMPANULATA.—*Bell-flowered Squill.*

A VIGOROUS species, long cultivated in England, one of the finest ornaments among early-summer-flowering bulbs, and, though a more southern species than most of the others, the most robust of the family. It is easily known by its strong pyramidal raceme of pendent, short-stalked, large, bell-shaped flowers, usually of a clear light blue. A variety known as *S. campanulata major* is larger in all its parts, and a noble early summer flower; and the white- and rose-coloured varieties, *S. campanulata alba* and *S. c. rosea*, are also excellent. It is never seen to greater advantage than when peeping here and there from the fringes of shrubberies and beds of evergreens, the shelter it receives in such positions protecting its very large leaves from strong winds. It is, however, sturdy enough to thrive in any position. Comes from the South of Europe, attains a height of from twelve to eighteen inches, and deserves to be naturalised alongside of wood walks, and in the semi-wild parts of every pleasure-ground.

SCILLA ITALICA.—*Italian Squill.*

A NATIVE not only of Italy but of Southern France and Southern Europe generally. This Squill, with its pale blue flowers, intensely blue stamens, and delicious odour, is one of the most interesting and distinct, if not the most brilliant, of cultivated kinds. It grows from five to ten inches high, the leaves somewhat shorter, slightly keeled, and oblique; the flowers small, spreading in short conical racemes, opening in May. It is perfectly hardy, living in almost any soil, but thriving best in sandy and warm ones. Increased by division, which had better be performed only every three or four years, when the bulbs should be planted in fresh positions. It is worthy of a sheltered sunny spot on rockwork, particularly as it does not seem to thrive so freely in this country as some of the other species.

SCILLA NUTANS.—*Bluebell.*

A WELL-KNOWN and much admired native plant, abounding in almost every wood and copse; the flowers always arranged in a gracefully drooping fashion on one side of the stem. The Bluebell is so very common that I should not have mentioned it here but for the sake of its several beautiful varieties, which are not so much known as they deserve to be, although fitted to be great ornaments of the early summer garden. I particularly allude to the white variety, *S. nutans alba;* the rose-coloured one, *S. nutans rosea;* the pale blue, *S. nutans cærulea*, and a pleasing "French white" variety, which is not dignified by a Latin name. These are all highly suitable for planting here and there in tufts along the margins of shrubberies, near rockwork, for borders, the spring garden, and for naturalisation in woods, among the common blue kind.

SCILLA PATULA.—*Spreading Squill.*

THIS is rather closely allied to the Bluebell, with flowers of a pleasing violet-blue, not sweet, like those of that species, nor arranged on one side, but larger, and more open, with narrow bracts. It is easy of culture on almost any soil; blooms late in spring, and is suitable for the same purposes as the varieties of the Bluebell. A native of France and Southern Europe generally.

SCILLA PERUVIANA.—*Pyramidal Squill.*

THE Peruvian Squill, which, however, is not a native of Peru, is a very noble plant where it thrives, and it does so perfectly in many mild parts of these islands, though it suffers on cold soils. The flowers are of a fine blue, very numerous, arranged in a superb, regular, umbel-like pyramid, which lengthens during the flowering period. The white stamens contrast charmingly with the blue of the flowers. In all but the warmest parts of the country, this fine plant should have a somewhat elevated, warm, and sheltered position, a deep, light, and well-drained soil, and the large pear-shaped bulbs should be planted six inches under the surface, which will better enable them to withstand the cold. A native of Southern Europe and North Africa, grows from six to eighteen inches high, flowers in May and June, and deserves a place in a sheltered, sunny nook on every rockwork, and on every warm raised bed or border devoted to choice hardy bulbs. There is a white variety, *S. peruviana alba*, which is not quite so beautiful as the ordinary form. Tufts of the Peruvian Squill should be taken up, when at rest, every three or four years, the bulbs divided, and immediately replanted.

SCILLA SIBIRICA.—*Siberian Squill.*

THIS beautiful and minute gem among the flowers of earliest spring is happily becoming very popular, and many will have had an opportunity of concluding for themselves that no rockwork, spring garden, or garden of bulbous plants, can be complete without the striking and peculiar shade of porcelain blue which quite distinguishes it from the other species. It has had a great number of synonyms, but, unlike *S. bifolia*, has sported into few varieties, *S. amœnula* being the only one worth mentioning, and it is not really distinct. Varieties with larger blossoms and with one instead of from two to five on a stem are preserved in herbariums and occasionally cultivated, but these are only trifling variations, often arising from the conditions in which they are placed.

There appears to be some doubt as to whether the plant is really a native of Siberia, but it is known to be widely distributed in Asia Minor and Persia, and I received a specimen found "growing among the Snowdrops," by a gentleman in the Caucasus. It is perfectly hardy in this country, and, like most other bulbs, thrives best in a good sandy soil. Bulbs of it that have been used for forcing should never be thrown away; if allowed to fully develop their leaves and go to rest in a pit or frame, and afterwards planted out in open spots, in warm soil, they will thrive well. It is needless to disturb the tufts except every two or three years for the sake of dividing them when they grow vigorously. It comes in flower in very early spring a little later than *S. bifolia*, but withstands the storms better than that plant, and remains much longer in bloom. In places where it does not thrive very freely, from the cold nature of the soil or other causes, it would be well, in placing tufts of it on rockwork or on borders, to put it in sheltered positions, so that the leaves may not be injured by the wind, and the plant thereby weakened. It may be used with good effect as an edging to beds of spring flowers, or choice alpine shrubs.

Of other cultivated Squills, the British ones, *S. verna* and *S. autumnalis*, are certainly not worthy of cultivation except in botanical collections ; the plant usually sold by the Dutch and by our seedsmen as *S. hyacinthoides* is generally *S. campanulata*, and occasionally *S. patula*. The true *S. hyacinthoides* of Southern Europe is scarcely worthy of cultivation ; *S. cernua* is not sufficiently distinct from *S. patula*, and one or two southern species allied to *S. peruviana* have not been proved sufficiently hardy for general cultivation.

SCUTELLARIA ALPINA.—*Alpine Skullcap.*

A SPREADING plant with all the vigour of the coarsest weeds of its natural order, but withal neat in habit, and ornamental in flower. The pubescent stems are prostrate, but so abundantly produced that they rise into a full round tuft, a foot high or more in the centre, and falling low to the sides ; the leaves are ovate-roundish or heart-shaped at the base, very shortly stalked, and notched, and the flowers in terminal heads, at first short, afterwards elongating, and purplish or with the lower lip white or yellow. The form with the upper lip purplish and lower pure white is very pretty. The variety *lutea* (*S. lupulina*) is a very ornamental kind, with yellow flowers. Both plants are admirably suited for borders, the margins of shrubberies, and the rougher parts of rockwork. A native of the Pyrenees, Swiss and Tyrolese Alps, and many other parts of Europe and Asia ; readily increased by division, and flowering freely in summer.

Of other kinds of Scutellaria in cultivation, *S. japonica, orientalis, scordifolia*, and the British *S. minor*, an interesting little plant for the artificial bog, are among the best, but it is doubtful if they are worth a place in any but a very large collection.

SEDUM ACRE.—*Stonecrop.*

GROWING on walls, thatched houses, rocks, and sandy places in almost all parts of Britain, this little plant, with its small, thick, bright green leaves and brilliant yellow flowers, is as well known as the common Houseleek. Like the Daisy, it is so very abundant in an uncultivated state that there is rarely occasion to introduce it to gardens, though one of the most brilliant and distinct of its very large family. Sheets of it in bloom look very gay, and it may well be used with dwarf alpine plants in forming carpets of living mosaic-work in gardens. The fact that it runs wild on comparatively new brick walls round London does away with the necessity of speaking of its cultivation or propagation. There is a variegated or yellow-tipped variety, *S. acre variegatum;* the tips of the shoots of this become of a yellow tone in early spring, so that the tufts or flakes look quite showy at that season. It is suitable for use in the spring garden, on the rockwork, and for the same purposes as the ordinary form.

SEDUM ALBUM.—*White Stonecrop.*

A BRITISH plant, with crowded fleshy leaves of a brownish green, and in summer a profusion of white or pinkish flowers in elegant corymbs. Like the common Stonecrop, this occurs on old roofs and rocky places in many parts of Europe, and may be cultivated with the same facility as that well-known plant. It is worthy of naturalisation on walls or old ruins, in places where it does not occur naturally, and also on the margins of the pathways or the less important surfaces of the rock-garden.

SEDUM ANACAMPSEROS.—*Evergreen Orpine.*

A SPECIES easily recognised by its very obtuse and entire glaucous leaves, closely arranged in pyramidal rosettes on the prostrate branches that do not flower. The rose-coloured flowers are in corymbs, not very ornamental, but the distinct aspect of the plant will secure it a place on the rockwork, or among very dwarf border-plants. A native of the Alps, Pyrenees, and mountains of Dauphiny, flowering in summer, easily propagated by division, and thriving in any soil.

SEDUM BREVIFOLIUM.—*Mealy Stonecrop.*

ONE of the most fragile and interesting of alpine plants, very nearly allied to *S. dasyphyllum,* but recognised at a glance by its pleasing, pinkish, mealy tone, without reference to the botanical characters which divide them. A native of the Southern Pyrenees and Corsica, in dry places, and somewhat too delicate for general planting in the open air; but it may be grown in dryish soil in sunny well-drained parts of rockwork. In small pans or pots it may be grown to great perfection in pit or frame, or the airy shelf of a cool greenhouse, the higher degree of warmth causing the plant to attain a fuller habit than it does out of doors in England. A very interesting subject for naturalisation on old ruins and walls, and nothing can be more pleasing than healthy tufts of it, though it is to its leaves rather than its flowers that it owes its attractiveness. *S. farinosum* resembles this in appearance, but so far as my experience goes, it is tender.

SEDUM DASYPHYLLUM.—*Thick-leaved S.*

A SINGULARLY pretty species, of a pleasing glaucous colour; indeed, not unfrequently the plant is of an amethystine blue tone. The leaves are usually perfectly smooth, very thick, and fat—in fact, quite swollen at the back—and very densely packed. The flowers are not ornamental, being of a dull white, tinged with rose, but the peculiarly neat habit and attractive hue of the plant, when not in flower, will always make it a favourite in collections of dwarf plants. It occurs abundantly on rocks, old walls, and humid stony places, in Southern and South-western Europe, and is found in some places in the South of England. Although hardy on walls and rocks, it has not the vigour and constitution of many of the other Stonecrops, and it is desirable to establish it on an old wall or dry stony part of rockwork, so as to secure a stock in case the plant perishes in winter on low ground, as I have seen it do occasionally. It is very suitable for association with such plants as the Cobweb Houseleek, and is an interesting subject for naturalisation on old ruins.

SEDUM EWERSII.—*Ewers's S.*

AN exceedingly neat, distinct, and diminutive species, with smooth, opposite, glaucous, and broad leaves, and purplish flowers in terminal corymbs, the whole plant being of a pleasing silvery tone and rather delicate appearance, but quite hardy, easily increased by division, and flowering in summer. A native of the Altai Mountains; merits a place on every rockwork and in collections of the dwarfest hardy succulent plants, rarely rising above two or three inches high.

SEDUM GLAUCUM.—*Glaucous S.*

A MINUTE species of a greyish tone, forming dense spreading tufts of short stems, densely clothed with fat leaves, and rather sparsely producing somewhat inconspicuous flowers. The very neat habit of the plant has caused it to become quite popular in our gardens of late years as a minute edging or surfacing plant; for edging purposes it is perhaps better divided every spring; thin but regular lines planted at that season forming neat, swelling, chubby-looking edgings four or five inches across in autumn. On the rockwork it may be used in any spot that is to spare, either to form a turf under other plants or for its own sake. Various other Sedums are very nearly allied to this, and all are probably but forms of one kind. A native of Hungary.

SEDUM KAMTSCHATICUM.—*Orange Stonecrop.*

A BROAD-LEAVED species, not unlike *Sedum spurium* in habit, but at once distinguished by its dark orange-yellow flowers. It is a prostrate plant, quite hardy, succeeding in almost any soil, but best in a warm rich loam, and flowering profusely in summer. Highly suitable for the rougher parts of the rock-garden, where it will take care of itself, and is a capital plant for the margin of the mixed border. It and *S. spurium* are much more worthy of being employed as edging-plants than the dull-coloured *S. denticulatum* and *S. oppositifolium*, frequently grown for that purpose.

SEDUM POPULIFOLIUM.—*Shrubby Stonecrop.*

DISTINCT from all its race, and forming a small, much-branched shrub, from six to ten inches high, with flat coarsely toothed leaves, and whitish flowers with red anthers. Not an ornamental plant, but being so different in habit to the other members of the family, it is worthy of a place in large and botanical collections. It grows in any soil, blooms rather late in summer, and comes from Siberia.

SEDUM PULCHELLUM.—*Purple American Stonecrop.*

A VERY neat species, at once distinguished by its purplish flowers arranged in several spreading and recurved branchlets, bird's-foot fashion, with numerous spreading stems densely clothed with alternate obtuse leaves. It is abundant in North America, and at present very rarely seen in our gardens, though far more worthy of cultivation than many commonly grown. In

France I have seen it a good deal used as an edging-plant, for which purpose, as well as for rockwork, it is well suited; and it is also a highly appropriate plant for the front margin of a mixed border, flowering in summer, growing in any soil, and easily increased by division.

SEDUM RUPESTRE.—*Rock Stonecrop.*

A GLAUCOUS densely-tufted plant, with numerous spreading shoots, these shoots generally rooting at the base and erect at the apex. It has rather loose corymbs of yellow flowers, and is frequently grown as an edging and border plant in gardens, though not so ornamental as some rarely grown kinds. There are several varieties or sub-species, notably the British *S. elegans*, and the green-leaved *S. Forsterianum*. A native of Britain and various parts of Europe, and of the easiest culture.

SEDUM SIEBOLDII.—*Siebold's S.*

A WELL-KNOWN and elegant species, frequently cultivated in pots, with roundish leaves, bluntly toothed in their upper part, of a pleasing glaucous tone, in whorls of three on the numerous stems that in autumn bear the soft rosy flowers in small round bouquets. At first the boldly-ascending stems form neat tufts, but as they lengthen, they bend outwards with the weight of the buds and flowers at the points, making the plant a graceful object for pots, small baskets, or vases. It is hardy, and merits a place on the rockwork, especially in positions where its graceful habit may be seen to advantage—that is to say, where its branches may fall without touching the earth; but except in favoured places, it does not make such a strong and satisfactory growth as most of the other Sedums, and is perhaps seen to greatest advantage as a frame or greenhouse plant. A native of Japan; easily propagated by division. In late autumn the leaves often assume a lovely rosy-coral hue. There is a variegated variety, but it is not so good as the ordinary form.

SEDUM SPECTABILE.—*Showy Stonecrop.*

THIS is one of the finest autumn-flowering plants introduced of late years—being at once distinct, perfectly hardy, fine when its delicate rose-coloured flowers, in very large heads, are in bloom, and pretty long before it flowers, from its dense bush of glaucous leaves; it has been gradually making way in our gardens under the names of *S. Fabarium*, *Fabaria*, and *Fabarinum*, all of which are wrong. It was very properly named *S. spectabile* by M. Boreau, curator of the Botanic Gardens at Angers, who has paid much attention to the group to which it belongs. Most valuable for association with hardy plants in beds, for use around shrubberies, as a pot-plant, a first-class border-plant, and also for banks, or grouped with the most vigorous subjects in the rougher parts of the rock-garden. It begins to push up its fleshy glaucous shoots in the very dawn of spring, keeps growing on all through the early summer, opens its flowers in early autumn, and continues in full perfection till the end of that season, worthily associating with such fine, autumn-flowering, hardy plants as the Tritomas, white Japan Anemone, and the broad-leaved Sea Lavender. The plant is one of the easiest to propagate and grow that has been introduced to this country, and forms round, sturdy, bush-like tufts of vegetation, eighteen inches or more high when well established in the full sun. A native of Japan.

SEDUM SPURIUM.—*Purple Stonecrop.*

SEVERAL kinds of Sedum, with large, flat, crenate leaves, occur in our gardens, of which this is much the best, its rosy-purple corymbs of flowers being handsome compared to the dull whitish flowers of allied kinds. A native of the Caucasus; exceedingly well suited for forming edgings, the margin of a mixed border, or the rockwork, and not at all sufficiently grown in gardens. It is of the easiest culture and propagation, and blooms late in summer, and often through the autumn. The leaves are slightly ciliated.

The preceding are the most distinct and ornamental kinds in cultivation. The pretty *S. cœruleum* is an annual, and *S. carneum variegatum* not hardy enough to stand our winters. Several Sedums with a monstrous development of stem, or what in botanical language is called *fasciation*, are in our gardens: *S. monstrosum*, *cristatum*, and *reflexum monstrosum*, to wit. The following is an enumeration of other species, or reputed species, now in cultivation in this country, the most desirable being marked with an asterisk. They are almost without exception of the easiest culture and rapid increase in ordinary soil.

Sedum aizoides	Sedum elongatum	Sedum orientale
Aizoon	Fabaria	pallens
albescens	*farinosum; rather	*pallidum
altaicum	tender	pruinosum
altissimum	Forsterianum	pulchrum
anglicum	grandifolium	reflexum
angulatum	*hispanicum	*sexangulare
arboreum	hispidum	*sexfidum
asiaticum	ibericum	*speciosum
aureum	involucratum	stellatum
Beyrichianum	Jacquini	Stephani
Brauni	libanoticum	telephioides
cæruleum	littoreum	Telephium
*corsicum	Lydium	teretifolium
cruciatum	Maximowiczii	ternatum
*cruentum	maximum	triangulare
*cyaneum	*monregalense	*Verloti
dentatum	*multiceps	villosum
denticulatum	neglectum	virens
*elegans	ochroleucum	Wallichianum

SEMPERVIVUM ARACHNOIDEUM.—*Cobweb Houseleek.*

ONE of the most singular of alpine plants, its tiny rosettes of fleshy leaves being covered at the top with a thick white down, which intertwines itself all over each plant like a spider's web. Widely distributed over the Alps and Pyrenees, this plant is perfectly hardy in our gardens, in which, however, it is rarely seen, except as a frame-plant. It thrives in exposed spots, in sunny arid parts of rockwork, forming sheets of whitish rosettes, which look as if a thousand fine-spinning spiders had been at work upon them, and send up pretty rose-coloured flowers in summer. About London it sometimes suffers from the sparrows plundering the "down." It should be on every rockwork; is easily increased by division, and thrives in moist sandy loam.

SEMPERVIVUM CILIATUM.—*Fringed Houseleek.*

THE margins of the leaves of this species are edged with transparent hair-like bodies, which give the whole plant a distinct appearance. The leaves are barred lengthways with brown and deep-green stripes. Flowers freely in summer, in close corymbs of many fine golden-yellow flowers, each scarcely half an inch across when fully expanded. It is, when grown at all in this country, usually kept in the greenhouse, but is hardy in warm well-drained spots on rockwork. It ought to be placed in some dry spot under a ledge of rock, and might be tried with advantage on the top of an old wall or on a ruin. A native of the Canary Islands; like the other Houseleeks, easily increased by division or cuttings.

SEMPERVIVUM MONTANUM.—*Mountain Houseleek.*

A DARK-GREEN kind, smaller than the common Houseleek, with a very pleasing, almost geometrical, arrangement of leaves, which are pubescent and glandular on both sides, ciliated, forming neat rosettes, from which spring dull rosy flowers in summer. It is very suitable for forming edgings or for rockwork; like all the others, grows in any soil, and, like all its fellows, is very easily propagated. A native of the Alps. When masses of it are in flower, they are visited by great numbers of bees.

SEMPERVIVUM SOBOLIFERUM.—*Hen-and-Chicken S.*

ONE of the neatest and most distinct in appearance of the family, particularly distinguished by growing in firm dense tufts, and throwing off little round offsets so abundantly that these are pushed clear above the tufts, and lie rootless, small, brownish-green balls on the surface. The full-grown rosettes are of a peculiarly light green, and of a decided chocolate brown at the tips of the under side of the leaves, for nearly one-third of their length. The small leaves of the young rosettes all turning inward, they appear of a purplish-brown colour. The rosettes are usually not more than one inch and a half in diameter, but I have seen them in France more than three inches; however, whether they were the rosettes of a form larger naturally than the common one, or the result of a higher culture, I cannot say. The plant, which I have not seen in flower, is admirably suited for forming wide tufts on rockwork, on banks beneath the eye. It grows freely in any soil.

SEMPERVIVUM TECTORUM.—*Common Houseleek.*

A NATIVE of rocky places, in the great mountain ranges of Europe and Asia, but which, having been cultivated from time immemorial on housetops and old walls, is well known to everybody. It is needless to describe the culture of a plant which thrives on bare stones, slates, and in the most arid position. It, like some less known species, may be used in flower-gardening for forming dwarf borders, &c., though it would be better to give a position in gardens to somewhat rarer species. It varies somewhat, a glaucous form called *rusticum* being one of the most distinct.

SEMPERVIVUM CALCAREUM.—*Glaucous Houseleek.*

THE Sempervivum now becoming very common in cultivation, under the garden name of *S. californicum*, is by some considered a very glaucous variety of *S. tectorum*, by others the same as the French *S. calcareum*; it is probably the last, but not having as yet met with the plant in flower, I am not able to determine this point with certainty. Of one thing, however, we may be assured—that no finer Houseleek has been introduced, and that, if not very nearly related to our common one, it is certainly as easily grown and as hardy as that much-enduring old plant. Planted singly, the rosettes of *S. calcareum* sometimes attain a diameter of nearly five inches, and as the leaves are of a decided glaucous tone, distinctly tipped at the points with chocolate, it is deservedly very popular for forming edgings in the flower-garden. It is also admirable for the rockwork, is easily increased by division, and thrives in any soil.

In addition to the preceding, which are among the most distinct and ornamental of the Houseleek race, there are a great number of species, or so-called species, wild in Europe, many of which are often cultivated in botanic gardens. In the following list the more ornamental kinds are marked with an asterisk.

Sempervivum acuminatum	*Sempervivum glaucum	*Sempervivum Pomelli
*anomalum	*globiferum	*Requieni
*arenarium	grandiflorum	ruthenicum
assimile	*Heuffelii	*sediforme
Braunii	juratum	stenopetalum
canescens	Mettenianum	urbicum
Cotyledon	molle	velutinum
dioicum	Neilreichii	villosum
*Funckii	*piliferum	

The under-mentioned kinds I first observed in cultivation in the Jardin des Plantes, at Paris. They are mostly sorts desirable for cultivation.

Sempervivum affine	Sempervivum Dœllianum	Sempervivum Schle-
albidum	Fauconetti	ani
barbatulum	fimbriatum	* Verloti
* Boutignianum	* Pseudo-arachnoi-	violaceum
Comollii	deum	

Umbilicus chrysanthus is frequently associated with the Sempervivums in botanic gardens, and much resembles a small Houseleek, but has spikes of golden flowers. It is worthy of a place in large collections, and is suitable for rockwork. The small reddish and thick-leaved sedum-like *U. sedoides* of the Pyrenees is also an interesting kind, and *U. spinosus* a remarkable-looking one. I have not seen it in cultivation in the open air except in the botanic garden at Geneva. Dry sunny parts of the rockgarden will be found most congenial to these plants.

SENECIO ARGENTEUS.—*Silvery Groundsel.*

A STURDY but minute silvery plant, almost like a diminutive of the popular *Centaurea ragusina*. The leaves are quite silvery, and vary from half an inch to one inch and a half long, the footstalk of the leaf channeled, and the blade cut into rounded lobes. The whole plant is not more than two inches high when fully developed and established; it stands any weather, and will live everywhere in sandy soil in well-drained borders. It will prove valuable for rockwork or borders, and, being well fitted to form beautiful dwarf edgings, will probably become very popular. The flower is not attractive, but, like the *Centaurea* and *Cineraria maritima*, the plant is valuable for the effect of its foliage. A native of the Pyrenees; increased by division.

SENECIO UNIFLORUS.—*One-flowered Groundsel.*

A VERY silvery dwarf species, growing little more than an inch high, very suitable for rockwork, but scarcely equal to the preceding, and not so easily grown. The flowers are poor, and should be removed, as tending to weaken and disfigure the plant. Increased by seed and division. A native of Switzerland, and perfectly hardy. *S. incanus* is another pretty dwarf alpine species, but hardly so fine as either of the preceding, or so easily grown.

SILENE ACAULIS.—*Cushion Pink.*

TUFTED into dwarf light-green masses like a wide-spreading moss, but quite firm, this plant safely defies the fiercest storms, snows, and arctic cold of numerous mountain climes in northern regions of the globe, from the White Mountains of New Hampshire to the Pic du Midi in the Pyrenees, and always excites admiration in the alpine traveller, covering, as it does, the most dreary positions with glistening and refreshing verdure at all times. In summer the Cushion Pink, or alpine Moss Campion, becomes a mass of pink-rose or crimson flowers barely peeping above the leaves and forming beautiful objects where nearly every other living thing fails from the earth, and making lovely carpets where all else is branded with desolation. Many places on the mountains of Scotland, Northern Ireland, North Wales, and the mountains in the Lake District of England, are quite sheeted over with its firm flat tufts of verdure often several feet in diameter. It is in cultivation as beautiful and distinct as in a wild state, and grows freely in almost any soil on rockwork, or in pots and pans. It may also be grown as a diminutive spreading border-plant, where borders are made with a view of growing such fairy plants. In a small state it would not be so easily seen as many we now grow, though, when spread out into wide tufts, it is visible enough. This plant is indispensable for rockwork, and those who have the opportunity would do well to carefully transfer several old established patches from the mountains to humid but sunny slopes on the rockwork, in peaty or sandy soil. It is, however, not a slow grower, and is easily increased by division. There are several varieties: *alba*, the white one; *exscapa*, with the flower-stems even less developed than in the usual form, and *muscoides*, dwarfer still; but none of them are far removed from the common plant or of greater importance either from a horticultural or botanical point of view.

SILENE ALPESTRIS.—*Alpine Catchfly.*

POSSESSES every quality that renders an alpine plant worthy of extended garden culture—great beauty of bloom, perfect hardiness, very dwarf and compact habit, growing only from four to six inches high, and a constitution that enables it to flourish in any soil. It flowers in May, the flowers being of a pure and polished whiteness, with the petals notched, and abundantly produced over the shining green masses of leaves, and is one of those plants which should be used in abundance on every well-made rockwork, with the Aubrietias, and tiny shrubs like *Menziesia empetriformis*. To secure a perfect bloom, it should not be disturbed, but allowed to spread forth into established tufts, and, not being fastidious as to soil, will require no attention except removing the flower-stems when the starry blooms have passed away. Like most high-mountain plants, it should have perfect exposure to the full sun; it should never be elevated amongst burrs or stones in such a position that a dry wind may parch the life out of the tiny roots, so unwisely cut off from revelling in the deep moist earth, and it should be protected from being overrun by coarser plants. I once regretted to see a colony of ants take up their abode under a tuft of this plant, and begin to elevate the soil amongst its tiny leaves; but as the ants built their hill, the Silene expanded its leaves, and finally grew to be quite a little pile of starry snow, finer than any of its neighbours. A native of the Alps of Europe; very readily increased by seed or by division. Some individuals of this species are quite sticky from viscid matter, and others perfectly free from it.

SILENE ELISABETHÆ.—*Elizabeth's Catchfly.*

A REMARKABLY beautiful, and as yet very rare, alpine plant, quite distinct from all its brethren, the flowers looking more like those of some handsome but diminutive Clarkia than those of the commonly grown Silenes. They are very large, of a bright rose colour, and with the claws or bases of the petals white, from one to seven being borne on stems three or four inches high, springing from tufts of acutely pointed and shining, slightly viscid, and pubescent leaves, half an inch broad. This is usually considered one of the plants difficult to cultivate, but when we have once secured strong plants, it will be found as easily grown as the Cushion Pink. It is rare in a wild state, occurring in the Tyrol and Italy, where I had the pleasure of gathering it on the top of Monte Campione, growing amidst shattered fragments of rock, and in one case in a flaky rock without any soil. It grows freely enough in sandy soil in a warm nook on rockwork, as I observed in M. Boissier's garden, in Switzerland. Flowers in summer, rather late, and is easily increased by seeds.

SILENE MARITIMA.—*Sea Catchfly.*

A BRITISH plant, not uncommon on sand, shingle, or rocks by the sea, or on wet rocks on mountains, forming level carpets of smooth glaucous leaves, from which spring generally solitary flowers about an inch across, and white, with purple inflated calyces. The handsome double variety of this plant, *S. maritima fl. pl.*, is well worthy of culture not only for its flowers but for the dense, sea-green, spreading carpet of leaves which it forms, and which make it particularly suitable for the margins of raised borders, for hanging over the faces of stones in the rougher parts of rockwork, or for the front edge of the mixed border. The flowers appear in June, and, in the case of the double variety, rarely rise more than a couple of inches above the leaves, which form a turf about two inches deep. Mr. Bentham unites this plant with the tall, ugly, and straggling Bladder Campion (*S. inflata*), but they are distinct, emphatically so considered from a gardening point of view, one being quite a weed and the other an ornamental plant.

SILENE PENNSYLVANICA.—*Wild Pink.*

THE wild Pink of the Americans is a dwarf and handsome plant, with narrow spoon-shaped and nearly smooth root-leaves, those on the stems lance-shaped, forming dense patches, and producing clusters of six or eight purplish-rose flowers, about an inch across, notched, and borne on stems from four to seven inches high, somewhat sticky, and hairy. A native of many parts of North America, in sandy, rocky, or gravelly places, flowering from April to June, and growing very freely in deep sandy soil. This plant is a fine ornament to rockwork, and will probably prove very useful for borders, in both positions requiring a certain degree of shade. It has only recently been introduced to cultivation, and is increased freely by seeds or cuttings.

SILENE PUMILIO.—*Pigmy Catchfly.*

A RARE and interesting species from the Tyrol, resembling the Cushion Pink of our own mountains in its dwarf firm tufts of shining green leaves, which are, however, a little more succulent and obtuse, and bearing much larger and handsomer rose-coloured flowers, rising taller than those of *Silene acaulis*, and yet scarcely more than an inch above the flat mass of leaves, so that the whole plant seldom attains a height of more than between two and three inches. This plant has been but recently re-introduced to cultivation by Messrs. Backhouse, of York, and will be found to thrive as well on our rockworks as *Silene acaulis*. It should be planted in deep sandy loam on a well-drained and thoroughly exposed spot, sufficiently moist in summer, facing the south, a few stones being placed round the neck of the young plant to keep it firm and prevent evaporation. Once it begins to spread, it will take care of itself. There is a white variety, but it is not in cultivation.

SILENE SCHAFTA.—*Late Catchfly.*

A MUCH branched plant, not compressed into hard cushions like the alpine, stemless, or dwarf Silenes, but withal forming very neat tufts, from four to six inches high, and becoming covered with large purplish-rose flowers from July to September, and even later. It comes from the Caucasus, is perfectly hardy, and a fine ornament for the front margin of the mixed border, but is particularly suitable for almost any position on rockwork. In planting it, it may be as well to bear in mind its late-flowering habit; it should not be used where a spring or early summer bloom is chiefly sought, but it may be employed in the summer flower garden in edgings to permanent beds, in the small circles round standard roses, &c., with better effect than most alpine plants, and is easily raised from seed or increased by division of established tufts.

SILENE VIRGINICA.—*Fire Pink.*

A BRILLIANT perennial, with flowers of the richest and brightest scarlet, nearly or quite two inches across, and sometimes more; the petals long, rather narrow, and with a deep notch at the end of each dividing it into two lobes; the lower leaves of a dark brown tone, and spoon-shaped, higher leaves lanceolate; stems, a chocolate brown, very brittle, thinly furnished with short hairs, and slightly viscid. It seems somewhat straggling in habit, is hardy and perennial, and, as the colour is as fine as that of the old scarlet Lobelia, will prove a very popular and excellent border and rock plant. A native of America, increased by seeds and division, growing from one to two feet high, and therefore most suited for association with the Aquilegias and taller alpine plants, and not with the dwarf or delicate sorts

Having in cultivation such brilliant and distinct plants as the preceding Catchflies, we must consider *Silene Zawadskii*, dwarf and with white flowers, the diminutive soft-tufted *S. quadridentata* (for which *S. alpestris* is often mistaken), the woody *Silene arborescens*, a dwarf, shrubby, evergreen species with rose-coloured flowers, and the dirty-white *Silene Saxifraga* —only worthy of a place in very large collections or in botanic gardens. *Silene rupestris*, a sparkling-looking, dwarf, white species, little more than three inches high when in bloom, and reminding one of a dwarf *S. alpestris*, is better worthy of a place. *Silene pendula*, a handsome, rose-coloured, biennial plant, now much used in the spring flower garden, is well worthy of being naturalised in stony or bare sandy places.

SMILACINA BIFOLIA.—*Two-leaved S.*

A SMALL plant, allied to the Lily-of-the-valley, but with smaller whitish flowers in close erect racemes or spikes, from one inch to one inch and a half long, on stems three to six inches high, each flowering-stem having usually two leaves, heart-shaped, but with the lobes and points somewhat elongated; the radical leaves larger, rounder, and heart-shaped. A very graceful and easily-grown little plant, found in woody places, and common on the Continent and in America in moist and mountain woods. It is found very plentifully near Scarborough, and occurs in Caen Wood, at Hampstead, and one or two other places in Britain, but probably is not truly wild. In cultivation it thrives either in shaded places, or, when fully exposed, forms crowded tufts of smooth leaves, a few inches high, freely spiked with flowers in early summer. It is most ornamental when it has spread into tufts a foot or eighteen inches in diameter, the chief care required being to keep the spot in which it grows free from coarser plants. It may be tastefully used in the rock-garden or in bare mossy spots near the hardy fernery, and is easily increased by division of its creeping white root-stalk.

SOLDANELLA ALPINA.—*Alpine S.*

ONE of the most interesting plants that live near the snow-line on many of the great mountain-chains of Europe—not brilliant, but withal beautiful in its pendent, pale-bluish, open, bell-shaped flowers, cut into numerous, narrow, linear strips, three or four being borne on a stem from two to six inches in height, and springing from a dwarf carpet of leathery, shining, roundish, or kidney-shaped leaves, nearly entire or obscurely indented. It is comparatively rare in gardens, is usually grown in pots, included among subjects considered very difficult to cultivate, and kept in frames, but if healthy young plants of it are placed out of doors on the rockwork or a raised border, in a little bed of deep and very sandy loam, they will be found to succeed perfectly well, especially in all moist districts, and in dry ones it will be easy to prevent evaporation by covering the ground near the young plants with some cocoa-fibre mixed with sand to give it weight. I have seen a perfect carpet, several feet square, of this plant growing on a bed of fine moist sandy earth on a flat spot in an old rockwork, in this country, and no specimens I saw in the Alps equalled it in luxuriance. The most suitable position for the plant is a level spot on the rockwork near the eye.

S. montana is very nearly allied to the preceding; in fact, except that it is usually somewhat larger in all its parts than *alpina*, and the flowers are of a bluer purple, there is no great difference in its character. It also inhabits several of the great continental chains, and will be found to thrive under the same treatment as the preceding. Both are readily increased by division, though, as they are usually starved and delicate from being confined in small worm-defiled pots, exposed to daily vicissitudes, they are rarely strong enough to be pulled in pieces. *S. pusilla*, with kidney-shaped leaves, heart-shaped at the base, and the corolla not nearly so deeply cut into fringes, and the very small *S. minima*, with minute round leaves and one flower fringed only for a portion of its length, are also in cultivation, though rare. They will thrive under the same conditions, but, being much smaller, especially the last, require more care in planting, and should be associated with the most minute alpine plants, in a mixture of peat and good loam with plenty of sharp sand, and get abundance of water in summer, especially in dry districts.

SPIGELIA MARILANDICA.—*Wormgrass.*

A MOST distinct and beautiful plant; the flowers being tubular, an inch and a half long, crimson outside and yellow within, from three to eight borne on a stem from six to fifteen inches high, and as, when the plant is well grown, these stems come up very thickly and form close erect tufts, the effect, when in bloom, is very brilliant. A native of rich woods in North America, from Pennsylvania to Florida and Mississippi, flowering in summer, and increased by careful division of the root. I have not seen it grown to perfection except in deep and moist sandy peat. It is very rare in our gardens, in which it should be placed on the warm side of the rockwork in the soil above mentioned.

STATICE TATARICA.—*Tartarian Sea Lavender.*

A BEAUTIFUL, dwarf, hardy species, with rosy or reddish flowers, contrasting prettily with the white membranaceous bracts; the leaves leathery, smooth, of a deep green, oblong, lance-shaped, and pointed; the flower-stems nine to twelve inches high, much branched, forming a wide-spreading inflorescence. This is the prettiest and most distinct of the dwarf Sea Lavenders I have seen in cultivation, its rose-tinted flowers quite removing it from the often-seen blue sorts, some of which are scarcely ornamental. There is a variety with narrow leaves and reddish flowers, *S. tatarica angustifolia.* Both bloom in autumn, thrive best in deep, well-drained, sandy loam, and are admirable for low sunny ledges or banks on or near rockwork. Increased by careful division or by seeds sown in spring, and comes from Tartary. *S. oleæfolia, globulariæfolia, eximia,* and several other species, are also suitable rock-plants.

SYMPHYANDRA PENDULA.—*Pendulous S.*

A CAMPANULA-LIKE plant, with branched pendulous stems, velvety, toothed, and ovate leaves, and very large, cream-coloured, drooping flowers (which are almost hidden amongst the leaves), bell-like and velvety at the throat. It is a very hardy dwarf plant, rarely reaching a foot high; a native of rocky places in the Caucasus, and easily increased by seed. In consequence of its pendulous habit, it is seen to best effect when elevated to the level of the eye in the rock-garden, but it is also a first-rate border-plant, and thrives in ordinary garden soil.

TEUCRIUM MARUM.—*Cat Thyme.*

I SHOULD no more have thought of including this in the present selection than I should the British Oak, previous to one beautiful afternoon in July, 1868. Sailing to one of the islands on Lago Maggiore, I noticed a charming mass of lilac flowers, on some plant that, from the great profusion of its bloom, appeared to be a dwarf well-grown Heath. I was pleasingly surprised to find it our old friend the Cat Thyme, which, flowerless and neglected, used occasionally to be seen in old greenhouses. Here in this dry old wall it had found a most congenial home, and become a perfect mass of flowers. This suggested that its true garden home was not in the greenhouse, but on some dry old sunny wall, or in a chalk pit or very dry spot on the southern face of rock-work. And indeed wherever there are cats the wall would seem to be the only way of preserving it, for they are desperately fond of it. I once placed a bushy old plant of it in the open air in early summer, and in passing a few days afterwards noticed it had disappeared, but, on looking closely, observed a stout stump about two inches high arising from one of the pots and quite covered with cats' hairs, just like a stake in a sheep-gap, and this was all they had left of our pungent Cat Thyme. Therefore a precipitous spot on a wall or very dry bank in the sunny and warmer districts is what is wanted for it, for two reasons. It is somewhat like the common Thyme, but quite grey, and more wiry and taller ; but when grown out of doors, as I suggest, it is dwarf and neat in habit. Hitherto it has in this country been grown as a greenhouse under-shrub, as which it has no merit except as a curiosity. A native of Spain ; readily increased by cuttings.

THALICTRUM ANEMONOIDES.—*Rue Anemone.*

A DELICATE, diminutive, and interesting species, with the "habit and frondescence of Isopyrum, the inflorescence of Anemone, and the fruit of Thalictrum." These qualities, in addition to its dwarf habit, usually only a few inches high, make it worthy of cultivation. The flowers are white, nearly an inch in diameter, open in April and May, the flower-stem bearing a few leaves near the summit, so as to form a sort of whorl round the flowers. A native of many parts of North America, increased by seed or by the division of its tuberous roots. There is a pretty double variety, *T. anemonoides fl. pl.*, with the flowers somewhat smaller than those of the single one, and very neat. Being small and fragile in its parts, it requires a little more care than most of its brethren, should have a light peaty and moist soil, and be associated with other delicate growers, or placed in a position where it is not liable to be overrun by coarse neighbours. = *Anemone thalictroides*.

THALICTRUM MINUS.—*Maidenhair Meadow Rue*.

PERHAPS of all the flowerless plants introduced to this country, none has given so much pleasure as the Maidenhair fern (*Adiantum cuneatum*), found in every stove and fernery, and which is in such great demand in every garden for mingling with cut flowers. I cannot give a plan by which our Brazilian friend may be grown in the open air in Britain, but have a substitute to recommend, which is as hardy as the common Crowfoot. *Thalictrum minus* is the plant, a native of Britain, but also found on the Continent and in Russian Asia. By pinching off the small, weak, and inconspicuous blooms that appear in summer, the plant presents a good resemblance in outline to the Maidenhair fern—looks, in fact, like a well-grown plant of *A. cuneatum* brought out of the stove and plunged in the open border. Singular to say, the finely dissected and elegant leaves are equally well adapted for mingling with cut flowers, and better in one respect, as they are of a pretty firm and wiry consistency, and do not fade quickly like those of the fern. It would form an excellent subject to plant in the mixed border, on the rockwork, or indeed in the flower-garden as a green edging. It has scarcely ever been grown in our gardens for these purposes, and for that reason I, having ascertained its merits, here speak so favourably of it. It will thrive in any soil, and requires no trouble whatever after planting, unless pinching off a few flowers may be so considered.

THLASPI LATIFOLIUM.—*Showy Bastard Cress*.

A DWARF but strong-growing plant, with large indented root-leaves and corymbs of pretty white flowers, somewhat like those of *Arabis albida*, but a little larger, and of a pure paper-white, appearing early in March. It is worth growing with the earlier, hardier, and more vigorous spring flowers such as the Aubrietias and yellow Alyssum, and is more ornamental than either *Arabis procurrens* or *A. alpina*, though not nearly so much known or grown. It comes from the Caucasian and Iberian mountains, and is easily increased by division. Being an early, vigorous, and showy plant, it is well suited for rough rockwork, or for naturalisation in rocky places and by wood walks.

THYMUS LANUGINOSUS.—*Downy Thyme.*

OF the various sorts of Thyme, this is, I am inclined to think, the most worthy of cultivation. It is usually considered a very woolly variety of *T. Serpyllum*, our common British Thyme, but, placed under the same conditions, it is a far more ornamental plant, pleasing at all seasons, and forming wide cushions in any soil, provided it be thoroughly exposed to the sun. Few plants are more suited for the most arid parts of rockwork, and for those in which, from various causes, many other plants will not thrive, though it spreads so quickly into wide dense cushions that it ought not to be placed near any delicate or very minute alpine plants.

Various other kinds of Thyme are worthy of a place on the dry arid slopes of the large rock-garden and on old ruins, but space forbids any more than the enumeration of them here. The minute, creeping, and strongly peppermint-scented *Thymus corsicus*, with flowers so small that they are almost invisible, should be planted on every rockwork, where it will soon become one of the welcome weeds. There is a neatly variegated form of the common garden Thyme, which makes a pretty tufted bush, and many subjects are grown and sold in collections of alpine plants not having half the merits of the Lemon Thyme as rock-plants. Other species or reputed species in cultivation are—*T. azoricus, azureus, bracteosus, Zygis,* and *thuriferus.*

TRIENTALIS EUROPÆUS.—*Starflower.*

A DELICATE and graceful inhabitant of shady, woody, and mossy places, with erect slender stems, rarely more than six inches high, bearing a whorl of five or six leaves, the largest nearly two inches long; from the centre of the whorl arise from one to four slender flower-stems, each supporting a star-shaped white or pink-tipped flower. A native of Northern and Arctic Asia, America, and Europe, and found in the Scotch Highlands and North of England. With healthy well-rooted plants to begin with, it is not difficult to establish among bog shrubs in some half-shady part of the rock-garden, or in the shade of Rhododendrons, &c., in peat soil. It is very suitable for association with the Linnæa, the Pyrolas, and Pinguiculas, among mossy rocks. Flowers in early summer, and is increased by division of the creeping root-stocks.

TRILLIUM GRANDIFLORUM.—*White Wood Lily.*

THIS, one of the most singular and beautiful of all hardy plants, belongs to a well-known American family, deriving its name from the larger parts being usually arranged in threes. When in good health, each stem bears a lovely, white, three-petalled flower, fairer than the white Lily, and almost as large when the plant is strong; but much depends on the vigour of the specimens. It seems to thrive under almost any kind of treatment, and blooms tolerably well even in small pots in frames. But what a difference between it in that state and when its leaves get large and fleshy, and the plant assumes its natural proportions and becomes a free-growing herb of goodly size in the open air! There can be no doubt as to its requirements— a free deep soil full of vegetable matter, and a shady position either in the hardy fernery or some depressed nook, or, failing such, among the Rhododendrons in peat beds. If placed in a sunny or exposed position, the large soft green leaves are not sufficiently developed, and consequently the plant fails to become strong. In a position much exposed to both sun and wind, I have grown it to perfection by planting it in peat, and keeping it covered with a clouded hand-glass so long as the leaves were above ground. At Biddulph Grange I first saw it in its true glory, forming bushes of the healthiest green, more than two feet high, and spreading out as freely as any border-plant. Every stem bore traces of flower, and it may easily be imagined what pictures of beauty these plants must have been in spring. They were planted in a moist spot, very much shaded by highly-raised root and rock work and shrubs, and perfectly sheltered by the same. In like positions it may be grown as well as in its native woods. Depressed shady nooks in the rock-garden or hardy fernery will suit it admirably. It is rather an uncommon plant; but I once saw it selling in the Nottingham market as cheaply as any common little border-plant.

There are several other species in cultivation—*T. atropurpureum*, *sessile*, and *pendulum*, none of them equal to *T. grandiflorum*, but some of them pretty, and all interesting.

TRITELEIA UNIFLORA.—*Spring Starflower*.

A NATIVE of Mendoza, in South America, with strap-shaped, spreading leaves, above which the flowers stand clear. They are on stems from about six to ten inches high, and are nearly an inch and a half across when the plants are not grown too thickly; colour white, with delicate descending bars of pale blue on the inside. The leaves, when bruised, smell exactly like those of an Onion, the flowers like those of the Persian-Iris—a delicate and grateful perfume. They open with the morning sun, and are conspicuously beautiful on bright days, precisely those in early spring when we are most disposed to visit and admire them, and close in dull and sunless weather. It comes into flower with or before *Scilla sibirica*, and remains during the last days of April still in effective bloom, when the vivid blue of the Squill has been long replaced by green leaves. An exposed position is usually a very bad one for hardy bulbs, as the leaves get lacerated, and the bulbs suffer in consequence; but from lying nearly flat on the ground, the leaves of Triteleia appear to escape injury from this cause; however, the warmer and more sheltered the position and better the soil, the better the bloom. Its bulbs increase as fast as those of Garlic, so that it may be propagated to any extent by division. A few years ago it used to be seen flowering most profusely in pots; but having put a small plant out in pure clay in a most unfavourable situation, I was surprised to find my single weak root flowering boldly after two hard winters in succession, and during the past few years it has become a general favourite. Associated with the best Scillas, *Leucojum vernum*, *Iris reticulata*, dwarf Daffodils, and the like, it forms a charming addition to the select spring garden, and is equally useful for rockwork, borders, or edgings. *Triteleia* (*Leucocoryne*) *alliacea* is a nearly allied plant, but scarcely so ornamental; it will thrive under similar circumstances.

TROPÆOLUM POLYPHYLLUM.—*Yellow Rock T.*

A VERY distinct-looking subject, whether in or out of flower; the leaves glaucous, almost rue-like in tone, and orbicular in outline, but cut into ten or eleven divisions or leaflets, which overlap each other. These leaves are densely crowded on a stem a quarter of an inch thick, at least when the plant is well grown; and when planted on a warm sunny rockwork, the stems creep about, snake-like, through the vegetation around, some to three or four feet in length. The flowers are a deep yellow, and produced as freely as the leaves. It is a tuberous-rooted kind, quite hardy in dry situations on rockwork and sunny banks, where it should not be often disturbed; springs up early, and dies down at the end of summer. It is very well grown by Mr. James Atkins, of Painswick. A native of the Cordilleras of Chili.

TROPÆOLUM SPECIOSUM.—*Brilliant Nasturtium.*

A SPLENDID creeping plant, with long and elegant annual shoots, gracefully clothed with six-lobed leaves, from the axils of which spring such brilliant vermilion flowers that a long shoot of the plant is startlingly effective, especially if seen wandering alone among Ivy leaves, or among verdure of any kind. It has been introduced a considerable time from South America, but, notwithstanding its graceful beauty and perfect hardiness, is but very little grown or known, especially in Southern England. It is impossible to find anything more worthy of a position in which its shoots may fall over or climb up the face of some high rock or bank in the rock-garden—or some open spot in the hardy fernery, or of any other position in which its peculiar beauty may be seen to full advantage. I never saw it more beautiful than when it was clambering through evergreen shrubs, nailed against terrace walls in Scotland. It enjoys a deep, rich, and rather moist soil, apparently flourishing best in cool moist places, or in those near the sea, and not so well in a dry atmosphere. No pains should be spared to establish this plant in a vigorous condition. When a position is selected for it, the soil should be made light, and deep, and free, by the addition of leaf-mould, peat, fibry loam, and sand, as the nature of the ground may require, and the surface should be mulched in summer with an inch or two of decomposed manure or leaf-mould, to prevent excessive evaporation. It will also enjoy a deep bed of manure beneath the roots, and put below the soil in which the young plants are first placed, and is best planted in spring, the roots inserted six or eight inches in the soil, and the young plants well watered. It is best planted where the shoots may ramble among the spray of shrubs, or ferns, or trailers; but, as it must in the first instance be placed on a cleared spot, it is well to put a few branchlets over the roots so that the young shoots may crawl over them when they begin to grow. When established, they may be allowed to take care of themselves, and it is much better to let them have their own wild way than to resort to any kind of staking or support, except that afforded by other subjects growing near.

Mr. W. Smythe, of Elmham Gardens, who grows the plant remarkably well, writes to me:—"It increases almost as freely from its thin white tuberous roots as Bindweed. It should be planted in light sandy mould, is quite hardy, and likes a half-shady situation. My plants do well on a fernery opening to the south, having trees in front of about eight feet or ten feet in height; they will not thrive under trees. It seeds freely, and the seeds, when ripe, are berry-like and of a beautiful blue; but they soon drop, and come up the next spring about the fernery round the old plant. It increases and grows most luxuriantly over large Box-trees six feet high, and over large ferns and logs of wood, forming one mass of bloom, and all who see it stop to admire it and ask its name. It ripens about August and September, and the seeds come up the next spring if sown in light sandy mould in pots, and placed in a greenhouse or pit."

TULIPA CELSIANA.—*Dwarf Yellow Tulip*.

A SPECIES having slightly concave glaucous leaves, the largest nearly an inch across, and bright yellow flowers, much smaller than those of the common bedding Tulips, and, when in clumps and fully open, sometimes reminding one of a yellow Crocus; the outside of the petals is tinted with reddish brown and green.

It begins to flower about the first of May, and usually attains a height of six to eight and sometimes twelve inches. The bulbs emit stolons after flowering. Comes from Southern Europe and the shores of the Mediterranean, and is well suited for rockwork or choice borders, in well-drained sandy soil.

TULIPA CLUSIANA.—*Clusius's Tulip.*

USUALLY our Tulips are great, bold, showy flowers, but in this species we have one delicate in tone, humble in stature, and modestly pretty in appearance. The bulbs are very small, the stem reaching from six to nine inches high, seldom more, and sometimes flowering when little more than three inches high. The flower is small, with a purplish spot at the base of each petal; the three outer divisions of the petals stained with a pleasing rose, the three inner ones of a pure transparent white. A native of the South of Europe, a little more delicate than most of its family, and requiring to be planted in good light vegetable earth in a warm, sheltered, and well-drained position to succeed to perfection. Although so small, it will be the better of being planted rather deeply, say at from six to nine inches, and of being placed in some snug spot, where it need not be disturbed too often. Readily known from other species by the peculiarity of its colouring, and well adapted for the rock-garden or the collection of hardy bulbs.

The two preceding Tulips are among the dwarfest and neatest known, but the many beautiful and brilliant varieties of our florists' and of our early-flowering bedding Tulips should be extensively used in every garden.

TUNICA SAXIFRAGA.—*Rock T.*

A SMALL plant, with narrow leaves, and a profusion of wiry stems, bearing elegant rosy flowers, small, but very numerous, thriving without particular care on most soils, and forming tufts a few inches high. A native of arid stony places on the Pyrenees and Alps, often descending into the low country, where I have found it on the tops of walls. There can be no doubt that it will grow in like positions in this country, and also on ruins, while it is a neat plant for the rock-garden or the margin of the mixed border. It is not unlike a Gypsophila in appearance, is easily raised from seed, and thrives in poor soil.

VACCINIUM VITIS-IDÆA.—*Red Whortleberry.*

A DWARF British evergreen, with box-like foliage, but of a paler green, and with clusters of small white, or pale rose-coloured, flowers, which appear in summer, and are followed by berries about the size of red Currants, very like those of the Cranberry, on wiry stems from three to nine inches high. It forms a neat little bush on rockwork, or in beds in peat soil.

The Marsh Cranberry (*V. Oxycoccos*), a native of wet bogs in Britain, with very slender creeping shoots and drooping dark-rose flowers, requiring wetter soil than the preceding, is also worthy of a place where bog-plants are admired. The American Cranberry (*V. macrocarpum*), a much larger plant, distinguished from the preceding by its oblong-obtuse leaves, and very much larger fruit, is also worthy of a place in moist sandy peat, associated with bog shrubs.

VERONICA CHAMÆDRYS.—*Germander Speedwell.*

A WELL-KNOWN and much-admired little native plant, with ovate, or heart-shaped, hairy leaves, and with hairs curiously arranged in two opposite lines down the stem, while the other portions are bare. The flowers are bright blue, produced in great numbers. It is abundant in nearly all parts of Britain, and may be allowed to crawl about here and there in the less important parts of rock- or root-work. Easily increased by seed or division.

VERONICA PROSTRATA.—*Prostrate Speedwell.*

A DWARF spreading plant, forming dark-green tufts, under six inches high, the leaves lance-shaped or linear; the stems covered with a short down, forming circular tufts, and nearly woody at the base; flowers of a deep blue, but varying a good deal, there being several varieties with rose-coloured and white blooms, appearing in early summer, somewhat earlier than *V. Teucrium*. A hardy and pretty plant, flowering so freely that, when in full perfection, the leaves are often quite obscured by the flowers. A native of France, Central and Southern Europe, occurring on stony hills and in dry grassy places, and, in cultivation, succeeding perfectly in dry sandy soil, though by no means fastidious, and easily increased by seeds or division. An admirable plant for rockwork, its prostrate habit fitting it best for sloping positions or fissures on vertical faces of rocks. It also thrives perfectly well as a border-plant, at least in well-drained sandy soils. It is nearly allied to *V. Teucrium*, but differs by flowering earlier, by having the divisions of the calyx smooth and not ciliated, the lobes of the corolla obtuse at the summit, and the fruit smooth and somewhat smaller.

VERONICA SAXATILIS.—*Rock V.*

A BRILLIANT, blue-flowering, dwarf, bush-like plant, a native of alpine rocks in various parts of Europe, and also in a few places in the Highlands of Scotland, forming very dressy tufts, six or eight inches high. The flowers are a little more than half an inch across, and of a very pretty blue, striped with violet, with a narrow but decided ring of crimson near the bottom of the cup, its base being pure white. Blooms in May and June abundantly, is easily increased by seed or cuttings, grows in ordinary soil, and should be in every rock-garden or collection of dwarf alpine plants.

VERONICA TAURICA.—*Taurian Speedwell.*

A VERY dwarf, wiry, and almost woody species, forming neat dark-green tufts, under three inches high; the leaves crowded, the upper ones distinctly toothed; the flowers a fine gentian-blue, abundantly produced. Perhaps the neatest of all rock Veronicas for forming spreading tufts in level spots, or tufts drooping from chinks, and admirable also for the margin of the mixed border, thoroughly hardy, growing in ordinary well-drained garden soil, and flowering in early summer. Suitable for association with the dwarfest alpine plants and mountain shrubs, being itself indeed a tiny compact prostrate shrub. A native of Tauria; increased by division or by cuttings.

VERONICA TEUCRIUM.—*Teucrium Speedwell.*

A CONTINENTAL plant, somewhat pubescent, with opposite leaves, the upper ones nearly linear, the lower oval-oblong, deeply and unequally incised and toothed; the stems forming spreading masses from eight inches to a foot high, and covered with flowers of an intense blue in early summer. The flowers are at first in dense racemes, which afterwards become much longer, the fine blue corolla having oval segments, the three lower ones pointed. It is an excellent plant either for the rockwork or borders, easily increased by seeds or division, and growing freely in ordinary garden soil.

V. Nummularia, of the Pyrenees, *V. aphylla*, the neat little bushy *V. fruticulosa*, *V. satureifolia*, and *V. candida*, with silvery-white leaves, very suitable for bedding out and edgings, are also worthy of a place.

VESICARIA UTRICULATA.—*Bladder-podded V.*

A PLANT with large yellow flowers, not unlike the alpine Wall-flower in habit and general appearance, but at once distinguished by its bladder-like pods. It usually grows from ten inches to a foot high, and has a vigorous constitution, though I have observed it perish in winter on cold soils. A native of mountains in France, Italy, and Southern Europe generally, usually on calcareous rocks, and most likely to flourish and endure on dry sunny parts of our rockworks in dryish soil. It is very easily increased from seed.

VICIA ARGENTEA.—*Silvery Vetch.*

A SPECIES with silvery and downy leaves, composed of from four to ten pairs of leaflets, and of prostrate habit, but without tendrils, and rarely more than eight inches high, spreading about, however, pretty freely in light, warm, and well-drained soil; the rather large whitish flowers are veined with violet in the upper, and spotted with purple in the lower part. It, however, is not a brilliant plant in flower, but the elegant silvery leaves make it worthy of a place in the rock-garden, in dry warm soil. A Pyrenean plant, rare in gardens; easily increased by division or seed.

VINCA HERBACEA.—*Herbaceous Periwinkle.*

A PLANT much less frequent than our common Periwinkles, and more worthy of culture on rocks, as it is not rampant in habit. A native of Hungary, flowering in spring and early summer, the stems dying down every year unlike those of its more familiar relatives; it thrives best in an open position.

The well-known *Vinca major* is very useful on large rockwork, on masses of rootwork, near cascades, &c., and also in rocky places or banks in the wilder parts of pleasure-grounds or by wood walks. There is a variety called *elegantissima*, finely blotched and variegated with creamy white, and several other variegated varieties, all exceedingly useful in the positions above named. The lesser Periwinkle (*V. minor*), a much smaller plant than the preceding, is also useful for like positions: there are several varieties of it well worthy of cultivation, a white-flowered one (*V. minor alba*), one with reddish flowers, one or two double varieties, and also, as of the larger, several variegated forms.

VIOLA BIFLORA.—*Two-flowered Yellow Violet.*

THIS is a bright little Violet, very widely distributed through Europe, Asia, and America, at present usually seen in such a delicate condition in gardens that few would suspect what a lovely little ornament it is on the Alps, in many parts of which every chink between the moist rocks is densely clothed with it. It even crawls far under the great boulders and rocks, and lines shallow caves with its fresh verdure and little golden stars. It is readily known from any other cultivated species by its very small but bright-yellow flowers, the lips streaked with black, being usually borne in pairs, and by its kidney- or heart-shaped leaves. In our gardens its home will be on the rockwork, running about among such plants as the yellow annual Saxifrage, the Dog Violet, *Arenaria balearica*, &c., in moist and half neglected spots. It will be found especially useful on large rockworks, where rude flights of stone are constructed to give one or more winding pathways over the mass, as it will run through every chink between the steps, and tend to make them, as well as the most select spots, replete with life and interest. If obtained in a small or weakly condition, it may seem difficult to establish, but this is not by any means the case; once fairly started in a moist and half-shady spot, it soons begins to creep about rapidly, and may then be readily increased by division. When well established on suitable rockwork, it is able to take care of itself.

VIOLA CALCARATA.—*Spurred Violet.*

This plant comes very near the well-known *Viola cornuta* in the flower, including the spur, but is easily known by the stipules, which are deeply divided into three lobes at the top, whereas in *V. cornuta* they are broad, leafy, and toothed. The stipule of *V. calcarata* is never toothed, nor wide like a leaf, and the plant is even more readily known by its habit of increasing by runners under the earth, somewhat after the manner of *Campanula pulla*, instead of forming strong leafy tufts like *V. cornuta*. It is a very pretty plant on the Alps, usually in very high situations, amidst very dwarf flowers, sometimes so plentiful that its large purple flowers form sheets of colour, the leaves being scarcely seen amidst the other dwarf plants that form the turf. I have not seen it in cultivation, but have no doubt it would form as charming a plant in the rock-garden as it does in its native wilds. There is a yellow variety, *flava* (*V. Zoysii*).

VIOLA CORNUTA.—*Horned Pansy.*

This fine Pyrenean and Alpine Violet is now to be seen in almost every flower-garden, its pale blue or mauve-coloured and sweet-scented flowers, so abundantly produced, making it very valuable in lines, borders, and mixtures. It has been cultivated for ages in our gardens as a rockwork and border plant, but its value as a continuous bloomer, and consequent capacity for bedding, only came to be noticed a couple of years ago. Generally speaking, it does poorly on dry soils and in warm districts, and exceedingly well in wet places. I have rarely seen anything to equal its appearance in the cold wet climate of East Lancashire, while it looks poor indeed in many gardens in the South. In long lines or ribbons, or large beds, it looks very pretty, the colour being of a quiet though decided tone, but it is in mixtures that it will prove truly beautiful. One of the most beautiful bits of colouring I have ever seen was produced by a mixture of Beaton's variegated Nosegay Geranium and *Viola cornuta*. In many cold and stormy districts, the blue Lobelia, so fine in the South, grows quite to grass instead of flower; that which spoils the Lobelia will highly improve this Violet. It is quite easily propagated by division, cuttings, or seeds.

VIOLA LUTEA.—*Mountain Violet.*

This is one of our native Violets classed by Bentham as a variety of *V. tricolor*, but considered distinct by other botanists, and certainly quite distinct enough for garden purposes. Being called *lutea*, one is surprised to find the flowers of nearly every wild plant of it a fine purple, with a yellow spot at the base of the lower petal. Both forms make very pretty rockwork ornaments, but the yellow one has lately become deservedly popular as a dwarf bedding and edging plant, and it is also a first-class plant for the front margin of the mixed border. In cultivation the yellow form is a very neat and compact plant, rising from two to six inches high, and flowering abundantly from the month of April onwards. The flowers are of a peculiarly rich and handsome yellow, the three lower petals striped with thin lines of rich black. It possesses first-class qualities as a bedding plant, and is less uncertain in its growth on the majority of soils than *V. cornuta*, while it is dwarf, neat, and of a colour much to be desired for the flower-garden. Hitherto we have had no very dwarf yellow plant that could be depended upon, or was at all satisfactory. The Calceolaria, which is of greater height, has of late become most precarious on many soils, so much so that many have given up its culture to a great extent; and therefore this Viola is all the more acceptable to flower-gardeners. Much dwarfer and more compact than *V. cornuta*, it may be planted in front of it with the best taste. More tasteful uses than that, however, must soon be found for it in the margining of choice beds, in forming low and rich mixtures with bright-leaved plants like Amaranthus or Coleus kept very dwarf, and in not a few other ways which will in due time suggest themselves to the flower-gardening reader.

VIOLA ODORATA.—*Sweet Violet.*

This well-known plant is in a wild state widely spread over Europe and Russian Asia, and common in various parts of Britain, but best known from its occurring in almost every garden, and from enormous quantities of it being sold in London, Paris, and many other cities. It is as needless to describe it as the Daisy; besides, its delicious odour distinguishes it immediately from the numerous other Violets. It is too well known to require praise, but it is very seldom used in the best way. The Sweet Violet and most of its varieties may be used in many places where few other things but weeds succeed; it will form carpets for open groves or the fringes of woods, or in open parts of copses, or on hedgebanks, demanding in such positions no care, and rewarding the planter by filling the cold March air with unrivalled sweetness; and in the garden, instead of confining it to a solitary bed for cutting from, as is often the case, it should be permitted to fringe the margins of shrubberies, or the margins of rockwork, or ferneries, or any like places where it may be allowed to exist and take care of itself. It will grow in almost any soil, but succeeds best in free sandy loams, and should be put in such when there is any choice. It is well to naturalise the plant on sunny banks, and fringes of woods, and on the warmer sides of bushy places, to encourage a very early bloom.

The cultivation of the Sweet Violet in gardens is of great importance, not only for the supply of private wants, but also for the vast supply required in large cities. About Paris the cultivation of the Violet for the markets is carried on to a great extent, and in some places near that city three or four acres may be seen covered with them, all in the possession of one cultivator, the ground being well exposed to the mid-day sun and of a rich, free, and warm nature, the plantations being made in spring, and those required during winter being grown in frames. The most successful cultivation of Violets I have observed in this country was at Bicton, carried out by Mr. Barnes. The plants, raised from seed, and having passed the summer in a slightly shady position, were transferred in early autumn to a little temporary border on the sunny side of a long glass-house, sheltered also by other structures near at hand; the border was formed of a few inches of fibrous loam on a row of flag-stones enclosed by an edging of brick, and was made every year afresh. The additional warmth and perfect drainage obtained in this position caused the plants to flower throughout the autumn, winter, and early spring months, and doubtless a similar plan would be very desirable in milder parts of the country. In cold dry parts and gardens in which Violets do not succeed well, and where they are required in mid-winter, it is better to raise a number of healthy plants every year and put them in a light frame in a sunny position in autumn. It is almost needless to say that they may be propagated to any extent by division, but strong, healthy, free-flowering plants are also easily raised from seed; and this is also better sown as soon as convenient after being gathered.

The Neapolitan Violet, a much admired variety of the Sweet Violet, is a little tenderer, and is usually grown in frames in winter. Mr. Barnes recommends the following mode of culture as the best. The plants, after flowering in spring, about the beginning of April, are encouraged to throw out runners by spreading some open sandy soil between the old plants, and afterwards giving a good watering if the weather be dry; a few weeks afterwards the strongest and healthiest of these runners are selected and planted on well prepared ground in a north or shady aspect; during summer the ground is kept clear from weeds, all side runners are cut off, and they are thoroughly watered should drought set in. By October these have formed strong plants, and they are then transferred to cold frames or pits in a sunny position, the plants being carefully taken up with balls, and placed in the pits so that they nearly touch each other. When it is desired to have the blooms early, or to force them, eighteen inches of slightly fermenting material are placed in the bottom of a turf pit, on that six or eight inches of sweet fibrous loam, in which the plants are placed. In fine weather the lights are always taken off, and as much air as possible given at other times.

The Sweet Violet varies a great deal; thus we have the single white and single rose, double white and double rose, the small Russian, the Czar, a very large and sweet variety; the Queen of Violets, with flowers almost as large as those of the double white Cherry; and the perpetual-blooming Violet, well known in France as the *Violette des quatre saisons*. This last differs but slightly from the common Sweet Violet, but is valuable for flowering long and continuously in autumn, winter, and spring. It is the variety used by the cultivators round Paris.

VIOLA PEDATA.—*Bird-foot Violet.*

THE most beautiful of the American Violets, with handsome flowers, an inch across, pale or deep lilac, purple or blue, the two upper petals sometimes deep violet, and velvety like a Pansy; the leaves deeply divided, like the foot of a bird, and the plant very dwarf and compact in habit. In a wild state it inhabits sandy or gravelly soil in the Northern States of America, and is of easy culture in this country, flowering in summer, and increased by seeds or division. It is best adapted for the choice rock-garden, but may also be grown in borders where the soil is sandy and moist. It does freely in pots where alpines are grown in cold frames, and should be amongst those that are grown for exhibition.

VIOLA TRICOLOR.—*Heartsease*.

THE Pansy is usually included under the head of *V. tricolor*, though it is more likely to have descended from *V. altaica*; in any case, a good many kinds seem very nearly allied to that species. But the kinds are so numerous, so varied, and, withal, so distinct from any really wild species of Violet in cultivation, that little can be traced of their origin. Of one thing we may be certain: the parents of this precious race were true mountaineers. Only alpines could give birth to such rich and brilliant colour and noble amplitude of bloom considering the size of the plant. Its season never ends, it blooms often cheerfully enough at Christmas, and is sheeted with delightful gold and purple when the Hawthorn is whitened with blossoms. Such a flower must not be ignored on our rock-gardens, even though it thrive in almost any soil and position. It may be treated as an annual, biennial, or perennial, according to climate, position, and soil. Good varieties are quickly and easily raised from seed, while the plant may be raised freely from cuttings or by division. It is, however, so well known that it is needless to describe its culture, and space forbids the shortest enumeration of its numberless varieties, which are, however, given in detail in many catalogues. In addition to the "florists' varieties," so richly and deeply stained, there is a race of fancy or Belgian Pansies now in cultivation; and there are also what are called bedding Pansies, many of them very fine, with simple colours, yellow, blue, or white. Of these, such fine varieties as Imperial Blue merit a place on the rougher slopes of the rockwork, where a free and fine effect of bloom is desired, as well as any wild species of rock-plant. Although in some soils the Pansy becomes perennial, the flowers on old plants are smaller and less beautifully coloured than those from young ones, and therefore, where a perfect yearly bloom is desired, it is necessary to increase the stock annually from seed or cuttings.

In addition to the Violets here described, other species are worthy of cultivation in large collections, for example: *V. striata, V. canadensis, V. obliqua, V. palmata, V. blanda, V. pennata, V. palmaensis,* and *V. cucullata;* but these are all exceeded in size and beauty of flower by those described, and all surpassed in odour by the Sweet Violet.

VITTADENIA TRILOBA.—*New Holland Daisy.*

A PRETTY Australian composite plant, bearing an abundance of flowers with yellowish disks and rosy-white rays, somewhat like those of a Daisy ; but the plant has a spreading diffuse habit, and forms neat little bushes nearly or quite a foot high. The seed is commonly sold, and the plant may be raised as freely as any annual, sown in frames or on a gentle hot-bed, in March or early in April ; when put out in April in free sandy soil in a sunny position, it flowers abundantly from early summer to late autumn. Even better results are obtained by sowing it in August, keeping it in pots over the winter. I probably should not have mentioned it in this book had I not met with it in North Italy beautifully embellishing rockwork on which it had become naturalised, and I am confident it will do the same in well-drained sandy loam on rockwork, and banks in the southern and milder parts of England and Ireland. Although frequently treated as an annual, it is really a perennial on soils and in positions where not destroyed by wet and frost.

WALDSTEINIA TRIFOLIA.—*Three-leaved W.*

A DWARF but vigorous plant, spreading about with stout but stubby strawberry-like runners. The trifoliate leaves are very deeply cut, and the flowers rich golden-yellow, on dwarf stems, with a dense brush of golden filaments and stamens in the centre. A thoroughly hardy and vigorous-growing subject, good for any kind of rockwork or the margin of the mixed border or shrubbery. Flowers in April, and is as readily propagated as any common weed.

Waldsteinia geoides is also worthy of a place on bare banks, and occasionally as an edging among spring flowers, but it is not so showy as the preceding.

ZEPHYRANTHES ATAMASCO.—*Atamasco Lily.*

A BEAUTIFUL, dwarf, lily-like plant, bearing handsome white flowers tinged with purple, three inches and a half across, on stems from six to twelve inches high. Although growing abundantly in North America, this fine plant is very rare in our gardens, where it is well worthy of culture on every rockwork, or in every collection of hardy bulbs, thriving freely in light, rich, sandy soil, and flowering in early summer. Dotted over a turf formed of some carpet-plant like the Lawn Pearlwort, it would be seen to great advantage when its great bell-like flower opened. The leaves are linear, concave, and fleshy, and appear at the same time as the flowers; the bulb small. It flowers in summer, and is increased by seeds, or division of established tufts.

Isolated rocks on plains eastward of the Rocky Mountains.

AN APPENDIX

OF

New or Rare Alpine Plants, or such as have been omitted in the preceding enumeration in Part II.

ACHILLEA AGERATIFOLIA (*Anthemis Aizoon*) Everlasting Achillea.—A beautiful silvery-leaved plant from the sub-alpine districts of Northern Greece, 4 to 7 inches high, with pure white flowers (resembling perfectly white Daisies), which appear early in summer. The leaves are narrow, tongue-shaped, beautifully crimped, and covered with white down, the lower ones crowded. This is a very neat and distinct-looking plant, and easy of cultivation in light soil on rockwork or in warm borders.

ACHILLEA AUREA (Golden Achillea).—One of the showiest of all the genus, and most useful for border purposes. Grows about 12 inches high; leaves finely cut; flowers bright golden; produced freely on upright stalks. A native of the Caucasus. This plant, which is also known under the name of *Pyrethrum achillæfolium*, is not often met with now-a-days.

ÆTHIONEMA GRANDIFLORUM.—This is of larger and more sturdy habit than *Æ. saxatile*, less spreading and prostrate in habit, less glaucous in tone, and with much larger flowers of a purplish rose colour, in elongated corymbs. It is a valuable plant for the rock-garden, thriving freely in sandy loam, and exceedingly well suited for margins and slightly-elevated rocky banks. As a border plant, it will also thrive where the soil is free and well drained. Being somewhat impatient of transplantation, it is desirable to allow some plants to ripen seed on sunny edges or borders. Seedlings in pots will transplant easily. It deserves a place in every collection of alpine and herbaceous plants.

AJUGA ALPINA (Alpine Bugle).—The true species is by far the finest of the genus. The flowers have the rich, pure blue of Salvia patens, and continue in perfection for a considerable time. As an alpine plant it comes next to Dracocephalum grandiflorum, and succeeds best in bog soil, where its roots have free room to ramble. Under such circumstances it spreads and increases rapidly. *Ajuga*

Moldavica, which grows only 2 inches high, is sometimes wrongly named *A. alpina*. It produces white flowers, and is a nice compact species worthy of cultivation.

ALYSSUM OLYMPICUM (Dwarf Alyssum).—With tiny, almost orbicular, greyish leaves. This is the dwarfest of all the Alyssums, and though not conspicuous in blossom, should have a place in every alpine collection. It likes full exposure, and will only thrive in a sandy well-drained soil. A native of Northern Greece.

ANDROSACE GLACIALIS (Glacier Androsace).—A rather rare and beautiful species, from Switzerland, growing in compact sheets, about 2 inches high, and bearing bright pink, or purplish rose-coloured, solitary flowers with a yellow throat and tube. The leaves are small, tongue-shaped, closely crowded, and forming small rosettes at the ends of the slender, red-tinged stems. Same treatment and positions as for *A. pubescens*.

ANEMONE RIVULARIS (River-side Anemone).—A native of the Himalayas, somewhat resembling *A. Narcissiflora*, growing about 18 inches high, and producing its whitish flowers in abundance. A distinct and good perennial species, readily increased by seeds, which are freely produced.

ANOMATHECA CRUENTA (Crimson-flowered Anomatheca).—A pretty and distinct little bulbous plant, from the South of Africa, growing 6 to 12 inches high. Flowers in summer, of a rich carmine-crimson, the three lower segments marked at the base with a dark spot; tube of the flower long and whitish. Leaves two-ranked, narrow, sword-shaped, and spreading above. Bulb ovate, rather large. It should be planted on warm slopes of rockwork, in very sandy dry soil, or in warm borders among the smaller and choicer bulbous plants: the bulbs to be set rather deep in the soil. It is easily increased by separation of the bulbs or from seed.

ANTIRRHINUM ASARINA (Heart-leaved Snapdragon).—A pretty trailing plant, admirably adapted to hang over a projecting ledge of rockery, where, when kept dry, it will stand our ordinary winters. The flowers are yellowish, and about the size of those of the ordinary Snapdragon. Closely allied to it is *A. molle*. Native of S. Europe.

AQUILEGIA LEPTOCERAS LUTEA (Yellow Long-spurred Columbine). —A new species from North America, and one of the finest perennials ever introduced. It grows in dense tufts, and produces a great abundance of large, golden-yellow, long-spurred flowers in a long succession of bloom, which commences in June. The flowers are much larger than those of *A. cærulea*, and have long, straight horns or spurs. This species is not to be confounded with *A. aurea* of Roetzl, of which the flower is hardly half the size, and of a sulphur-yellow, shaded with green. Same culture and positions as for *A. cærulea*.

AQUILEGIA PYRENAICA (Pyrenean Columbine).—A very dwarf species from the Pyrenees. 6 to 9 inches high, allied to *A. alpina*, but smaller in all its parts. It flowers early in summer, producing from 1 to 3 small blue flowers on each of the almost leafless stems. The leaves at the base are 1 or 2-ternate, with linear segments. It is a very suitable subject for rockwork, the margin of the mixed border, or for cultivation in pots in moist sandy loam.

ARENARIA MULTICAULIS (Many-stemmed Sandwort).—From the south of Spain, resembles *A. balearica*, but has its leaves more ovate and ciliate, and its flowers elevated higher above the foliage and larger than that species; recently re-introduced by Mr. Maw, of Broseley. *Arenaria grandiflora* is a somewhat larger flowered species than *A. montana*, and but rarely seen true in cultivation.

ARNEBIA ECHIOIDES (Russian Bugloss).—A charming plant from the Ural Mountains, 6 to 10 inches high, forming a hardy stem, from which its annual growths are made. Its flowers are yellow on their first expansion, but on the second day five dark spots appear near the throat of the corolla, ultimately assuming almost a black appearance. It is exceedingly rare in this country; in fact, we believe is only to be seen at Comely Bank Nursery in Edinburgh. It is very difficult to increase, as it rarely seeds, and cuttings will not strike.

ASTER BESSARABICUS (Dwarf Amellus Aster).—This variety of *A. Amellus* is the finest of all the Asters as a dwarf border plant, and grows about 12 to 15 inches high; flowers large, bright blue, produced in October; habit good, requires no staking.

ASTER RAMOSUS (Branching Aster).—A neat rock-plant, from the Caucasus. An abundant flowerer. Flowers rosy lilac, nearly as large as those of *A. alpinus*.

ASTER REEVESI (Reeves's Aster).—A very charming dwarf Aster, with slender branching stems 9 to 12 inches high, and very small, linear, acute, Heath-like leaves. It flowers in autumn, producing a great abundance of small white yellow-centered blossoms in a dense pyramidal panicle. Well suited for rockwork or borders, in ordinary soil. N. America.

ASTER SCORZONERIFOLIUS.—A new species introduced by M. Roetzl from the Sierra Nevada of California, resembling *A. Alpinus* in habit, but differing from all other known species in its radical leaves. These are very long, ribbon-like, smooth, entire, channelled, usually 5-ribbed, erect, or nearly so, and longer than the stem, which is simple and about 10 inches high. Stem-leaves sessile, linear-lance-shaped. Flowers terminal, solitary (seldom axillary), large, 'of a pale violet colour with yellow disk. The plant is quite hardy, and flowers in July and August.

ASTRAGALUS ALPINUS (Alpine Astragalus).—This plant, which is considered by some to be a variety of *A. Onobrychis*, is a native of various parts of northern Europe and Siberia, and occasionally, though rarely, found in Britain. It is a prostrate hairy herb, with branching stems varying from a few inches to a foot in length, and producing, in summer, short racemes of bluish-purple (sometimes whitish), drooping flowers. The leaves consist of from eight to twelve pairs of ovate or oblong leaflets, with an odd one. Same positions and culture as for *A. Onobrychis*.

ASTRAGALUS DASYGLOTTIS (Clover Astragalus).—A dwarf and pleasing species, nearly allied to *A. Hypoglottis*, but somewhat taller. It produces an abundance of bright purplish-lilac flowers, resembling those of Clover, the flower-stems extending a little beyond the leaves. It is one of the finest and freest-flowering of the genus. Ural and Altai Mountains.

ASTRAGALUS VAGINATUS (Sheathed Astragalus).—A handsome species with erect pubescent stems, about one foot high. Flowers in summer, of a rosy-purple colour, with white-tipped wings, large (each flower about an inch long), and arranged in dense spikes on stalks longer than the leaves. The leaflets, which are usually in seven or eight pairs, with an odd one, are of an elongated-oblong shape, and covered on both sides with short, silvery, adpressed hairs. A native of Siberia and N. America, and a good plant for either rockwork or borders, in sandy loam.

BERBERIS EMPETRIFOLIA (Fuegian Berberis).—A dwarf, shrubby, or trailing species, from the Straits of Magellan, well adapted for rock cultivation, provided a good depth of peaty soil be given it for its underground shoots to ramble in. Its flowers are of a bright orange colour, produced singly along the whole length of the previous year's growth. They would be admirably adapted for bouquets were it not that each leaf is terminated with a sharp mucro, which somewhat rudely admonishes a person wishing to gratify a second sense, that he must be content with that of sight. This Berberis possesses a very delicate fragrance.

BRODLÆA COCCINEA (Crimson-flowered Brodiæa).—A very beautiful Californian bulbous plant of the Lily family, with a flower-scape from 2 to 3 feet in height, bearing in summer a dense terminal umbel of rich magenta-crimson pendent tubular flowers; each flower being from 1½ to 2 inches long, yellow at the extremity, and with reflexed green tips; the umbel consisting of five to twenty blossoms, according to the vigour of the plant. The leaves are linear, channelled, lax, and nearly as long as the flower-scape. This very charming plant should be planted in a warm position in the rock-garden or choice mixed border among select plants, and in warm sandy loam.

BRYANTHUS BREWERI (Brewer's Bryanthus).—A handsome dwarf shrub of compact habit, nearly a foot high, from the lofty sierras of

California. The rigid ascending stem and branches are thickly covered with smooth linear leaves about half an inch long, and narrowly revolute at the margin. The flowers are half an inch across, cleft to, or rather beyond, the middle, of a bright rose-violet or red-purple colour, the upper part of the petals being of a darker shade. The beauty of the flowers is enhanced by the great length of their ten light-coloured stamens, tipped with violet anthers. The inflorescence is at first in dense roundish clusters at the ends of the erect branches, becoming corymbose as the flowering advances. This is a valuable addition to the rockwork or choice border, and should be planted in sandy peat soil.

CALTHA PALUSTRIS PLENO (Double Marsh Marigold) is a remarkably showy plant, and though properly, as its name, "Marsh Marigold," would imply, a marsh plant, yet it flourishes well and flowers freely in any ordinary garden soil, more especially if of a heavy clay texture.

CAMPANULA RAINERI (Rainer's Campanula).—An exceedingly dwarf, pretty, and rare species from the mountains of Styria and Carinthia. It blooms early in summer, producing erect, light purplish-blue, funnel-shaped flowers, over an inch across, on stems 1 to 2 inches high, each stem bearing from one to three flowers, and issuing from a rosette of small, roundish, hairy leaves. This is one of the most interesting Campanulas in cultivation, and admirably suited for warm ledges of rockwork, or the margin of the choice mixed border, where its creeping roots will spread rapidly in fine sandy soil.

CAMPANULA SOLDANELLÆFLORA (Soldanella-flowered Campanula). —Sometimes named *C. rhomboidea pleno*; but the former name is more appropriate, as the flower has all the appearance and beauty of its specific foster-parent, associated with the light graceful character that is inseparably connected with the Harebell family. It is a plant of slender habit, with linear leaves. When growing strong, it attains a height of 15 inches. The flowers, being double, have a much longer duration than were they single. We have, however, never met with a single-flowered form that we could identify as the normal type of this plant.

CAMPANULA WARNERI (Warner's Campanula).—A distinct and handsome species, 6 to 10 inches high, with drooping, dark-blue, tubular bell-shaped flowers, each 1½ inch long, appearing in May on long one-flowered axillary and terminal peduncles. Leaves lance-shaped, unequally toothed, the lower ones decurrent on long leaf-stalks. A native of Transylvania and the Banat Alps. At present it is rather scarce, and until more plentiful should be confined to the rockwork; but it will probably prove an excellent border plant. Ordinary free or sandy soil.

CHEIRANTHUS MARSHALLI (*Marshall's Wallflower*).—This, which is said to be a hybrid between *Cheiranthus ochroleucus* and *Erysimum*

Peroffskianum, is a half shrubby plant, 1 to 1½ foot high, with erect angular branches. The flowers appear in spring or early summer, are nearly three-quarters of an inch across, of a deep clear orange at first, afterwards becoming somewhat paler, and are freely produced in terminal racemes. The leaves, which are more or less spoon-shaped, are crowded together at the lower part of the stems; the upper ones are narrowly lance-shaped, and more distant from each other. The fine orange-colour of the flowers of this plant renders it a very pleasing subject for the rock-garden, or borders in light, well-drained soil. It is increased by cuttings, and a young stock should be kept up, as it is not perennial, and is apt to perish in winter.

CHRYSOSPLENIUM OPPOSITIFOLIUM and ALTERNIFOLIUM.—The two Burnet Saxifrages are lovely plants, and are admirably adapted for peculiar situations, that will always occur in well-formed artificial rockeries—namely, for the margin of rivulets of water, or for a soft, boggy place. Under these circumstances they are perfectly at home, whether in shade or sunshine. The latter species is the rarer of the two, and adapted for a drier locality than the former.

CLAYTONIA VIRGINICA (Spring Beauty).—A handsome American plant of the Portulacaceæ or Purslane family, sending up in March and April simple stems bearing a pair of opposite linear lance-shaped leaves from 3 to 6 inches long, and a loose raceme of pretty rose-coloured flowers marked with deeper veins, which, unlike the flowers of most of the species of this family, remain open for more than one day. Suited for the rock-garden or borders, in loam and leaf-mould. *C. sibirica* and *C. alsinoides*, although only biennials, or perhaps little better than annuals, are so much at home on rockwork, and sow themselves freely in such impossible crevices, that they must not be omitted.

CORONILLA IBERICA (Caucasian Coronilla).—A plant with glaucous foliage and decumbent habit, not rising 4 inches from the ground, and producing freely umbels of yellow blossoms. Somewhat similar in appearance, but much larger than our own familiar Lotus corniculatus. It flourishes admirably with its woody roots well bedded in rockery, and will cover completely 2 or 3 square feet of rock surface, when so placed. The Caucasus.

CROCUS IMPERATI (Imperati's Crocus).—A very early spring-blooming species, nearly allied to *C. versicolor*, but much handsomer, 3 to 6 inches high. The flowers are sweet-scented, of a lilac-purple on the inside, while the outside is of a creamy-white, marked with three longitudinal dark-purple lines, of which the two outer ones and the end of the middle one are curiously feathered or fringed with short lines of the same colour. This very charming Crocus was found on the mountains of Calabria, in Southern Italy, at an altitude of from 3,000 to 6,000 feet.

We may here mention also another fine species of recent introduction from the mountains of Greece, viz., Crocus Aucheri, a dwarf, very early spring-flowering species, with deep orange-yellow flowers. Syn. *C. chrysanthus*.

CROCUS BORYANUS (White Autumn Crocus).—A very pretty autumn-flowering kind, from Asia Minor, the Morea, and the Greek Islands. The flowers, which do not appear until late in autumn, are of a creamy-white, with an orange-yellow throat, the base of the segments sometimes marked externally with dull purple lines. This species, which is as yet rare in gardens, is a very pleasing addition to the list of our autumn-flowering plants.

CYANANTHUS LOBATUS (Lobed-leaved Cyananthus).—A brilliant and remarkable Himalayan rock-plant, spreading loosely about, with prostrate habit and procumbent or ascending stems about 4 inches high. The flowers, which appear in August and September, are purplish-blue, with a whitish centre, solitary, usually terminal, about an inch across, funnel shaped, with five spoon-shaped lobes, and are finely fringed on the throat with numerous long, soft, whitish hairs. The leaves are lozenge-shaped, small, fleshy, alternate, deeply and irregularly lobed, greyish underneath. The best position for this plant is on a sunny ledge of rockwork, where its overhanging flower-laden stems will produce a fine effect. It grows best in a mixture of sandy peat and leaf-mould, with plenty of moisture during the growing season. It is easily and freely increased by cuttings. The seed requires a dry favourable season to ripen it; in wet weather the large, erect, persistent calyx becomes filled with water, which remains and rots the included seed-vessel.

CYPRIPEDIUM GUTTATUM (Spotted Lady's Slipper).—A very charming Siberian species, which has been described as "one of the most beautiful of vegetable productions," although at present seldom seen in gardens. It grows from 6 to 9 inches high, and flowers in June, producing solitary, rather small but beautiful snow-white flowers, heavily blotched or marbled with deep rosy-purple. The flower-stem rises from a single pair of broadly-ovate downy leaves. It requires a shady position on the rockwork or mixed border, in leaf-mould, moss, and sand, and should be kept rather dry in winter.

DALIBARDA VIOLOIDES (Violet-leaved Dalibarda).—A plant rarely met with, even in the most select collections. It grows about 2 inches in height, with somewhat reniform leaves, and white blossoms shaded with the most delicate rose-colour. It loves a deep, peaty soil; and, though a permanent and long-lasting plant, it is by no means of rapid growth. Canada.

DAPHNE RUPESTRIS (Rock Daphne).—A neat and diminutive shrub with shining and fleshy spoon-shaped leaves, from a third to one-half an inch long, channelled above, and with a conspicuous and large blunt keel below, the spaces between the keel and margins being of a greyish tone. Flowers somewhat like those of *Daphne indica*, and very sweet-scented, appearing in early spring. New Zealand. Syn. *Gaultheria rupestris*.

DAPHNE STRIATA (Striated Daphne).—A sweet-scented hardy trailing species, resembling *D. Cneorum*, but somewhat more shrubby. It forms dense, twiggy, spreading masses, 1 to 3 feet across, which, in June and July, are covered with a profusion of rosy-purple, carnation-scented, tubular flowers, growing in clusters. The trailing and freely-spreading habit of this plant recommends it as an excellent subject for covering bare parts of rockwork. A native of France.

DIANTHUS ARENARIUS (Sand Pink).—A neat, compact rock-plant, about 6 inches high, with very dense foliage, and white, fimbriated or fringed flowers. It blooms in May and June, and should have a dry, sunny position, either on the rockwork or in the border. North Europe.

DIANTHUS CRUENTUS AND VAGINATUS.—Both belong to the fasciculated or clustered flowered section of this genus, and are beautiful plants, well worthy of a place in all collections. The flowers in the former, elevated on a stem from 9 to 12 inches high, are of a crimson scarlet, rich and intense; those of the latter are double the size, of carmine colour, nearly equal to *Calandrinia*, and only 6 inches high; it is a new and rare species, continuing in bloom for nearly two months.

DIANTHUS FISCHERI (Fischer's Dianthus).—A very beautiful, and as yet rare, species from Russia, 3 to 4 inches high; blooms in summer; flowers numerous, of a light rose-colour, with the petals not much cut; flowers produced solitary like *D. Alpinus*; leaves lance-shaped, stiff, with a single vein on the upper side. Deserves a good position in the rock-garden, in moist, sandy, or gritty loam.

DIANTHUS TYMPHRESTEUS.—A free and continuous blooming species from Northern Greece, growing from 15 to 18 inches high, with deep rosy flowers; makes a good perennial and showy border plant.

DONDIA EPIPACTIS (Syn. HACQUETIA).—A most unusual form of the umbel-bearing plants; this is amongst our earliest bloomers. It grows only some 3 or 4 inches high, and though the blossoms individually are small, they are surrounded with a bright golden involucre that reminds one of the Burnet Saxifrages (Chrysosplenium), retaining its brilliant colour for nearly two months of the spring. It is a strong-rooted plant, likes a good stiff loam, and is perfectly hardy. Carinthia and Carniola.

DRABA CILIATA (Eye-lashed Whitlow Grass).—This is really a good white Draba, not unlike a diminutive specimen of *Arabis albida*. The leaves are broadly spoon-shaped, sparsely but distinctly ciliated, in loose rosettes. Flowers in early spring; pure white, about eight on a stem; the whole plant when in bloom not being more than two inches high. Mountains of Croatia and Carniola.

DRABA GLACIALIS (Glacier Draba).—A very dwarf species, forming dense little cushions 1 to 2 inches high, which in April are covered with bright golden-yellow flowers. Leaves linear, smooth, ciliated, forming small rosettes closely packed in pincushion-like masses. The plant very much resembles a small specimen of *D. aizoides*, and is considered by Koch to be merely a variety of that species, growing at a higher elevation; but it differs from it by having a few-flowered stem, pedicels shorter than the pod, and a short style. It is found on the granitic Alps of Switzerland, and is suited for exposed spots in the rock-garden, in moist and very gritty or sandy soil, and associated with the dwarfest alpine plants.

EDRAIANTHUS PUMILIO (Silvery Harebell).—A singularly pretty and minute rock-plant, with foliage resembling that of a dwarf tufted Pink, hoary and silvery, with adpressed hairs on the upper side, very finely ciliated at the edges, the under sides quite smooth, shining, and dark green. The flowers, about an inch long, are almost seated among the leaves (the whole plant not being more than 2 inches high), of a pure purplish blue, vasiform, cut into segments for about one-third of their length, and opening in May. A gem for the rock-garden, thriving in moist loam with abundance of sand or grit, and well-suited for association with the alpine Forget-me-not and other dwarf and choice May-flowering rock-plants. Easily raised from seed. Croatia.

EPILOBIUM OBCORDATUM (Dwarf Californian Willow Herb).—This, which is by far the finest of the alpine Willow-herbs, has been recently introduced from the Rocky Mountains of North America. It forms handsome little shrub-like tufts, 3 or 4 inches high, and flowers late in summer, producing large rosy-crimson blossoms, resembling those of *Clarkia integripetala*. The leaves are small, smooth, and roundish, closely set on the erect stems, and the plant, even when not in flower, has a remarkably neat and distinct appearance from the foliage alone. Coming from the summit of the Sierra Nevada, it is perfectly hardy, and will prove one of the most valuable and attractive of rock-plants. It is also an exquisite pot-plant, and thrives to perfection in ordinary peaty loam.

ERODIUM CARVIFOLIUM (Caraway-leaved Erodium). — A good perennial species, 6 to 10 inches high, producing red flowers larger than those of *E. romanum*, the whole plant being more vigorous, and more decidedly perennial than that species. A native of Spain.

ERYTHRONIUM AMERICANUM (Yellow-flowered Dog's-tooth Violet). —A species from North America, as its name indicates; narrower in foliage, with bright yellow pendent flowers; thrives freely in deep peaty soil, but is rarely met with in cultivation. It stands in a somewhat similar relationship to *E. Dens-canis* that *Anemone ranunculoides* does to *Anemone nemorosa*.

ERYTHRONIUM GIGANTEUM (Gigantic Dog's-tooth Violet).—A very large species from North America, growing from 9 to 18 inches high, and bearing from three to ten large flowers on one stalk, arranged in a branching and somewhat confluent spike, each flower 3 inches or more across, and usually of a creamy white, shaded with delicate pink or purple. There is great variety, however, in the colour. In different districts plants are found with pure white, with light lemon-yellow, and with clear red-purple flowers. The plant continues in bloom through February, March, and April, and the flowers are very handsome, the petals being broad and well expanded. The leaves are blotched and marbled with purplish-brown. Same culture and positions as for the common Dog's-tooth Violet.

GALANTHUS IMPERATI (Imperati's Snowdrop). This is a very large species of Snowdrop, found on mountains in that part of Italy which was lately called the Kingdom of Naples. It resembles our common Snowdrop, but is double the size in all its parts, and flowers later; one of the external divisions of the flower, also, is larger than the others. The leaves are very broad and long, but not plicate at the margin, like those of the great Crimean Snowdrop, *G. plicatus*. It flowers ten days to a fortnight earlier than the latter. This very desirable species is at present scarce in this country; but we have no doubt that a demand for it would stimulate our nurserymen to provide themselves with a supply.

GALAX APHYLLA (White Wand Plant).—A very distinct and interesting evergreen herb from N. America, forming a thick matted tuft of scaly creeping root-stocks, thickly set with fibrous red roots, from which it sends up a number of roundish heart-shaped, veiny, shining leaves (about 2 inches wide) on slender stalks. The flowers appear in June, and form a wand-like spike of small and minutely-bracted white flowers, on the summit of a slender, naked stem, from 1 to 2 feet high. A very striking plant for the rock-garden, or choice mixed border in loam and leaf-mould.

GERANIUM CRISTATUM (Crested Geranium).—A trailing species, with peculiar crested appendages to its seed carpels. It is admirably adapted for rockery, covering fully 18 inches square; but is liable to smother more delicate plants, unless room be given to it. Flowers, lilac striated, very abundant, produced in May, and again in August or September. Native of Albania in Northern Greece.

GLOBULARIA TRICHOSANTHA (Hair-flowered Globularia).—This species is distinguished by its glaucescent foliage and finely-divided petals. It grows from 6 to 8 inches high, and flowers in summer, producing large sky-blue many-flowered heads, very similar in shape to those of *Erigeron alpinus*. The flower-stems are herbaceous and leafy, and bear one flower-head each. The whole plant is very smooth and glaucescent. Native of Asia Minor. Same positions and culture as for G. *nana*.

GYPSOPHILA PANICULATA (Panicled Gypsophila).—A very handsome herbaceous plant, forming a dense compact globe-like bush, 2 to 3 feet in height, and as much across. Flowers from Midsummer to the end of August; small, white, exceedingly numerous, arranged on thread-like stalks in many-branched panicles, with the light, airy, graceful effect of certain ornamental grasses. It seeds but very sparsely, but forms a thick, woody, root-like stem, and the young shoots removed therefrom, when about two inches long, will strike freely. A native of Siberia and Sicily. Admirably adapted for bouquets.

IBERIS JUCUNDA (Glaucous Iberis).—A little novelty in leaf and flower, from the mountains of Asia Minor; about two and a half inches high. The leaves are ob-lanceolate, about three quarters of an inch long by one-eighth of an inch broad, somewhat glaucous, much attenuated at the base, and with a minute point at the apex. The flowers are in small clusters, and of a pleasing flesh-colour, prettily veined with rose in early summer. This does not as yet appear to possess the rude vigour of our common evergreen Iberises, but it is none the less valuable as a rock-plant for being unlike them, and is fitted for association with a dwarfer and more select set of subjects. Should be planted on warm and sunny parts of the rock-garden, in well-drained sandy loam. Increased by cuttings or seeds. Syn., *Æthionema coridifolium*.

IRIS IBERICA (Iberian Iris).—A remarkable striking Iris, reminding one of *I. susiana*, but quite distinct in leaf and flower. Grows from 4 to 16 inches high, and blooms in summer, producing solitary flowers, the external divisions of which are of a dull red, marked with tawny streaks, and an oval, black, purple-edged spot in the middle, while the internal divisions are of a very pale purple, with streaks of a darker hue, and veined and spotted about the base. The leaves are linear, arched, almost curled, and folded lengthwise.

LEWISIA REDIVIVA (Bitter Root Plant).—A very singular and ornamental plant, allied to the Mesembryanthemums, resembling a very fleshy-leaved Thrift, and forming rosettes of leaves, each 2 to 3 inches long, on a thick, woody, branching root-stalk. After the leaves attain their full growth in spring or early summer, a profusion of beautiful flowers issue from the rosettes, nearly hiding the whole plant. Each blossom is from 3 to 4 inches across, and consists of eight or twelve shaded pink petals, the centre being nearly white and the tips rose-colour, the whole having a very pleasing satiny lustre. The calyx is elegantly veined with red, and is of a consistency like paper. The flowers open only during sunshine. Native of the west parts of N. America, particularly in Washington Territory and Oregon. Should have a warm position in the rock-garden, in dryish soil.

LILIUM PULCHELLUM (Russian Lily),—From South-western Russia, is even a more beautiful species than *L. tenuifolium*; slender in habit, the stem rises to a height of 12 to 15 inches, crowned at the top with upright (not pendent) deep orange blossoms, slightly marked with black dots; not by any means common in gardens, but a distinct and most desirable species.

LINARIA PILOSA and HEPATICÆFOLIA.—These are somewhat allied species, although perfectly distinct. They are both so dwarf as to defy their height being characterised by inches. In fact they grow flat on the ground, and produce their lilac blossoms just raised above the foliage very freely. The first is a native of France; the other is found in Corsica.

LINUM SALSOLOIDES (Heath-like Linum).—A hardy, dwarf, half-shrubby species, somewhat like a dwarf Heath or Leschenaultia, with the stem twisted at the base. It grows from 3 to 6 inches high, and blooms in June and July, producing an abundance of flowers which are each an inch across, and white with a purple centre or eye. The leaves are linear, smooth, and scattered; the lower ones shorter and almost imbricated. A native of the South of Europe, this plant is not only well adapted for rockwork, but is also a very pretty object on the common border, in well drained sandy soil.

LINUM VISCOSUM (Viscid Linum).—A fine herbaceous species from the Pyrenees, with slightly branching downy stems from 1 foot to 1½ foot high. Blooms from May to August. Flowers nearly an inch across, of a rich, rosy lilac colour, in an erect corymb. Leaves alternate, lance-shaped, covered with feeble whitish hairs and viscid glands. A fine subject for the rock-garden or choice border, in well-drained, moist, sandy loam.

LITHOSPERMUM CANESCENS (Hoary Lithospermum).—A species distinguished by its hoary appearance, the whole plant being covered with soft white hairs. Grows from 6 to 15 inches high, and flowers in May, producing large, sessile, salver-shaped flowers, of a deep orange-yellow colour. The leaves are narrowly oblong, obtuse, more or less downy beneath, and roughish above, with close adpressed hairs. Northern States of America. Suited for the rockwork or borders, in sandy or peaty loam and grit. Syn. *Batschia canescens*.

LITHOSPERMUM GASTONI (Gaston's Lithospermum).—A rare and beautiful species from the Pyrenees, with erect herbaceous stems about a foot high, which from May to August bear terminal clusters of large bright sky-blue flowers, about twice the size of those of *L. prostratum* (commonly sold under the erroneous name of *L. fruticosum*). Leaves obovate lance-shaped. Grows freely on rockwork, or in any open border in rich well-drained loam.

MERTENSIA DAHURICA (Dahurian Gromwell).—This plant, which was formerly known under the name of *Pulmonaria dahurica*, although of a very slender habit, and liable to be broken by high winds, is perfectly hardy, and was formerly often to be met with in British gardens. It grows from 6 inches to a foot high, with erect branching stems, which are angular and furrowed, and clothed with decumbent white hairs. It blooms in June, producing handsome bright azure-blue drooping flowers in racemose panicles. The leaves are ovate, roughish, slightly glaucous, and clothed with small decumbent hairs. It is a very pretty plant, and suited either for the rock-garden or borders, where it should be planted in a sheltered nook in a mixture of peat and loam. It is easily propagated by division or from seed.

MITCHELLA REPENS (Variegated Partridge Berry).—This is one of the most interesting of the pretty woodland plants that accompany the Ground Laurel (*Epigæa*), the tree Lycopodium, the Rattlesnake Plantain (*Goodyera*), etc., in the Pine woods of North America. It is a smooth and trailing little evergreen herb, with roundish shining leaves, and minute stipules. The flowers are white, sometimes tinged with purple, and bear pretty scarlet berries in autumn. I saw it in Long Island, running about in the Moss, etc., at the bottom of Pine trees, and it occurred to me at the time that it would be a charming addition to shady parts of our rock-gardens, hardy ferneries, etc., in which it would thrive under the same conditions as the Pyrolas, the Linnæa, etc. It is named after Dr. John Mitchell, an old Virginian botanist and correspondent of Linnæus.

MUSCARI ARMENIACUM (Armenian Grape Hyacinth).—A strikingly beautiful and scarce species, growing about 6 inches high and flowering in May, when it produces a dense spike of flowers of a fine cobalt-blue colour, with three small yellow dots near the mouth of the corolla. The spike is 2½ inches or more in length, and the flowers are most agreeably fragrant. The leaves are about 9 inches long, ribbon-like, concave, and pointed. An exquisite plant for level spots on rockwork, for the bulb-garden, or for borders in light soil.

MUSCARI HELDREICHII (Greek Grape Hyacinth).—A beautiful, and as yet rare, kind from Greece, about 14½ inches high, with flowers of a fine deep sky-blue with white mouth, somewhat like those of *M. botryoides*, but nearly twice as large, and arranged in a longer spike. Leaves about 4 inches long, flat, like those of *M. commutatum*, but not open at the top, like those of that species. A fine subject for the rock-garden or the choice border, in deep sandy soil. Comes into bloom in the end of March.

ŒNOTHERA PUMILA (Dwarf Œnothera), as the name would indicate. This plant is of diminutive size, growing about 4 or 5 inches high, forming dense, compact tufts, flowering abundantly from May to September. The flowers are small, and of a bright yellow colour. The plant possesses a modest simplicity of character, which renders it very attractive. N. America.

ŒNOTHERA RIPARIA (Rock Œnothera).—Considerably more vigorous in growth, but with a somewhat similar habit to the foregoing. This species, flowering freely and continuously, has been introduced into our flower-gardens, and forms a beautiful rock or border plant. Its flowers are canary yellow, and nearly 1¼ inch in diameter. The plant grows from 9 to 12 inches high, and is very showy when in bloom. N. America.

ONOSMA TAURICA (Golden Drop).—A Boraginaceous evergreen plant, 6 to 8 inches high, producing an abundance of tubular, somewhat ventricose, canary-coloured flowers, in long cymes, early in summer; it is a noble rock-plant. It should be planted on an elevated knoll, so as to allow its branches to hang over the surrounding rocks. Though perfectly hardy, it is somewhat impatient of water during winter, and if growing very vigorously, is liable to rot off. Caucasus.

OPUNTIA RAFINESQUII (Hardy Opuntia).—A dwarf spreading Cactus, forming clusters of thick, ovate, very green joints, each 3 or 4 inches long and about 3 inches broad, studded with small tufts of minute, sharp-pointed, reddish, hair-like spines. Flowers in summer, of a bright sulphur yellow. The fruit is said to be edible and "like a Gooseberry." North America. This Opuntia is a very free bloomer, and its dwarf, branching habit makes it better suited for out-of-door vase or rockwork culture than for a border. For several years past, the hardiness of *O. Rafinesquii* in the climate of London and Paris has been a subject of remark, and various persons in England and northern France have testified to its hardiness. The fact, however, that it stands and grows well in a London back-garden, deprived to a great extent of the sun, is as much proof as we need in that respect. This hardy species resists much greater cold than we ever have in Britain, and it is probable we shall find that half a dozen or more species of Cactus are quite as hardy. Along the line of the Pacific Railway, Cacti are abundant in some places—in districts frosty and silvered with snow when I passed over them in November, 1870, and on the flanks of the Wassatch Mountain, near Salt Lake City, deeply covered with snow during the winter. It is desirable, in gathering the small mountain plants there, and in sitting down on the ground, to look well for a small, poignantly prickly Cactus, with round stems, which abounds there, and which communicates a peculiarly acrid sting to all soft, fleshy parts that touch it. I gathered this in company with Astragali and other plants, which are usually termed alpine. In the northern States of America, which are very cold in winter, as everybody knows, there are three species of hardy Opuntias—*O. vulgaris* (the common Prickly Pear), which goes as far north as New England ; *O. Rafinesquii*, in Wisconsin and Kentucky ; and *O. missouriensis*, in Wisconsin and towards the great plains. And from what one sees along the Pacific route, it is very likely a greater number of Cacti go north along the Rocky Mountains' dry plains and sierras than we find on the eastern side of the continent. It is very likely we shall some day have quite a group of dwarf hardy Cacti keeping company with the Houseleeks on our rock-gardens, and rivalling them in hardiness. They should be planted on the drier parts of the rock-garden, on dry sunny banks, on the edges of old walls, old bridges, ruins, etc. They will also thrive on borders, but are most appropriately placed in the positions above named. Mr. James Atkins, of Painswick, informs me that he has grown a variety of *Opuntia vulgaris*, which he found in Piedmont, for upwards of 18 years without the slightest protection, on a part of his rockwork, which has a S.W. aspect.

ORCHIS FOLIOSA (Leafy Orchis).—A very handsome and showy Orchis, found on rocky banks in the island of Madeira, and growing from 1½ to 2½ feet high. It flowers in May, when it produces a very great number of purplish flowers in a large ovate or oblong-ovate spike, about 9 inches long and 3 inches across. The leaves are oblong, unspotted, the lower ones obtuse. May be grown in sheltered nooks in the rock-garden, or in the mixed border in deep light soil.

OROBUS HIRSUTUS (Hairy Orobus).—This is a distinct and desirable dwarf species, growing about 9 inches high. It has somewhat broad leaflets of a light green colour, and lively blue and rose-coloured flowers. It grows freely.

OROBUS LATHYROIDES (Lathyrus-like Orobus).—Though too tall in growth for association with the more truly alpine plants, this is a lovely border plant. Its height, when growing vigorously, is from 18 inches to 2 feet ; its flowers bright blue, produced in dense racemes, and its general habit all that can be desired ; it is increased freely by seeds, which it produces abundantly. In continental gardens it is known under the name of Vicia unijuga.

OROBUS VICIOIDES PLENO is a double form, a good deal like *O. vernus* in habit, but a more compact grower, and like all double flowers retaining its beauty for a long period. Well worthy of cultivation.

O. TAURICUS and AURANTIACUS.— Nearly allied species ; both have orange-coloured flowers, and require to be well established plants before they will bloom freely, and be seen in perfection.

OURISIA COCCINEA (Scarlet Ourisia).—A dwarf Scrophulariaceous creeping plant, producing racemes of crimson flowers ; is of recent introduction, and has a good habit, though, as we have found it, somewhat shy in blooming. A native of the Andes of Chili.

OXALIS LASIANDRA (Woolly-stamened Oxalis).—A singular and handsome Mexican species, growing from 9 to 18 inches high, and producing umbels of large crimson flowers in summer—about twenty flowers in each umbel. The leaves are all radical and digitate; the leaflets (7 to 9 in number) being 3 inches long and 1 inch broad, of an oval spoon-shape, dark green above, pale underneath, and spotted with crimson. It is a fine plant either for the rock-garden or borders, and should have a warm position in well-drained sandy soil.

OXALIS LOBATA (Lobed Oxalis).—This is often grown under the synonym of *O. granulata*; it is well worth growing in pots for greenhouse decoration, as it produces its bright yellow flowers in November, a period of the year when such a colour, other than among the Chrysanthemums, is of rare occurrence.

OXYTROPIS URALENSIS (Purple Oxytropis).—An elegant little perennial about 6 inches high, thickly clothed with soft silky silvery hairs in every part, and bearing dense round heads of bright purple flowers in summer. Leaves composed of 10 to 15 pairs of ovate, acute leaflets with an odd one, more or less white, with adpressed hairs. Native of Scotland and Northern Europe. An excellent plant for either rockwork or border in rich loam. Closely allied to it, and similar in size and growth, is a species grown for years by Mr. Niven, Botanic Gardens, Hull, and provisionally named by him Oxytropis cyanea; the flowers are larger, and of a lovely amethyst blue. The original plant has now been growing in a pot for fifteen years, and blooms freely every season; it is, however, very shy to increase otherwise than by seed, the production of which is very exceptional.

PAPAVER PILOSUM (Pilose Poppy).—A showy species, distinguished by its very hairy stems and pale green foliage; 1 to 2 feet high. Flowers in summer, brick-red or deep orange, with a whitish spot at the base of each petal. Leaves of the root oblong; stem-leaves ovate oblong, with serrated lobes, pilose on both sides. Asia Minor.—*Papaver spicatum* resembles the above, but is more woolly in the leaf, and more compact in growth, owing to the shorter peduncles; it is well adapted for rockwork; its flowers are of an exceedingly delicate texture, and of a yellowish orange. The plant grows about 18 inches high.

PHLOX NIVALIS (White-flowered Phlox).—A species now rarely met with in cultivation; much smaller in leaf and more rigid in habit than *Phlox subulata*; the flowers are pure white. Doubtless it and *subulata* constitute the origin of the pink-eyed *Phlox Nelsoni*; this origin appears to be well indicated by the shorter leaves and the white flowers—characters which it obtains from *P. nivalis*.—*Phlox frondosa* belongs to the same section as *subulata* and *setacea*; is a much stronger and freer grower than either, with larger flowers; admirably adapted for such a position as will enable it to trail its procumbent branches over the face of a mass of rock.

PINGUICULA LUTEA (Yellow-flowered Pinguicula).—This is a native of the Southern States of America, and ought to have a very sheltered corner; its yellow blossoms, associated with the large blue *P. grandiflora*, would have a very pretty effect, but it is rare in cultivation.

POLEMONIUM HUMILE (Dwarf Polemonium).—This is a dwarf species, with small circular leaflets, and the whole plant is covered with pubescence; the flowers are of a slaty blue, produced on a short foot-stalk; the general appearance of the plant recommends itself for cultivation, more, perhaps, than the flowers themselves. Columbia. Syn. *P. pulcherrimum*.

POTENTILLA ARGENTEA (Silvery Potentilla).—As the name would imply, this plant is covered over with silvery down; it is of a creeping habit, not exceeding 6 inches in height; and though scarcely definite enough in its argent character to give it a status in the gaudy ranks of the flower-garden, it is yet a very desirable plant to place as a variety among dark-leaved plants in a rockery.

PRIMULA AURICULATA (Auricled Primula).—This Caucasian species is closely allied to *P. farinosa* in outward appearance, but has larger flowers—and, indeed, is altogether of a more vigorous habit. It grows from 6 to 8 inches high, and flowers in April and May, producing an umbel of numerous rosy-violet flowers. The leaves are oval lance-shaped, smooth, rather thick, irregularly toothed on the margin, and gradually narrowed at the base into a broadly-winged leaf-stalk. They are not nearly like those of *P. farinosa*. A good plant for rockwork. Sometimes called *P. Magellanica*.

PRIMULA JAPONICA (Japanese Primrose).—A very noble and perfectly hardy species, recently introduced from Japan, and not inaptly named the "King of the Primulas." Its flower-stems rise from 1 to 2½ feet in height, and in the month of May begin to exhibit the brilliant magenta or deep purplish-crimson flowers, which are very numerous and arranged in from three to six whorls on the top of the stem, each flower being from 1 to 2¼ inches across. The leaves resemble those of *P. denticulata*, and are from 6 to 10 inches long and 3 to 4½ inches broad, forming a rosette 1½ foot, or more, across. If planted in the rock-garden, it should not be among minute alpine species, but grouped with subjects growing a foot or more high, and in sheltered positions, where its fine foliage would not be injured by harsh winds. It is a magnificent plant for a warm sheltered border, or for a position among dwarf shrubs, in sandy loam and leaf-mould. Propagated from seed, which it yields freely. The seed should be sown in light soil and mixed with silver sand, in pans placed in cold frames, where they should be allowed to remain until it germinates. Two years may elapse before the young plants appear. Many persons, ignorant of this, have thrown out the contents of their seed-pans in despair, because no sign of germination appeared during the first year, and a great deal of good seed has doubtless been lost in this way. Another cause of failure is placing the seed-pans in heat; they should be kept cool. After the young plants are pricked off and potted, they are sometimes benefited by a little heat, just to start them afresh; but they must afterwards be carefully hardened off.

PRIMULA PARRYI (Parry's Primula).—This Primula is spoken of by American botanists as the finest of all the alpines of the Rocky Mountains. It grows from 9 to 12 inches high, and flowers from April to June, bearing heads of the richest purple blossoms. The leaves are erect, and resemble those of *P. longiflora* on a large scale. It requires an abundance of moisture, as in its native habitats it occurs "in shallow streams springing from the melting snows at an elevation of 7,000 feet or more." This fine species has only recently been reintroduced into this country by Messrs. Backhouse and Son, York.

PRUNELLA GRANDIFLORA (var. LACINIATA).—This is a very distinct variety, producing, along with the large flowers of the parent species, leaves divided into linear lobes. It is considered by some to possess sufficient character to render it specifically distinct; but as there is a tendency in seedlings raised from it to revert to the normal type, it is better to retain it as a variety, and a very distinct one, of *grandiflora*.

PULMONARIA SIBIRICA and PULMONARIA ANGUSTIFOLIA.—These are both perfectly distinct species from *P. officinalis*. The former has the leaves beautifully mottled with white—so much so as to render it worthy of cultivation for its foliage only. The latter has narrow leaves, and both form dense crowns, from which the flowering branches arise in bright masses to a height of 9 to 12 inches.

RHODODENDRON PRÆCOX (Early Rhododendron).—In this we have a perfect gem for the rock-garden, the margins of choice dwarf beds of American plants, or any position in which a very beautiful dwarf shrub may be desired. It is a hybrid between the very early-blooming *R. dauricum atrovirens* and the fine *R. ciliatum*, and has the very early-flowering habit of the former with the large and handsome blooms of the latter. It is frequently in bloom early in March. The leaves are dark green, shining, oval, and slightly ciliated; the flowers, within an eighth or so of 2 inches across, of the most delicate satiny lilac, unmarked by spottings of any kind, and abundantly produced on the compact little bushes. It is a worthy companion to the Dog's-tooth Violet, the Siberian Squill, and *Sisyrinchium grandiflorum*, which are in full beauty at the same time.

SAPONARIA CÆSPITOSA (Tufted Soapwort).—A neat, dense-habited species from the Pyrenees, forming very rigid tufts like those of *Silene acaulis*, but much more robust, and bearing sub-umbellate heads of showy rose-coloured flowers, which first appear in June. The blooms are nearly as large as those of *Dianthus alpinus*, to which they also bear a considerable resemblance. It should be planted in deep loam, mixed with gravel and sand, in sunny fissures of rockwork.

SAXIFRAGA LONGIFOLIA, var. ELATIOR.—This plant was for years grown as the true type of *longifolia*. It bears some resemblance to *S. ligulata*, but is perfectly distinct when in flower, the panicle rising to 2 and even 3 feet high, somewhat lax in the arrangement of the flowers, but the individual flowers are double the size of those of *S. ligulata*, and beautifully dotted with tiny crimson spots; a plant known as *S. nepalensis* is doubtless this old form of *longifolia* under a new name.

SAXIFRAGA MUTATA (Changeable Saxifrage).—A yellow-flowered species, bearing considerable similitude to *S. ligulata*; its flowering panicle is about 18 inches high, and I presume it is rarely seen in cultivation, owing to the fact that it not infrequently exhausts all its vigour in producing blooms, and rarely matures seeds in this country; further, it rarely produces offsets, as most of this section do. It is a native of the Alps, but limited in its distribution.

SAXIFRAGA PENTADACTYLIS (The 5-fingered Saxifrage).—This species, so named from its linear 5-lobed leaves, is the most rigid of all the mossy section, dwarf and compact in habit; it has a further peculiarity of exuding a gummy secretion, in the form of a white powder, that gives the plant (especially in spring) the appearance of frosted silver. In this respect it is allied to *S. ladanifera*, but in the latter the exudation is brown-coloured. It produces pure white flowers, and ought to find a place in every collection of this extensive genus. Native of the Pyrenees.

SCUTELLARIA MACRANTHA (Long-flowered Scutellaria).—A native of Siberia, and the finest of all the species of this genus; it forms a hard woody root-stock, and is an excellent perennial alpine plant; it grows 9 inches high, producing an abundance of rich velvety dark blue flowers, much finer in colour than those of *S. japonica*.

SEMPERVIVUM ANOMALUM.—A native of Dauphiny, in the South of France; the rosettes are small, and covered with glandular hairs. It bears rich chocolate-coloured blossoms, on a dwarf flower-stalk. Is one of the prettiest species, is perfectly hardy, and of as free growth as our own common Houseleek.

S. arenarium, when grown in dense patches, has a lovely effect. It is much smaller than *S. globiferum*, to which it is allied, and is usually of a rich crimson colour; the leaves in the rosettes are not incurved, as in the latter species; the flowers are small, but crimson, and exceedingly pretty. *S. Heuffelli* is remarkable for the rich tints its foliage assumes in the autumn—almost a chocolate crimson. It is very distinct in the form of its flowers from all other species, its flowers being more tubular, and not expanded; they are of a yellow colour, and, as regards the value of the plant in point of beauty, may be readily dispensed with. When grown in a dry situation, the leaves are almost the colour of a Copper Beech. *S. hirtum, piliferum* and *fimbriatum* are all remarkable for possessing little dense tassels of hair at the extremity of each leaf; they are closely allied to one another, though bearing distinctive characters, not, perhaps, noticeable to the casual observer. *S. Pittoni.*—A somewhat tender-looking plant, but, nevertheless, perfectly hardy; it is a very distinct species, the leaves being covered with fleshy hairs that give the whole plant the appearance of being frosted over. *S. Reginæ Amaliæ.*—A fine vigorous grower, from Greece; scarce yet in cultivation, and likely to remain so for some time, on account of its shyness in producing offsets. It belongs to the same section as *S. Heuffelli*.

SILENE QUADRIFIDA and QUADRIDENTATA.—These species deserve more than mere nominal mention; in fact, among the dwarf rock-plants they have scarcely any equals. *Quadrifida* has narrow leaves, grows about 4 inches high, producing two or three flowers on each stem, of pure white, and beautifully notched. *Quadridentata* grows about 2 inches high, and for at least a month is covered with its small but exquisitely symmetrical white blossoms. They are both perfectly hardy when given a suitable locality on a rockwork, and will be found well worthy of that distinction. Natives of the Alps.

SISYRINCHIUM GRANDIFLORUM (Spring Satin-flower).—A beautiful early spring-flowering perennial, 6 to 10 inches high. The flowers are of a rich deep purple, with red style and filaments, and yellow anthers (or pure white, with transparent style and white filaments), issuing from a two-flowered spathe with a thin transparent margin. The leaves are narrow, sword-shaped, and sheathing at the base. A native of the N.W. regions of N. America. Suitable either for the rock-garden or for borders, in light peaty soil or very sandy loam, and in warm positions. It is also a charming subject for growing in pots in cold frames, from which it may be removed to the greenhouse when in flower in early spring. Multiplied by careful division.

STATICES.—*Statice minuta, bellidifolia, nana,* and *speciosa* are all species well worthy of cultivation, compact in habit, the flowers produced freely and of long continuance, added to which they are on a dry subsoil perfectly hardy. There are no flowers better adapted, in lightness and elegance, for giving a sort of neutral tint finish, if I may use the expression, to a bouquet of flowers.

TEUCRIUM PYRENAICUM (Pyrenean Teucrium).—A dwarf alpine plant with roundish leaves, somewhat woolly, and producing an abundance of yellowish flowers. It is well worthy of a place in all collections. Flowers from July to September.

THALICTRUM TUBEROSUM (Tuberous Meadow-Rue).—Grows about 9 inches high. Besides the usually graceful foliage which we find in all the dwarf forms of the genus *Thalictrum*, we have in this instance an additional beauty in the abundant mass of yellowish cream-coloured flowers which this plant produces. It is perfectly hardy, and thrives in deep peat soil. A native of Spain.

THYMUS SERPYLLUM ALBUM (White-flowered Thyme).—The white variety of our common wild Thyme is a lovely plant for a sunny bank. Nothing can be more charming than such a bank covered with a mixture of the common wild form and the white variety. The first plant cultivated in the Botanic Gardens, Hull, came from the Carpathian Mountains, and Mr. Niven believes the said plant is the parent of all now in cultivation.

TROLLIUS (Globe-flowers).—This genus, although more generally grown as a border plant, is a true alpine. Who, in rambling in the fields that border, as it were, the altitudinal zone of cultivation, has not been struck with the great contrast produced between the golden colour of the Buttercups, or still more golden of the Marsh Marigold (*Caltha palustris*), and the delicate lemon tint of our own *Trollius europæus*? And how many beautiful congeners has this *Trollius europæus*, ranging from *T. pumilus pallidus* (sometimes called *albus*), to the noble border plant *T. napellifolius*, with its large orange blossoms, almost the size of oranges; the former some 9 inches high, with cream-coloured flowers; the latter when growing vigorously full 30 inches high. Especially noticeable in this group are *T. sinensis*, in which the modified linear petals are exserted, and form a sort of rigid crown of intense gold, rising from amid the expanded petals; and *T. americanus*, rarely to be met with true, but when obtained to be cherished. Its petals almost approach the scarlet tint along their outer margin. There are many other species, but these are the most distinct.

UMBILICUS SPINOSUS (Spiny Umbilicus).—A very singular-looking plant, with somewhat the appearance of a small Apicra or Haworthia. The root-leaves are oblong in shape, convex towards the points, and form a rosette like a Sempervivum, each leaf bearing a spine at the apex. The flowers, which appear early in summer, are yellow, and form a terminal cylindrical spike on the top of the flower-stem, which is furnished with flat lance-shaped leaves. A native of Siberia, China, and Japan. A good plant for the rock-garden, in dry sunny spots. It is perfectly hardy, and my friend, Mr. James Atkins, of Painswick, informs me that on his rockwork it withstood the severe weather of January and February, 1871, without the slightest injury. Its chief enemies are the slugs, which destroy it whenever they have a chance.

VERONICAS.—*Veronica saturæifolia* is one of the very finest species; the flowers are about the size of *V. saxatilis*, of the same intense blue, and produced in abundant upright racemes. A somewhat rare plant, but when once seen growing in perfection never to be forgotten. Add

Veronica lactea (syn. *repens*), *verbenacea*, *pectinata*, *rupestris*, *bellidifolia*, *Daubenezi*, all species well worth cultivation, of good dwarf habit, and admirably adapted for rockery. Nor should the pink variety of *V. officinalis* be omitted—forming, as it does when established, dense patches of pink-coloured blossoms, elevated some 3 inches above the surface of the ground.

WALDSTEINIA FRAGARIOIDES (Strawberry Waldsteinia).—A showy plant from N. America, with creeping, bright-red, hairy stems. Grows about 6 inches high, and flowers in summer, when it produces numerous bright-yellow blooms about half an inch across. The leaves are ternate, on long channelled stalks, with obovate, smooth green leaflets, which fade to a lurid red. Suitable for rockwork, borders, fringes of shrubberies, etc., and thriving in ordinary soils. Syn. *Comaropsis*.

WULFENIA CARINTHIACA (Carinthian Wulfenia).—A remarkably dwarf, almost stemless evergreen herb, 12 to 18 inches high, bearing in summer showy spikes of purplish-blue, drooping, tubular flowers, issuing on short stalks from the axils of the bracts. The leaves are oblong, narrowed at the base, doubly crenated, stalked. Found only on one or two mountains in Carinthia. A very ornamental plant for rockwork or borders, in light, moist, sandy loam.

ZAPPANIA NODIFLORA (Creeping Vervain).—A pretty and modest-looking, compactly-spreading, trailing plant, with prostrate stems 2 or 3 feet or more in length. It blooms late in summer, producing small purplish flowers in small roundish heads, on long stalks springing from the axils of the leaves, which are spoon-shaped, coarsely and irregularly notched, with a wedge-shaped base, attenuated into a stalk. A native of Asia and America. Very suitable for the rock-garden, or for borders or edgings, in any rather warm soil. Syn. *Lippia nodiflora*.

PART III.

SELECTIONS

OF

ALPINE AND ROCK-PLANTS

FOR VARIOUS PURPOSES.

SELECTIONS OF ALPINE AND ROCK-PLANTS FOR VARIOUS PURPOSES.

A SELECTION OF DWARF ALPINE AND ROCK-PLANTS THAT WILL THRIVE IN ORDINARY SOIL ON THE LEVEL GROUND.

So much is said about the difficulties of growing alpine flowers, that I have made out the following list of kinds that will succeed in any soil. The greater number of these, if planted in the middle of a ploughed field of ordinary loam, will thrive as well as they do in their native homes. Association with rockwork is not required for any of them, though they may of course be arranged with good effect on it. A word is added to any kinds best enjoying any soil other than that of which our garden borders are usually composed. Nearly all the species enumerated (nearly 300) form a beautiful finish to raised borders and beds.

Acæna microphylla
Acantholimon glumaceum
Achillea tomentosa
    umbellata
Adonis vernalis
Æthionema coridifolium
Ajuga genevensis
Alyssum montanum
    saxatile
    spinosum
Anemone alba
    alpina
    angulosa
    apennina
    blanda
    coronaria
    fulgens
    Hepatica
    palmata
    pavonina
    Pulsatilla
    ranunculoides
    stellata
    sulphurea
    sylvestris
    trifolia
Antennaria dioica
Anthyllis montana
Antirrhinum rupestre
Aquilegia cærulea
    glandulosa
Arabis albida
    petræa
Arenaria laricifolia
    montana
    purpurascens
    verna
Armeria cephalotes
Artemisia frigida

Aster alpinus
    Reevesii
    versicolor
Astragalus hypoglottis
    monspessulanus
Aubrietias (all)
Bellium (dry soil), in variety
Bulbocodium vernum
Campanula alpina
    cæspitosa
    carpatica
    fragilis
    pulla
    Raineri
    turbinata, and nearly all the kinds
Cerastium, in variety
Convolvulus lineatus
    Soldanella
Cornus canadensis
Coronilla iberica
    minima
    montana
    varia
Corydalis lutea
    nobilis
    solida
Crocus Imperati
    nudiflorus
    Orphanidis
    reticulatus
    speciosus, and many other fine species and varieties
Cyclamen europæum
    hederæfolium
Daphne Cneorum
Dianthus alpinus

Dianthus cæsius
    deltoides
    dentosus
    neglectus
    petræus
    superbus, and many other species and varieties
Dielytra eximia
Diotis maritima
Dodecatheon integrifolium
    Jeffreyanum
    Meadia (cum grandiflorum
Dracocephalum austriacum
    Ruyschianum
Dryas Drummondi
    octopetala
Erica carnea
Erodium macradenum
    Manescavi
    petræum
    romanum
Erysimum ochroleucum
Erythronium Dens canis
Ficaria grandiflora
Fumaria bulbosa
Funkia, in variety
Galanthus plicatus
Gaultheria procumbens
Genista sagittalis
Gentiana acaulis
    Andrewsii
    asclepiadea
Geranium argenteum
    cinereum
    lancastriense
Globularia nudicaulis
    trichosantha

Gypsophila, in variety
Helianthemum, in variety
Helichrysum arenarium
Hippocrepis comosa
Hutchinsia alpina
Hyacinthus amethystinus
Iberis corifolia
    correæfolia
    gibraltarica
    sempervirens
    Tenoreana
Iris cristata
    nudicaulis
    pumila
    reticulata
Isopyrum thalictroides
Jasione perennis
Jeffersonia diphylla
Leucojum vernum
Leiophyllum buxifolium
Linaria alpina
Linum alpinum
    arboreum
    campanulatum
    narbonnense
    perenne
Lithospermum prostratum and others
Lychnis Lagascæ
    Viscaria
Melittis Melissophyllum
Mertensia virginica
Muscari, in variety
Myosotis alpestris
    dissitiflora
    sylvatica
Narcissus Bulbocodium
    juncifolius
    minor
    triandrus, and any other kinds
Nierembergia rivularis
Œnothera, all the dwarf kinds
Omphalodes Luciliæ
    verna
Ononis, in variety
Onosma taurica
Orobus cyaneus
    flaccidus
    variegatus
Orobus vernus, and many others
Oxalis Bowiei
    floribunda and various others
Papaver nudicaule
Paradisia Liliastrum
Pelargonium Endlicherianum
Pentstemon glaber
    procerus
    Scouleri, and other dwarf kinds
Phlox divaricata
    reptans
    subulata, and any other dwarf varieties
Plumbago Larpentæ
Polygonum vaccinifolium
Potentilla alpestris
    nitida
    pyrenaica
    verna, and all choice dwarf kinds
Primula altaica
    Auricula, in variety
    cortusoides
    ,, amœna
    marginata
    Palinuri
    viscosa
    vulgaris, fl. pl.
Prunella grandiflora
Puschkinia scilloides
Pyrethrum alpinum
Ranunculus aconitifolius
    amplexicaulis
    gramineus
    montanus
    parnassifolius
Sagina glabra
Sanguinaria canadensis
Santolina alpina
    incana
Saponaria cæspitosa
    ocymoides
Saxifraga, nearly all the kinds
Scabiosa graminifolia
    Parnassiæ
Scilla amœna
Scilla bifolia
    italica
    sibirica, and any hardy kinds
Scutellaria alpina
Sedum, any of the numerous kinds
Sempervivum, any of the numerous hardy kinds
Senecio argenteus
Silene acaulis
    alpestris
    Elisabethæ
    Pumilio
    Schafta
Sisyrinchium grandiflorum
Smilacina bifolia
Statice oleæfolia
    tatarica, and all dwarf kinds
Sternbergia lutea
Symphyandra pendula
Thalictrum anemonoides
Thlaspi latifolium
Thymus, in variety
Triteleia uniflora
Tunica Saxifraga
Veronica candida
    prostrata
    saxatilis
    taurica, and any dwarf kinds
Vesicaria utriculata
Vicia argentea
Vinca herbacea
Viola altaica
    calcarata
    cornuta
    lutea
    odorata
    pedata
    tricolor, and many others
Vittadenia triloba (dry soil)
Waldsteinia trifolia
Zapania nodiflora
Zauschneria californica
Zephyranthes Atamasco
Zietenia lavandulæfolia

A SELECTION OF ALPINE AND ROCK-PLANTS WORTHY OF BEING GROWN IN NURSERIES.

GENERALLY nurserymen do not now grow alpine plants, probably appalled by the display of vigorous weeds seen here and there under the name of "alpine plants," and when one does begin to collect them he usually errs by not selecting ornamental kinds. Most intelligent nurserymen would gladly grow a collection if they were confident of securing really ornamental kinds. I have therefore taken considerable pains in selecting the following; many of them will thrive in good ordinary soil, and all those requiring a special treatment are such as well repay for it. Nurserymen would also do well to consult the long list of plants that will thrive in ordinary soil in borders.

Acæna microphylla
Acantholimon glumaceum
Achillea Clavennæ
    tomentosa
Adonis vernalis
Æthionema coridifolium
Ajuga genevensis
Alyssum saxatile
Andromeda fastigiata
    tetragona
Androsace carnea
    Chamæjasme
    lanuginosa
    vitaliana
Anemone alba
    alpina
    angulosa
    apennina
    blanda
    coronaria
    fulgens
    Hepatica
    nemorosa, and varieties
    palmata
    stellata
    sylvestris
Antennaria dioica
    tomentosa
Anthyllis montana
Aquilegia alpina
    cærulea
    californica
    glandulosa
Arabis albida
    blepharophylla
    petræa
Arenaria balearica
    montana
    purpurascens
Armeria cephalotes
    vulgaris, and varieties
Aster alpinus
Astragalus monspessulanus
Aubrietia, in varieties
Bryanthus erectus
Bulbocodium vernum
Calandrinia umbellata
Campanula alpina
    cæspitosa
    carpatica
    fragilis
    pulla

Campanula Raineri
    turbinata
Cerastium, the finer silvery kinds
Colchicum, in varieties
Convolvulus lineatus
Cornus canadensis
Coronilla minima
    varia
Corydalis lutea
    nobilis
Crocus Imperati
    nudiflorus
    Orphanidis
    reticulatus
    Sieberi
    speciosus
    vernus
Cyclamen Coum
    europæum
    hederæfolium
    vernum
Cypripedium spectabile
Daphne Cneorum
Dianthus alpinus
    cæsius
    deltoides
    neglectus
    superbus
Dielytra eximia
Dodecatheon integrifolium
    Meadia
Draba aizoides
    Aizoon
    bœotica
Dryas Drummondi
    octopetala
Eranthis hyemalis
Erica carnea
Erinus alpinus
Erodium Manescavi
Erysimum ochroleucum
Erythronium Dens-canis
Funkia grandiflora
Galanthus plicatus
Gaultheria procumbens
Genista sagittalis
Gentiana acaulis
    Andrewsii
    asclepiadea
    verna
Geranium argenteum
    sanguineum

Geranium striatum
Geum coccineum
    montanum
Gypsophila prostrata
Hedysarum obscurum
Helianthemum, in variety
Helleborus niger minor
    trifolius
Hyacinthus amethystinus
Iberis correæfolia
    sempervirens
Iris cristata
    nudicaulis
    pumila
    reticulata
Leiophyllum buxifolium
Leontopodium alpinum
Leucojum vernum
Lilium tenuifolium
Linaria alpina
Linum alpinum
    arboreum
    narbonnense
    perenne
Lithospermum prostratum
Lychnis Lagascæ
    Viscaria, and varieties
Menziesia empetriformis
Muscari botryoides
Myosotis alpestris
    dissitiflora
    sylvatica
Narcissus Bulbocodium
    juncifolius
    minor
    triandrus
Nierembergia rivularis
Œnothera marginata
    missouriensis
    taraxacifolia
Omphalodes verna
Onobrychis montana
Onosma taurica
Orobus cyaneus
    vernus
Oxalis Bowiei
    floribunda
Paradisia Liliastrum
Pentstemon glaber
    procerus
Phlox divaricata
    reptans
    subulata

Polygala Chamæbuxus
Potentilla alpestris
    pyrenaica
Primula Auricula, in varieties
    cortusoides amœna
    integrifolia
    marginata
    nivea
    viscosa
Prunella grandiflora
Puschkinia scilloides
Ranunculus aconitifolius fl. pl.
    alpestris
    amplexicaulis
Rhexia virginica
Rhododendron Chamæcistus
Sanguinaria canadensis
Santolina incana
Saponaria ocymoides
Saxifraga Aizoon
    Andrewsii
    aretiodes
    biflora
    cæsia
    ceratophylla

Saxifraga cordifolia
    Cotyledon
    crassifolia
    diapensioides
    Geum
    granulata fl. pl.
    hypnoides
    juniperina
    ligulata
    longifolia
    oppositifolia
    retusa
    Rocheliana
    Stansfieldii
    umbrosa
Scilla bifolia
    sibirica
Sedum Ewersii
    glaucum
    kamtschaticum
    pulchellum
    rupestre
    Sieboldii
    spectabile
    spurium
Sempervivum arachnoideum
    calcareum

Sempervivum globiferum
    Heuffelli
    montanum
    rusticum
    tectorum
Senecio argenteus
Silene acaulis
    alpestris
    Elisabethæ
    Pumilio
    Schafta
Soldanella alpina
Spigelia marilandica
Symphyandra pendula
Thlaspi latifolium
Trillium grandiflorum
Triteleia uniflora
Tunica Saxifraga
Veronica candida
    fruticulosa
    prostrata
    saxatilis
    taurica
Viola cornuta
    lutea
    pedata
Waldsteinia trifolia
Zephyranthes Atamasco

Alpine and Rock-plants with White or Whitish Flowers.

Achillea Clavennæ
    umbellata
Allium neapolitanum
Andromeda tetragona
Androsace Chamæjasme
    cylindrica
    helvetica
    obtusifolia
    pubescens
    villosa
Anemone alba
    alpina
    nemorosa
    sylvestris
    trifolia
Aquilegia cærulea alba
Arabis albida
    petræa
    procurrens
Arenaria balearica
    montana
    verna
Asperula odorata
Aster alpinus albus
Astragalus hypoglottis albus
Calla palustris
Campanula cæspitosa alba
    carpatica alba
    rotundifolia alba
Cardamine trifolia
Cerastium Bieberstcinii

Cerastium grandiflorum
    lanuginosum
    tomentosum
Chimaphila maculata
Chrysanthemum alpinum
Convallaria majalis
Cornus canadensis
Crocus biflorus
Cyclamen Coum album
Diapensia lapponica
Dryas octopetala
Erinus alpinus albus
Funkia grandiflora
Galanthus nivalis
    plicatus
Helleborus niger minor
Hoteia japonica
Hutchinsia alpina
    petræa
Iberis corifolia
    corræfolia
    Garrexiana
    sempervirens
Isopyrum thalictroides
Jeffersonia diphylla
Koniga maritima
Leucanthemum alpinum
Leucojum vernum
Lychnis Viscaria alba
Menziesia polifolia alba
Muscari botryoides album
Myosotis sylvatica alba

Nierembergia rivularis
Œnothera marginata
    taraxacifolia
Ononis alba
Oxalis floribunda alba
Papaver alpinum albiflorum
Paradisia Liliastrum
Parnassia, in variety
Phlox subulata alba
Potentilla alba
Primula cortusoidesamœna alba
    Munroi
    nivea
Pyrola rotundifolia
Ranunculus aconitifolius
    alpestris
    amplexicaulis
    parnassifolius
    rutæfolius
Sagina glabra
Sanguinaria canadensis
Saxifraga affinis
    ceratophylla
    Cotyledon
    diapensioides
    granulata fl. pl.
    hypnoides
    longifolia
    oppositifolia alba
    palmata

Saxifraga Rocheliana
    Stansfieldii
Scilla bifolia alba
Sedum album
Silene alpestris
    maritima fl. pl.

Smilacina bifolia
Symphyandra pendula
Thlaspi latifolium
Thymus Serpyllum album
Trientalis europæus

Trillium grandiflorum
Vinca minor alba
Viola cornuta alba
    odorata alba
Zephyranthes Atamasco

ALPINE AND ROCK PLANTS WITH BLUE, BLUISH, OR PURPLISH FLOWERS.

Ajuga genevensis
Anemone angulosa
    apennina
    blanda
    Hepatica
    Pulsatilla
Aquilegia alpina
    cærulea
    glandulosa
    pyrenaica
    vulgaris
Aster alpinus
Astragalus hypoglottis
Aubrietia, in variety
Bulbocodium vernum
Campanula cæspitosa
    carpatica
    fragilis
    garganica
    isophylla
    muralis
    pulla
    Raineri
    turbinata, and many
      others
Clematis, fine new hybrid
    and purple varieties
Crocus nudiflorus
    speciosus
    vernus
Dracocephalum austria-
    cum, in variety
    grandiflorum
    Ruyschianum
    canescens
Erpetion reniforme
Gentiana acaulis
    alpina
    Andrewsii
    asclepiadea
    bavarica

Gentiana caucasica
    Pneumonanthe
    pyrenaica
    septemfida
    verna
Globularia cordifolia
    nana
    nudicaulis
    trichosantha
Hottonia palustris
Houstonia cærulea
Hyacinthus amethystinus
Ionopsidion acaule
Iris cristata
    nudicaulis
    pumila
    reticulata
Jasione humilis
    perennis
Linaria alpina
Linum alpinum
    narbonnense
    perenne
    viscosum
Lithospermum Gastoni
    petræum
    prostratum
    purpureo-cæruleum
Mazus Pumilio
Meconopsis aculeata
Mertensia virginica
Mesembryanthemum acin-
    aciforme
Muscari botryoides
    Heldreichii
    monstrosum
    racemosum
Myosotis alpestris
    azorica
    dissitiflora
    palustris

Myosotis sylvatica
Omphalodes verna
Orobus cyaneus
    vernus
Pentstemon glaber
    Jeffreyanus
    speciosus
Pinguicula grandiflora
    vulgaris
    longifolia
Plumbago Larpentæ
Polygala calcarea
Primula glutinosa
    latifolia
    purpurea
    scotica
Prunella grandiflora
Pulmonaria angustifolia
Puschkinia scilloides
Ramondia pyrenaica
Scilla amœna
    bifolia
    campanulata
    italica
    sibirica
Scutellaria, in variety
Sisyrinchium grandiflorum
Soldanella alpina
Veronica candida
    Chamædrys
    saxatilis
    taurica
Vicia Cracca
Vinca herbacea
    major
    minor
Viola calcarata
    cornuta
    cucullata
    pedata
Wulfenia carinthiaca

ALPINE AND ROCK-PLANTS WITH ROSY, CRIMSON, SCARLET, RED, AND PINKISH FLOWERS.

Acæna microphylla
Acantholimon glumaceum
Acis autumnalis
Æthionema saxatile
Androsace carnea
   ciliata
   lanuginosa
   Wulfenii
Anemone coronaria
   fulgens
   Hepatica
   hortensis
   pavonina
   stellata
Antennaria dioica-rosea
Anthyllis montana
Arabis blepharophylla
Armeria cephalotes
   vulgaris rosea
Astragalus alpinus
   monspessulanus
Bryanthus erectus
Calandrinia umbellata
Centaurea uniflora
Convolvulus arvensis
   Soldanella
Coronilla varia
Cortusa Matthioli
Cyclamen Coum
   europæum
   hederæfolium
   repandum
   vernum
Cypripedium acaule
Daphne alpina
   Cneorum
Dianthus alpinus
   cæsius
   Caryophyllus
   neglectus
   petræus

Dianthus plumarius
   superbus
Dielytra eximia
   spectabilis
Dodecatheon integrifolium
   Jeffreyanum
   Meadia
Erica carnea
   hibernica
   Mackieana
   Tetralix
Erinus alpinus
Erodium cicutarium
   Manescavi
   petræum
   romanum
Geranium argenteum
   sanguineum
Geum coccineum
Helianthemum venustum
Lathyrus grandiflorus
   tuberosus
Lilium longiflorum
   tenuifolium
Lychnis alpina
   Coronaria
   flos-Jovis
   fulgens
   Haageana
   Lagascæ
   Viscaria
Menziesia cærulea
   empetriformis
   polifolia
Ononis arvensis
   rotundifolia
Oxalis Bowiei
   floribunda
   speciosa
Pelargonium Endlicherianum

Phlox divaricata
   reptans
   subulata
Polygonum vaccinifolium
Primula cortusoides
   amœna
   denticulata
   farinosa
   Flœrkiana
   integrifolia
   marginata
   minima
   pedemontana
   pubescens
   viscosa
Rhexia virginica
Rhododendron amœnum
   Chamæcistus
Saponaria cæspitosa
   ocymoides
Saxifraga biflora
   cordifolia
   oppositifolia
   retusa
Sedum Ewersii
   Sieboldii
   spectabile
   spurium
Sempervivum arachnoideum
Silene acaulis
   Elisabethæ
   pennsylvanica
   Pumilio
   Schafta
   virginica
Spigelia marilandica
Tropæolum speciosum
Tunica Saxifraga
Zauschneria californica

ALPINE AND ROCK-PLANTS WITH YELLOW FLOWERS.

Achillea ægyptiaca
   tomentosa
Adonis vernalis
Alyssum alpestre
   montanum
   Wiersbeckii
   saxatile
Androsace Vitaliana
Anemone palmata
   ranunculoides
   sulphurea
Calceolaria Kellyana
Caltha palustris
Cheiranthus alpinus

Cheiranthus Cheiri
Coronilla minima
   montana
Corydalis lutea
   nobilis
Crocus luteus
Cypripedium Calceolus
Dondia Epipactis
Draba aizoides
   Aizoon
   alpina
   aurea
   bœotica
   ciliaris

Draba cuspidata
Dryas Drummondi
Eranthis hyemalis
Erysimum ochroleucum
   pumilum
Ficaria grandiflora
Galium verum
Genista prostrata
   sagittalis
   tinctoria
Geum montanum
Helianthemum formosum
   ocymoides
   Tuberaria

Helianthemum vulgare
Helichrysum arenarium
Hippocrepis comosa
Hypericum calycinum
   humifusum
   Nummularium
   Coris
Linum arboreum
   campanulatum
Lithospermum canescens
Lotus corniculatus
Lysimachia nemorum
   Nummularia
Meconopsis cambrica
Narcissus Bulbocodium
   Jonquilla
   juncifolius
   minor
Œnothera Drummondi
   missouriensis
   riparia

Onosma taurica
Othonna cheirifolia
Papaver nudicaule
Polygala Chamæbuxus
Potentilla alpestris
   calabra
   pyrenaica
   Tonguei
   verna
Primula Auricula lutea
   elatior
   Palinuri
   sikkimensis
   veris
   vulgaris
Ranunculus acris, fl. pl.
   bulbosus, fl. pl.
   bullatus, fl. pl.
   gramineus
   montanus
   spicatus

Ranunculus Thora
Santolina Chamæcyparissus
Saxifraga aizoides
   aretioides
   Cymbalaria
   Hirculus
   juniperina
Sedum acre
   kamtschaticum
   rupestre
Sternbergia lutea
Tropæolum polyphyllum
Vesicaria utriculata
Viola biflora
   lutea
   tricolor
   lutea
   Zoysii
Waldsteinia geoides
   trifolia

A Selection of Choice Dwarf Shrubs for the Rock-Garden, etc.

Alyssum spinosum
Andromeda fastigiata
   hypnoides
   tetragona
   floribunda
Arctostaphylos alpina
   Uva-ursi
Azalea amœna
Bryanthus erectus
Calluna vulgaris, finer varieties
Chimaphila maculata
   umbellata
Cistus algarvensis
   formosus, and others
Cornus canadensis
   suecica
Daphne alpina
   Cneorum
   collinum
Diapensia lapponica
Dryas Drummondi
   octopetala
Epigæa repens
Erica carnea
   ciliaris
   cinerea
   hibernica

Erica Mackieana
   mediterranea
   Tetralix
Euonymus radicans, variegated
Euphorbia Myrsinites
Gaultheria procumbens
Genista prostrata
   sagittalis
   tinctoria
Globularia cordifolia
Hedera, variegated and other interesting varieties
Helianthemum canum
   formosum
   ocymoides
   rosmarinifolium
   Tuberaria
Iberis correæfolia
   garrexiana
   gibraltarica
   sempervirens
Juniperus squamata
Kalmia latifolia nana
Leiophyllum buxifolium
Linnæa borealis

Linum arboreum
Lithospermum fruticosum
   petræum
   prostratum
Loiseleuria procumbens
Lycopodium dendroideum
Menziesia cærulea
   empetriformis
   polifolia
Polygala Chamæbuxus
Polygonum vaccinifolium
Rhododendron amœnum
   Chamæcistus
   ferrugineum
   hirsutum
   myrtifolium
   Torlonianum
Salix lanata
   reticulata
   serpyllifolia
Santolina incana
Scabiosa Webbiana
Skimmia, in variety
Vaccinium, in variety
Veronica pinguifolia
   fruticulosa
   saxatilis

A SELECTION OF ALPINE AND ROCK-PLANTS TO RAISE FROM SEED.

ALL flowering rock-plants, it need hardly be said, seed and grow from seeds in a wild state, but in selecting the kinds we wish to raise from seed in gardens, it is desirable to avoid—first, kinds more quickly obtained in other ways, as the kinds that run about very quickly; secondly, those very difficult to raise, as some Gentians; thirdly, such as it is difficult to obtain seed of, as the Epimediums. The following is compiled as a guide to those interested in alpine and rock-plants, so that they may know the direction in which to work. Nothing is more common than for amateurs to make mistakes in this way. Some kinds enumerated are not in seed catalogues, but many amateurs may gather these in their native wilds.

Achillea, the rare and smaller kinds
Æthionema, any kinds
Alyssum, any kinds
Andromeda, any kinds
Androsace, any kinds
Anemone, any kinds
Anthyllis montana, and any others
Aquilegia, the rarer alpine kinds
Arabis, any kinds
Arenaria, any kinds
Armeria, any kinds
Aster alpinus or any other dwarf kinds
Astragalus, in variety
Aubrietia, various kinds
Begonia Veitchii
Bellis, in variety
Bellium, in variety
Calandrinia umbellata
Callirhoe involucrata and pedata
Calystegia dahurica
Campanula, many dwarf kinds
Cardamine trifolia
Cheiranthus, rare mountain kinds
Cerastium, the rarer kinds
Cistus, in variety
Convolvulus, the dwarf kinds
Coptis trifolia
Cornus, small alpine kinds
Coronilla, ditto
Cortusa Matthioli
Corydalis, in variety
Crocus, rare kinds
Cyananthus lobatus
Cyclamen, all the kinds
Daphne, dwarf mountain kinds
Dianthus, any kinds
Dielytra, in variety
Diotis maritima
Dodecatheon, any kinds

Draba, any kinds
Dracocephalum, in variety
Dryas, in variety
Erigeron, in variety
Erinus, any kinds
Erodium, small and rare kinds
Erpetion reniforme
Erysimum ochroleucum
Erysimum pumilum
Funkia, in variety
Galanthus plicatus
Genista, dwarf alpine kinds
Gentiana, any kinds
Geranium, dwarf rock kinds
Geum, the finer kinds
Globularia, in variety
Gypsophila, in variety
Hedysarum obscurum
Helianthemum, in variety
Hieracium aurantiacum
Hippocrepis comosa
Houstonia cærulea
Hutchinsia alpina petræa
Iberidella rotundifolia
Iberis, any kinds
Ionopsidion acaule
Iris, rare and dwarf kinds
Isopyrum thalictroides
Jasione humilis
Jeffersonia diphylla
Linaria alpina, and dwarf or rare alpine kinds
Linum, any kinds
Lithospermum, any kinds
Lychnis, any kinds
Malva campanulata
Matthiola tristris
Mazus Pumilio
Meconopsis, in variety
Medicago falcata
Melittis Melissophyllum
Mertensia virginica
Mimulus, in variety

Myosotis, in variety
Narcissus, rare kinds only
Nierembergia, in variety
Nertera depressa
Œnothera, in variety
Omphalodes Luciliæ verna
Ononis, in variety
Onosma taurica
Orobus, any kinds
Oxytropis, in variety
Papaver, alpine kinds
Paradisia Liliastrum
Parnassia, in variety
Pelargonium Endlicherianum
Pentstemon, dwarf alpine kinds
Petrocallis pyrenaica
Phlox, dwarf species
Pinguicula, any kinds
Potentilla, alpine kinds
Primula, any kinds
Prunella grandiflora
Ramondia pyrenaica
Ranunculus, alpine kinds
Santolina, in variety
Saponaria, dwarf species
Saxifraga, any kinds
Scutellaria, in variety
Sedum, any kinds
Sempervivum, any kinds
Senecio, dwarf alpine kinds
Silene, in variety
Soldanella, in variety
Symphyandra pendula
Thymus, in variety
Tunica Saxifraga
Veronica, dwarf and alpine kinds
Vesicaria utriculata
Vicia argentea
Viola, in variety
Vittadenia triloba
Wulfenia carinthiaca
Zauschneria californica

A Selection of Alpine Plants, etc., suitable for Planting on the Margins of Beds and Masses of Rhododendrons and other American Shrubs.

The masses of peat in American plant-beds and borders are just those in which many interesting plants grown with difficulty in ordinary soil and borders will thrive apace. The appearance of the borders themselves, often bare and earthy towards the edge, will be greatly improved by the presence of the choice dwarf plants enumerated in the following list, some of which, thriving in the watery woods of the eastern side of North America, do little good with us unless both in peat and partial shade. Of course numbers of others besides those enumerated would grow in the above-named positions, which are, however, better reserved for rare and somewhat fastidious subjects.

Acis autumnalis
Andromeda fastigiata
    polifolia
    tetragona
Anemone angulosa
    blanda
Begonia Veitchii
Bryanthus erectus
Calandrinia umbellata
Campanula pulla
Chimaphila maculata
    umbellata
Coptis trifolia
Cornus canadensis
    suecica
Cortusa Matthioli
Crocus, rare kinds
Cyclamen, in variety
Cypripedium, hardy kinds
Daphne Cneorum
Dianthus alpinus
Diapensia lapponica
Dodecatheon, in variety
Epigæa repens
Epimedium, all the kinds
Erica, in variety
Erythronium Dens-canis
Galanthus plicatus
Gentiana, any herbaceous kind like G. Pneumonanthe
Helleborus, in variety

Hoteia japonica
Hyacinthus amethystinus
Ionopsidion acaule
Iris cristata
    reticulata
Jeffersonia diphylla
Leiophyllum buxifolium
Leucojum vernum
Lilium longiflorum
    tenuifolium
Linnæa borealis
Linum arboreum
Loiseleuria procumbens
Lycopodium dendroideum
Menziesia cærulea
    empetriformis
Mertensia virginica
Myosotis azorica
    dissitiflora
Narcissus Bulbocodium
    juncifolius
    triandrus
Nierembergia rivularis
Omphalodes Luciliæ
    verna
Orchis foliosa
    latifolia
    maculata
Orobus cyaneus
Parnassia asarifolia
    caroliniana
    palustris

Pinguicula grandiflora
Polygala Chamæbuxus
Primula Auricula
    cortusoides
    amœna
    farinosa
    integrifolia
    longiflora
    marginata
    Munroi
    Palinuri
    purpurea
    sikkimensis
    viscosa
Puschkinia scilloides
Pyrola, in variety
Ramondia pyrenaica
Rhexia virginica
    marilandica
Rhododendron Chamæcistus
    præcox
Sanguinaria canadensis
Sisyrinchium grandiflorum
Smilacina bifolia
Soldanella alpina
Spigelia marilandica
Thalictrum anemonoides
Trientalis europæus
Trillium grandiflorum, and any other kind

A Selection of Plants for forming "Carpets."

Very little reflection suffices to show that a much more beautiful effect may be obtained from a mingling of several distinct types and sizes of vegetation than from an array of any one species, or even plants of one size. On the mountain-sides Violets and Lilies-of-the-Valley bloom beneath Hazel and Mezereon, and below the golden showers of the Laburnum, while the forest vegetations reign over all. One of the most successful ways of getting like effects in the rock-garden and on the choice border is by covering the ground with small spreading plants, which heighten the effect of the taller objects placed among them, and indeed often benefit them by keeping the ground in a more open and natural condition. Besides, the highest effect is not possible in any garden where there are bare surfaces in spring or early summer. I should strongly advise the reader to "carpet" his choice mixed border as well as his rock-garden with any dwarf spreading plants he may think suitable. It is obvious that almost all herbaceous plants may be used in this way under certain conditions. A species that might seem a giant compared to some of those in the following list, would form a "carpet" beneath the branches of forest trees. This list, however, is confined to things of dwarf stature suitable for beds, borders, and the rock-garden. Those requiring plants to form "carpets" in woods may consult the list of plants that will grow in woods and copses. Subjects, however, as large as the Vincas, Common Forget-me-not, and Creeping Forget-me-not (*Omphalodes verna*) are included, as they would answer well for placing beneath Rhododendrons and other shrubs, and also in permanent arrangements of the stronger perennials. Not a few plants may be enjoyed in this way better than in any other, as, for example, such as Mentha Requieni and Ionopsidium acaule, which are not usually considered ornamental enough for cultivation, but used as "carpets" in the ground occupied by choice bulbs, or alpine plants, their effect will be of the happiest kind. Annuals are included in this selection in consequence of their freedom of growth and the facility with which they may be raised and grown, for at once forming a turf. Somewhat slow-growing things are also included, as, for example, the Æthionemas, but these could be used with the best effect as surface plants for groups or small beds of neat shrubs, or other subjects that thrive best when planted permanently.

Acæna mycrophylla
Æthionema, in variety
Agrostis nebulosa
Alyssum maritimum
    montanum
Anagallis, in variety
Anemone apennina
Antennaria alpina
    dioica
    hyperborea
    tomentosa
Anthyllis montana
Arabis albida
Arenaria balearica
    montana
    purpurascens
    verna
Artemisia frigida
Asperula odorata
Aubrietia, in variety
Brachycome, in variety
Calandrinia discolor
    grandiflora
Campanula cæspitosa
    fragilis
    garganica
    hederacea
    pulla
    Raineri
Cenia turbinata
Cerastium, in variety
Claytonia sibirica
Clintonia elegans
    pulchella
Collinsia, in variety
Convolvulus althæoides
    mauritanicus
    Soldanella
    tricolor
Cornus canadensis
    suecica
Coronilla varia
Cosmidium filifolium
Crucianella stylosa
Dianthus deltoides

Draba, in variety
Dryas Drummondi
    octopetala
Erinus alpinus
Erpetion reniforme
Fragaria, in variety
Galium verum
Genista prostrata
    sagittalis
Gilia, in variety
Godetia, in variety
Gypsophila dubia
    elegans
    muralis
    prostrata
    viscosa
Hedera, in variety
Helianthemum, in variety
Helichrysum arenarium
Hepatica, in variety
Hippocrepis comosa
Hutchinsia alpina
Hypericum humifusum
Ionopsidium acaule
Jasione humilis
Lamium maculatum album
Lathyrus tuberosus
Leptinella scariosa
Leptosiphon, in variety
Linaria alpina
    Cymbalaria
    Elatine
    repens
Linnæa borealis
Linum alpinum
Lithospermum prostratum
Lotus corniculatus
Lysimachia Nummularia
Malcolmia maritima
Malva campanulata
Medicago elegans
Mentha Requieni
Mesembryanthemum, in
    variety
Mimulus, in variety

Myosotis, in variety
Nemophila, in variety
Nertera depressa
Nierembergia rivularis
Nolana, in variety
Œnothera acaulis
    Bistorta
    marginata
    taraxacifolia
Oxalis Acetosella
    Bowiei
    corniculata, in variety
    floribunda
    Valdiviana
Parochetus communis
Paronychia serpyllifolia
Pentstemon procerus
Phlox canadensis
    Drummondi
    procumbens
    reptans
    subulata
Platystemon californicus
Polygonum vaccinifolium
Polypogon monspeliensis
Potentilla alba
    calabra
    gracilis
    reptans
    verna
Pyrethrum Tchiatchewi
Ranunculus repens
Reseda odorata
Rhodanthe, in variety
Sagina glabra, variety cor-
    sica
Santolina alpina
Sanvitalia procumbens
Saponaria calabrica
    ocymoides
Saxifraga, in great variety
Schizopetalon Walkeri
Sedum, in great variety
Selaginella denticulata
Sempervivum, in variety

Senecio argenteus
Sibthorpia europæa
Silene acaulis
   alpestris
   maritima
   pendula
   Schafta
Smilacina bifolia
Specularia pentagonia
Specularia speculum

Symphyandra pendula
Symphytum caucasicum
Thymus lanuginosus
Tropæolum, in variety
Tunica Saxifraga
Umbilicus chrysanthus
Venidium calendulaceum
Veronica alpina
   Chamædrys

Veronica fruticulosa
   repens
   saxatilis
   syriaca
Vicia argentea
Vinca, in variety
Viola, in variety
Waldsteinia fragarioides
Zapania nodiflora

Alpine Plants that will grow well in and near Cities.

Acantholimon glumaceum
Alyssum saxatile
Anemone apennina
   blanda
   fulgens
Anthyllis montana
Aquilegia alpina
   cærulea
Arabis albida
Armeria vulgaris-rosea
   cephalotes
Aster versicolor
Aubrietia deltoidea, and varieties
Calandrinia umbellata
Campanula cæspitosa
   fragilis
   garganica
   turbinata
Cerastium Biebersteinii
   grandiflorum
Dianthus cæsius (on walls)
   deltoides
   neglectus
Dryas octopetala
Erinus alpinus (on walls)
Erysimum ochroleucum
Gaultheria procumbens

Gentiana acaulis
   Andrewsii
Gypsophila prostrata
Helianthemum, many kinds
Hepatica angulosa
   triloba
Iberis corifolia
   corræfolia
   sempervirens
Lithospermum prostratum
Onosma taurica
Phlox reptans
   subulata
Plumbago Larpentæ
Polygonum vaccinifolium
Ranunculus amplexicaulis montanus
Saponaria ocymoides
Saxifraga Andrewsii
   capillaris
   ceratophylla
   coriophylla
   cristata
   crustata
   hypnoides
   incurvifolia
   juniperina

Saxifraga lævis
   longifolia
   muscoides
   oppositifolia
   pectinata
   pyramidalis
   recta
   Rocheliana
Sedum Ewersii
   hispanicum
   kamtschaticum
   monstrosum
   pulchellum
   reflexum
   rupestre
   Sieboldii
   spectabile
   spurium
Sempervivum calcareum
   glaucum
   globiferum
   hirtum
   montanum
   soboliferum
Silene alpestris
Tunica Saxifraga
Veronica candida
   saxatilis

Alpine Plants Green in Winter.

Alyssum montanum
Antennaria dioica
Arabis albida
   lucida
   procurrens
Arbutus Uva-ursi
Arenaria balearica
   verna
Aubrietia deltoidea, and all the varieties
Dianthus deltoides
Dryas octopetala
Gaultheria procumbens
Gentiana acaulis
Helianthemum, in variety
Hieracium aurantiacum
Hippocrepis comosa
Iberis corifolia
   sempervirens
Linum arboreum
Lysimachia Nummularia

Phlox reptans
   subulata
Sagina glabra
Santolina viridis
Saxifraga affinis
   cæspitosa
   contraversa
   decipiens
Geum
   hypnoides
   juniperina
   lævis
   muscoides
   palmata
   pedata
   Stansfieldii
Sedum album
   anglicum
   reflexum
   rupestre
   sexangulare

Sempervivum arenarium
   flagelliforme
   montanum
   soboliferum
Silene acaulis
   alpestris
   maritima
Thymus corsicus
   bracteosus
   lanuginosus
   Serpyllum
   micans
   Zygis
Tunica Saxifraga
Vaccinium Vitis-idæa
   Oxycoccos
Veronica gentianoides
   saxatilis
   taurica
Vinca major
   minor

A Selection of Alpine Plants suited for Culture in Pots and for Exhibition.

Acæna microphylla
Acantholimon glumaceum
Achillea ægyptiaca
    Clavennæ
    tomentosa
    umbellata
Æthionema coridifolium
Alyssum montanum
Andromeda fastigiata
    hypnoides
    tetragona
Androsace carnea
    Chamæjasme
    ciliata
    cylindrica
    glacialis
    helvetica
    imbricata
    lactea
    lanuginosa
    obtusifolia
    pyrenaica
    villosa
    Vitaliana
Anemone alba
    alpina
    angulosa
    blanda
    narcissiflora
    palmata
    Pavonina
    stellata
    sylvestris
Anthyllis montana
Antirrhinum rupestre
Aquilegia alpina
    cærulea
    pyrenaica
Arabis blepharophylla
Arenaria montana
Armeria cephalotes
Aster versicolor
Astragalus monspessulanus
    hypoglottis
Aubrietia deltoidea, and varieties
Begonia Veitchii
Calceolaria Kellyana
Callirhoe involucrata
Campanula carpatica
    fragilis
    garganica
    pulla
    turbinata
Convolvulus lineatus
    mauritanicus
Cornus canadensis
Coronilla minima
Cyclamen Coum
    hederæfolium
    vernum
Cypripedium acaule
    Calceolus

Cypripedium pubescens
    spectabile
Dianthus alpinus
    neglectus
Dielytra eximia
    spectabilis
Dodecatheon, all the species and varieties
Draba aizoides
    Aizoon
    cuspidata
Epimedium macranthum
    pinnatum elegans
Epipactis palustris
Erinus alpinus
Erodium Manescavi
    macradenum
    petræum
    Reichardi
Funkia grandiflora
Genista sagittalis
Gentiana bavarica
    verna
Geranium argenteum
    cinereum
    lancastriense
    sanguineum
Geum coccineum
    montanum
Goodyera pubescens
Gypsophila repens
Helianthemum venustum
Iberis, in variety
Iris cristata
    nudicaulis
    reticulata
Leucojum vernum
Lilium longiflorum
    tenuifolium
Linaria alpina
Linum alpinum
    arboreum
    flavum
Lithospermum prostratum
Lychnis alpina
    Lagascæ
Malva campanulata
Menziesia empetriformis
Mimulus repens
Myosotis alpestris
    azorica
    dissitiflora
Narcissus Bulbocodium
    juncifolius
    minor, and many others
Nierembergia rivularis
Œnothera marginata
    missouriensis
Omphalodes Luciliæ
    verna
Ophrys apifera
    arachnites

Orchis foliosa
    latifolia
    laxiflora
    nigra
Orobus cyaneus
    vernus
Oxalis Bowieana
    floribunda
    speciosa
Parnassia caroliniana
Pentstemon procerus
Phlox reptans
    subulata
Polemonium cæruleum
    variegatum
Primula amœna
    Auricula, in variety
    denticulata
    erosa
    farinosa
    integrifolia
    latifolia
    longiflora
    marginata
    minima
    Munroi
    nivea
    purpurea
    viscosa
Pulmonaria virginica
Pyrola media
    rotundifolia
Ramondia pyrenaica
Ranunculus amplexicaulis
    parnassifolius
Rhexia virginica
Santolina alpina
    incana
Saponaria ocymoides
Saxifraga biflora
    cæsia
    Cotyledon
    diapensioides
    juniperina
    oppositifolia
    retusa
    Rocheliana
    Stansfieldii
Sedum Ewersii
    Kamtschaticum
    pulchellum
    Sieboldii
    spectabile
    spurium
Sempervivum arachnoideum
    Laggeri
    montanum
    Pittoni
    Pomellii
Silene alpestris
    acaulis
    Elisabethæ

Silene Pumilio
    Schafta
    virginica
Sisyrinchium grandiflorum
Soldanella, in variety
Spigelia marilandica

Spiræa japonica
Statice oleæfolia
Symphyandra pendula
Trillium cernuum
    erectum
    grandiflorum

Veronica saxatilis
    taurica
Viola pedata
Wulfenia carinthiaca
Zauschneria californica

Alpine and Herbaceous Plants, suitable as Flowering Edgings for Beds or Borders.

Achillea tomentosa
Ajuga genevensis
Alyssum saxatile
Anemone apennina
    coronaria
Arabis albida
Armeria vulgaris-rosea
Aster versicolor
Aubrietia deltoidea
Bellis hortensis aucubæ-
    folia, and double
    varieties
Calandrinia umbellata
Campanula carpatica
    alba
    cæspitosa
        alba
    fragilis
    garganica
Cerastium Biebersteinii
    grandiflorum
    tomentosum
Convolvulus mauritanicus
Dianthus alpinus

Dianthus deltoides
    petræus
Dodecatheon Meadia
Dryas octopetala
Erodium Reichardi
Gentiana acaulis
Gypsophila repens
Helianthemum, in variety
Iberis sempervirens
    corifolia
    Tenoreana
Linaria alpina
Lithospermum prostrata
Myosotis dissitiflora
    sylvatica
Nierembergia rivularis
Œnothera marginata
    taraxacifolia
Omphalodes verna
Oxalis, all hardy species
Phlox reptans
    subulata
Pentstemon procerus

Plumbago Larpentæ
Polygonum vaccinifolium
Primula Auricula, in var.
    vulgaris, in variety
Saponaria ocymoides
Saxifraga hypnoides, and
    most of the mossy
    section
Sedum Kamtschaticum
    pulchellum
    Sieboldii
    spectabile, for large
    beds
    spurium
Silene alpestris
    Schafta
Spergula pilifera
Statice oleæfolia
Veronica fruticulosa
    saxatilis
    taurica
Viola cornuta, in variety
    lutea

A Selection of Ornamental Aquatic Plants.

Acorus Calamus
    gramineus
Actinocarpus Damasonium
Alisma, in variety
Aponogeton distachyon
Aster Tripolium
Butomus umbellatus
Calla palustris
Caltha palustris
Carex paniculata
    pendula
    Pseudo-cyperus
Cyperus longus
Epilobium hirsutum
Equisetum, in variety
Glyceria aquatica
Hippuris vulgaris
Hottonia palustris

Houttuynia cordata
Hydrocharis Morsus-ranæ
Iris Pseudacorus
    sibirica
Limnanthemum nymph-
    oides
Lobelia Dortmanna
Lysimachia thyrsiflora
Lythrum Salicaria
Menyanthes trifoliata
Myosotis palustris
Myriophyllum, in variety
Nuphar advena
    Kalmiana
    lutea
    pumila
Nymphæa alba
    minor

Nymphæa odorata
Œnanthe fistulosa
Orontium aquaticum
Osmunda, in variety
Phormium tenax
Polygonum amphibium
    Hydropiper
Pontederia cordata
Ranunculus aquaticus
    Lingua
Rumex Hydrolapathum
Sagittaria sagittifolia fl. pl.
Scirpus lacustris
Sparganium, in variety
Stratiotes aloides
Thalia dealbata
Typha, all the kinds

A Selection of Plants thriving in Marshy or Boggy Ground.

Butomus umbellatus
Calla palustris
Caltha, in variety
Carex pendula
Chrysobactron Hookeri
Coptis trifolia
Cornus canadensis
Crinum capense
Cypripedium spectabile
Drosera, in variety
Epilobium hirsutum
Epipactis palustris
Equisetum, in variety
Eriophorum, in variety
Eupatorium, in variety
Ficaria, in variety
Galax aphylla
Gentiana Pneumonanthe

Gunnera scabra
Helonias bullatus
Hibiscus, in variety
Hydrocotyle bonariensis
Iris graminea
    Monnieri
    ochroleuca
    Pseudacorus
    sibirica
Leucanthemum lacustre
Leucojum æstivum
    Hernandezii
Linnæa borealis
Lobelia syphilitica
Lycopodium, in variety
Lysimachia thyrsiflora
Lythrum, in variety

Narcissus, in variety
Nierembergia rivularis
Orchis, in variety
Petasites vulgaris
Phormium tenax
Pinguicula, in variety
Primula Munroi
    sikkimensis
Pyrethrum serotinum
Rhexia virginica
Sagittaria, in variety
Sarracenia purpurea
Spigelia marilandica
Swertia perennis
Symplocarpus fœtidus
Tofieldia, in variety
Tradescantia virginica

Trailers, Climbers, etc.

The selection of plants to cover bowers, trellises, railings, old trees, stumps, rootwork, etc. suitably is important. The plants fitted for these purposes are equally useful for the rough rockwork, precipitous banks, flanks of rustic bridges, riverbanks, ruins natural or artificial, covering cottages or outhouses, and many other uses in garden, pleasure-ground, or wilderness.

Vitis æstivalis
    amooriensis
    cordifolia
    heterophylla variegata
    Isabella
    Labrusca
    laciniosa
    riparia
    Sieboldii
    vinifera apiifolia
    vulpina
Hedera, all the finer varieties of Ivy, both green and variegated
Aristolochia Sipho
    tomentosa
Clematis azurea grandiflora
    campaniflora
    elliptica
    Flammula
    florida
        plena
        Standishi
    Fortunei

Clematis Francofurtensis
    Hendersoni
    insulensis
    Jackmani
    lanuginosa
    montana
    nivea
    patens Amelia
        Helena
        insignis
        Louisa
        monstrosa
        Sophia
        violacea
    pubescens
    rubro-violacea
    Shillingii
    Sieboldii
    tubulosa
    Viticella
        alba
        venosa
Calystegia dahurica
    pubescens plena
Wistaria sinensis

Asparagus Broussoneti
Cynanchum acutum
    monspeliacum
Apios tuberosa
Tamus communis
Periploca græca
Hablitzia tamnoides
Boussingaultia baselloides
Menispermum canadense
    virginicum
Cissus orientalis
    pubescens
Ampelopsis bipinnata
    cordata
    hederacea
    tricuspidata
Jasminum nudiflorum
    officinale
    revolutum
Passiflora cærulea
Lonicera Caprifolium
    confusa
    flava
    japonica
    Periclymenum

List of Ferns that may be Grown in the Rock-Garden.

Adiantum pedatum
Asplenium Adiantum nigrum
    Filix-fœmina and varieties
    fontanum
    germanicum
    Halleri
    lanceolatum
    monanthemum
    Ruta-muraria
    septentrionale
    Trichomanes
    viride
Ceterach officinarum
Cystopteris alpina
    fragilis
Cheilanthes odora
Cyrtomium caryotideum
    falcatum
Dennstædtia punctilobula
Diplazium thelypteroides

Lastrea Filix-mas and varieties
    Goldieana
        assurgens
    intermedia
    marginalis
    novæboracensis
    atrata
    erythrosora
    opaca
    Standishii
Lomaria magellanica (in warm districts)
Onoclea sensibilis
Osmunda cinnamomea
    Claytoniana
    gracilis
    regalis
        cristata
    spectabilis
Platyloma atropurpurea

Polypodium hexagonopterum
    Phegopteris
    vulgare
Polystichum acrostichoides
    aculeatum
    angulare
    vestitum venustum
Pteris aquilina
Scolopendrium vulgare and varieties
Struthiopteris germanica
    pennsylvanica
Woodsia hyperborea
    ilvensis
    polystichoides
Woodwardia areolata
    aspera
    japonica
    orientalis
    radicans

Selection of Alpine and Rock-plants for Growing on Old Walls, Ruins, Stony Banks, etc. (The most suitable kinds are marked *.)

*Corydalis lutea
*Cheiranthus Cheiri pleno, in variety
Arabis albida
    *arenosa
    lucida variegata
    petræa (old mossy walls)
    blepharophylla (do.)
Aubrietia, all the varieties
Hutchinsia petræa
Vesicaria utriculata
Schivereckia podolica
*Alyssum montanum
    saxatile
    spinosum
Koniga maritima
Petrocallis pyrenaica (mossy and moist old walls)
*Draba aizoides
    bœotica
Ionopsidion acaule (north side of old walls)
Thlaspi alpestre
Iberis, in variety
*Reseda odorata (sown in chinks in walls this sometimes becomes perennial)
Helianthemums (many of the varieties might be grown upon old ruins, stony banks, etc.)

*Gypsophila muralis
    prostrata
*Tunica Saxifraga
*Dianthus cæsius
    deltoides
    monspessulanus
    petræus
Saponaria ocymoides
Silene acaulis (moist walls, to be first carefully planted in a chink)
    alpestris
    rupestris
    Schafta
*Lychnis alpina
    lapponica
Sagina procumbens pleno
*Arenaria balearica
    cæspitosa
    ciliata
    graminifolia
    montana
    verna
Linum alpinum
Malva campanulata (ruins)
Erodium romanum (old walls)
    Reichardii
Ononis alba
Astragalus monspessulanus
Coronilla minima
    varia
Acæna Novæ Zealandiæ (moist mossy walls)
*Cotyledon Umbilicus

Umbilicus chrysanthus
*Sedum acre
    *acre variegatum
    Aizoon
    *album
    anglicum
    brevifolium
    cæruleum
    *dasyphyllum
    elegans
    *Ewersii
    farinosum
    hispanicum
    *kamtschaticum
    multiceps
    pulchellum
    *sempervivoides
    sexangulare
    sexfidum
    *spurium
*Sempervivum arachnoideum
    arenarium
    calcareum
    globiferum
    *Heuffelli
    hirtum
    *montanum
    piliferum
    *soboliferum
    *tectorum
Saxifraga bryoides
    cæsia
    crustata
    cuscutæformis

Saxifraga diapensioides
  *Hostii
  intacta
  *ligulata
  *longifolia
  pectinata
  pulchella
  retusa
  *rosularis
  Rocheliana
  *Rhei
  sarmentosa
Cesperula cynanchica
*Antranthus ruber
  albus
  coccineus
Antennaria dioica minima (mossy chinks, with a little soil)
Bellium bellidioides
  crassifolium

Bellium minutum
Santolina incana
Achillea tomentosa
Symphyandra pendula
Campanula Barrelieri
  fragilis
  *garganica
  cæspitosa
  alba
  rotundifolia
*Antirrhinum rupestre
  *majus
  Orontium
*Linaria Cymbalaria
  alba
  vulgaris
*Erinus alpinus
Veronica fruticulosa
  saxatilis
Thymus citriodorus (earthy chinks)

Iris germanica, and varieties
  pumila
Polypodium vulgare
Adiantum Capillus Veneris (on moist warm walls)
Asplenium Adiantum-nigrum
  fontanum
  *septentrionale
  *Ruta-muraria
  *germanicum
  *lanceolatum
  *Trichomanes, and varieties
  *viride
*Ceterach officinarum
*Matthiola tristis.

Dwarf Hardy Plants of a Silvery or Variegated Tone, and mostly suitable for Edgings.

Achillea ægyptiaca Clavennæ
Ajuga reptans rubra variegata
Alyssum saxatile variegatum
  spinosum
Andryala lanata
Antennaria hyperborea tomentosa
Arabis albida variegata
  lucida variegata
  procurrens variegata
Artemisia frigida
Aubrietia deltoidea variegata
Cerastium Biebersteinü

Cerastium grandiflorum tomentosum
Convolvulus lineatus
Euphorbia Myrsinites
Glechoma hederacea variegata
Hieracium Camerarii
Linaria Cymbalaria variegata
Origanum vulgare variegatum
Polemonium cœruleum variegatum
Salvia argentea
Santolina incana
Saxifraga Aizoon

Saxifraga Cotyledon
  crustata
  ligulata
  longifolia
  pectinata
  recta
Sedum glaucum
Sempervivum arachnoideum
  calcareum
Sideritis syriaca
Stachys lanata
Thymus lanuginosus
Veronica gentianoides variegata
  candida

Dwarf Shrubs suited for the rougher parts of Rock-gardens and for Intermingling with the Larger Alpine Plants, when Planted on the Margins of Shrubberies, etc.

Iberis, in variety
Helianthemum, in variety
Cistus, in variety
Polygala Chamæbuxus
Hypericum humifusum
Genista tinctoria
  sagittalis
  prostrata
Hedera, variegated and other curious vars.
Othonna cheirifolia
Erica carnea, and all hardy species and varieties
Arbutus Uva-ursi
Pernettya mucronata
Gaultheria procumbens
Andromeda hypnoides

Andromeda fastigiata
  tetragona
Bryanthus erectus
Menziesia cærulea
  empetriformis
  polifolia, and varieties
Daphne Cneorum
Lithospermum prostratum
Thymus Mastichina
Polygonum vaccinifolium
Veronica saxatilis
  taurica
Euphorbia Myrsinites
Salix lanata
  reticulata
  serpyllifolia
Empetrum nigrum

Santolina Chamæcyparissus
  incana
Euonymus radicans variegatus
Rhododendron hirsutum
  ferrugineum
  Chamæcistus, and others
Azalea amœna
Epigea repens
Skimmias, in variety
Vaccinium Myrtillus
  macrocarpum
  Oxycoccos
  Vitis-idæa
  uliginosum
Juniperus squamata.

List of Dwarf Alpine Shrubs, etc., for the Rock-Garden.

Andromeda floribunda
    tetragona
Astragalus Tragacantha
Azalea amœna
Betula nana
Bryanthus erectus
Calluna vulgaris, in var.
Cistus, various species
Clematis, numerous kinds
Cornus canadensis
Cotoneaster microphylla
    thymifolia
Cytisus sessilifolius
Daphne alpina
    Cneorum
    collina
    Mezereum
Deutzia, several kinds
Empetrum nigrum
    rubrum
Epigæa repens
Erica, all hardy species

Euonymus japonicus fol.
    argenteis
    nanus
    radicans fol. variegatis
Gaultheria procumbens
    Shallon
Genista anglica
    hispanica
    sagittalis
    tinctoria
Helianthemums, many
    kinds
Hydrangea, several kinds
Hypericum calycinum
Indigofera Dosua
Ivies, in great variety
Kalmia latifolia
    nana
Leiophyllum buxifolium
Menziesia, several kinds
Ononis fruticosa

Ononis rotundifolia
Pernettya, several kinds
Polygala Chamæbuxus
Potentilla floribunda
    fruticosa
Rhododendron, dwarf
    kinds
Ruscus Hypoglossum
    racemosus
Salix lanata
    reticulata
Santolina Chamæcyparis-
    sus
    viridis
Skimmia japonica
    laureola
    oblata
Spiræa, several dwarf kinds
Vaccinium, three or four
    kinds
Vinca, various kinds

Selection of Alpine and Rock-plants of Prostrate or Drooping Habit, suited for Placing so that they may Droop over the Brows of Rocks, and like Positions.

Arabis albida
    procurrens
Aubrietias
Alyssum montanum
    saxatile
Iberis corifolia
    sempervirens
    Tenoreana
Helianthemum, in variety
Gypsophilas, several
Dianthus deltoides, and
    others
Tunica Saxifraga
Saponaria ocymoides
Cerastium Biebersteinii
    grandiflorum
    tomentosum
Malva campanulata
Callirhoe involucrata
    pedata
Hypericum humifusum
Tropæolum speciosum
    polyphyllum
Genista prostrata
    tinctoria
    sagittalis
Ononis arvensis albus
Trifolium repens penta-
    phyllum
Lotus corniculatus
    pleno
Astragalus monspessulanus
Coronilla iberica
    varia

Hippocrepis comosa
Vicia argentea
Orobus roseus
Dryas octopetala
Fragaria indica
Rubus arcticus
Potentilla alpestris
    calabra
    Hopwoodiana
    M'Nabiana
    Tonguei
    verna, and numerous
        varieties and hybrids
Zauschneria californica
Œnothera acaulis
    missouriensis
    taraxacifolia
Sedum spurium
    Ewersii
    kamtschaticum
    reflexum
    Sieboldii
Saxifraga hypnoides
    ceratophylla
    biflora
    sarmentosa
    oppositifolia, and va-
        rieties
Linnæa borealis
Galium verum
Scabiosa graminifolia
    Webbiana
Santolina incana
Diotis maritima

Artemisia argentea
    frigida
Campanula Barrelieri
    rotundifolia
        alba
    fragilis
        hirsuta
    garganica
    muralis
    cæspitosa
        alba
Erica carnea
Cornus canadensis
Epigæa repens
Phlox subulata
    reptans
Convolvulus mauritanicus
Lithospermum prostratum
Antirrhinum rupestre
Linaria alpina
    Cymbalaria
Pentstemon procerus
Veronica taurica
    prostrata
Thymus lanuginosus
    Serpyllum, white vari-
        ety
Zietenia lavandulæfolia
Dracocephalum argunense
Zapania nodiflora
Androsace lanuginosa
Lysimachia Nummularia
    nemorum
Plumbago Larpentæ

Polygonum vaccinifolium
Euphorbia Myrsinites
Salix lanata
    reticulata
Empetrum nigrum
Polygonum complexum
Boussingaultia baselloides
Medicago falcata

Lathyrus grandiflorus
    latifolius
    albus
    tuberosus
Vicia Cracca
Convolvulus arvensis
Calystegia dahurica
    pubescens

Vinca major
    minor
    herbacea
Clematises, the new varieties of the lanuginosa section
Cyananthus lobatus.

THE END.

Watson & Hazell, Printers, London and Aylesbury.

www.ingramcontent.com/pod-product-compliance
Lightning Source LLC
Chambersburg PA
CBHW022134300426
44115CB00006B/175